Botany: Plant Identification and Classification

Botany: Plant Identification and Classification

Editor: Edward Ackles

Callisto Reference
www.callistoreference.com

Callisto Reference,
118-35 Queens Blvd., Suite 400,
Forest Hills, NY 11375, USA

Visit us on the World Wide Web at:
www.callistoreference.com

ISBN: 978-1-64116-257-9 (Hardback)

The publisher's policy is to use permanent paper from mills that operate a sustainable forestry policy. Furthermore, the publisher ensures that the text paper and cover boards used have met acceptable environmental accreditation standards.

Printed in China.

Cataloging-in-Publication Data

Botany : plant identification and classification / edited by Edward Ackles.
p. cm.
Includes bibliographical references and index.
ISBN 978-1-64116-257-9
1. Botany. 2. Plants--Identification. 3. Plants--Classification. I. Ackles, Edward.
QK97.5 .B68 2020
580--dc23

Table of Contents

Preface

Botany is the branch of biology associated with the study of plant life. It is also known as plant biology, phytology or plant science. Modern botany is a multidisciplinary subject with inputs from most other areas of science and technology. The topics for research in botany include the study of plant's structure, growth, classification, biochemical processes, primary metabolism, properties, evolutionary relationships, diseases and interaction with the environment. The branches of botany are divided into three groups. Organismal topics focus on groups of plants such as grasses, mosses and algae. Core topics are associated with the study of the classification and description of plant diversity, and the fundamental natural phenomena and processes of plant life. Applied topics study the ways in which plants may be used for economic benefit in forestry, horticulture and agriculture. This book contains some path-breaking studies in the field of botany. It attempts to understand the multiple branches that fall under this discipline and how such concepts have practical applications. It aims to serve as a resource guide for students and experts alike and contribute to the growth of this field.

After months of intensive research and writing, this book is the end result of all who devoted their time and efforts in the initiation and progress of this book. It will surely be a source of reference in enhancing the required knowledge of the new developments in the area. During the course of developing this book, certain measures such as accuracy, authenticity and research focused analytical studies were given preference in order to produce a comprehensive book in the area of study.

This book would not have been possible without the efforts of the authors and the publisher. I extend my sincere thanks to them. Secondly, I express my gratitude to my family and well-wishers. And most importantly, I thank my students for constantly expressing their willingness and curiosity in enhancing their knowledge in the field, which encourages me to take up further research projects for the advancement of the area.

Editor

1

THREE NEW SPECIES OF *ELATOSTEMA* J.R. FORST. & G. FORST. (URTICACEAE) IN LIMESTONE CAVES FROM GUANGXI AND GUIZHOU, CHINA

YI-GANG WEI, FANG WEN, LONG-FEI FU[1] AND WEN-TSAI WANG[2*]

Guangxi Institute of Botany, Guangxi Zhuang Autonomous Region and Chinese Academy of Sciences, Guilin 541006, China

Keywords: Elatostema; New species; China.

Abstract

Three new species of *Elatostema* J.R. Forst. & G. Forst. from Guangxi and Guizhou, China are described and illustrated. They are *E. atrostriatum* W.T. Wang & Y.G. Wei, which is most similar to *E. stewardii* Merr.; *E. jingxiense* W.T. Wang & Y.G. Wei, which is most similar to *E. malacotrichum* W.T. Wang & Y.G. Wei, and *E. schizodiscum* W.T. Wang & Y.G. Wei, which is most similar to *E. angulaticaule* W.T. Wang & Y.G. Wei.

Introduction

Elatostema J.R. Forst. & G. Forst. is a specialized genus for adapting to humid and dark habitats, e.g. dense forests, deep gorges and caves. It is one of the biggest genera in the family Urticaceae comprising at least 500 species all over the world, and mainly occur in tropical and subtropical Asia, Africa and islands of Oceania (Chen *et al.*, 2003). Accompanying with the further field investigations on South and South-west China, more and more new taxa of this genus were discovered and published, including new series, Series *Albopilosoides* Q. Lin & L.D. Duan (Lin and Duan, 2008), Series *Neriifolia* W.T. Wang & Zeng Y. Wu (Wu *et al.*, 2012), nearly 50 new species (Duan, 2010; Lin *et al.*, 2011; Wang, 2010a, b, 2011a, b, 2012; Wei *et al.*, 2011; Wei and Wang, 2011a, b; Wu *et al.*, 2011a, b, 2012) and 4 new varieties (Wang, 2010b, 2012; Wu *et al.*, 2011a). Bi *et al.* (2011) described new taxa of *Elatostema* from Thailand and India. Recently three unknown specimens belonging to the genus *Elatostema* were collected from Guangxi and Guizhou. After critical examination of these specimens and carefully consulting relevant literature, they were identified as new species.

1. Elatostema atrostriatum W.T. Wang & Y.G. Wei **sp. nov.** **(Fig. 1).**

Type: CHINA. Guangxi: Jiayou village, Lingyun County, 1 April 2009, Y.G. Wei 103 (*Holotype:* PE, *Isotype:* IBK).

Diagnosis: *Species nova haec est similis E. stewardii* Merr., *quod foliis majoribus, apice cuspidatis, captitulis pistillis receptacula majorbus, bracteis pistillatis triangulatibus, dense puberulis, apice corniculatis, non striatis differt.*

Herbs perennial, terrestrial. Stems c. 3 clumping, 100-135 × 1-2 mm, erect, simple, glabrous. Stipules subulate, 1-7 × 0.1-0.3 mm. Leaves distichous, alternate, short petioled, petioles 1-4 mm long, glabrous; leaf blades 28-105 × 10-34 mm, oblique, narrowly obovate, leathery, nerves pinnate, broader-half lateral nerves 3-6, 30-70° to the midrib, narrower-half lateral nerves 4-6, 30-70° to the midrib; upper surface cystoliths dense, conspicuous or inconspicuous, rod-shaped,

*Corresponding author. Email: wentsaiwang@yeah.net.

[1]College of Life Sciences, Guangxi Normal University, 541004, Guilin, China.

[2]State Key Laboratory of Systematic and Evolutionary Botany, Institute of Botany, The Chinese Academy of Sciences, Cn-100093 Beijing, China.

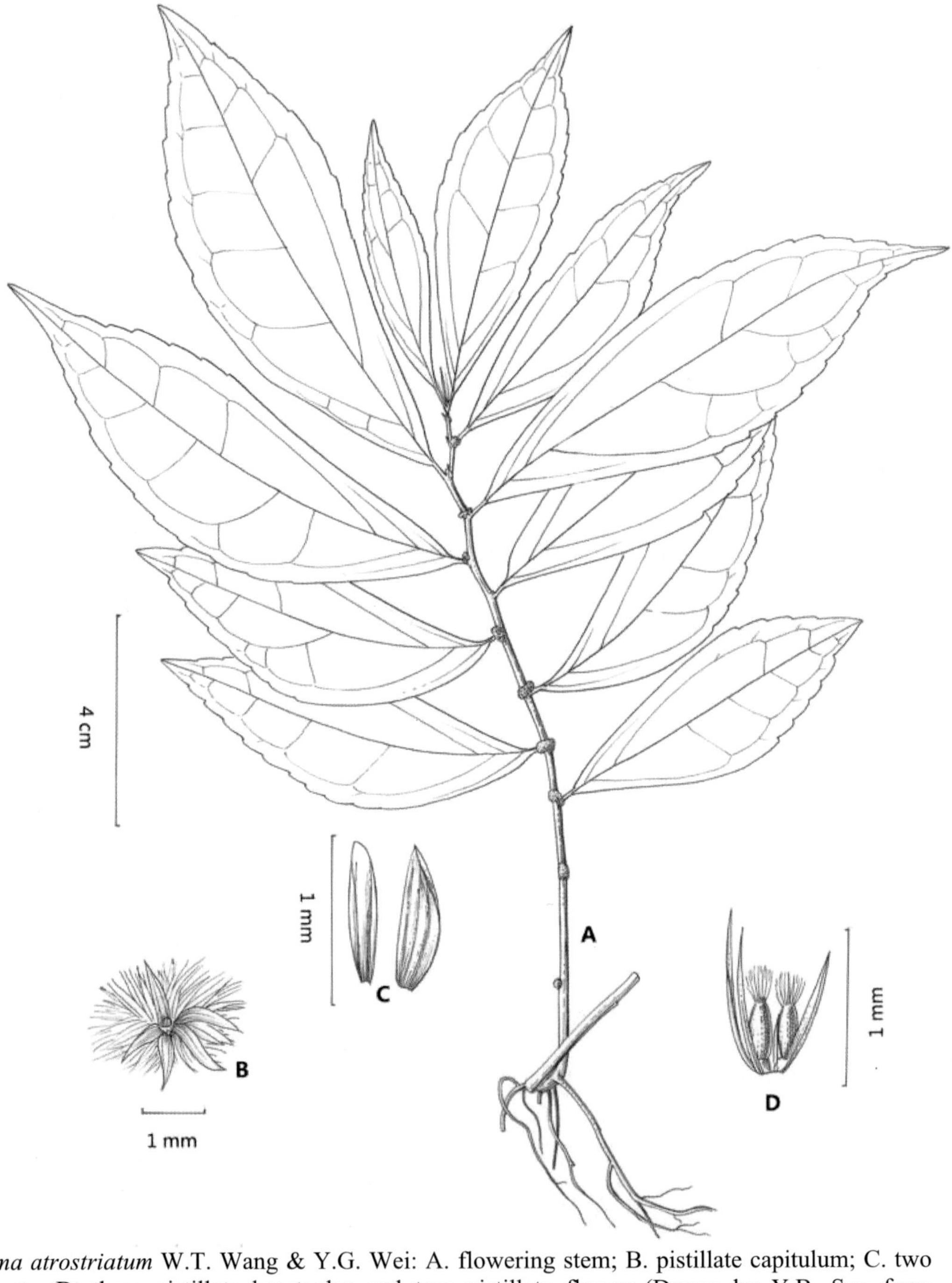

Fig. 1. *Elatostema atrostriatum* W.T. Wang & Y.G. Wei: A. flowering stem; B. pistillate capitulum; C. two pistillatae bracts; D. three pistillate bracteoles and two pistillate flower (Drawn by Y.B. Sun from holotype).

0.1-0.25 mm long, lower surface cystoliths absent; base asymmetrical, broader-half rounded and narrower-half cuneate, margin entire or denticulate, basal ≤1/2 entire, apex acuminate or blunt;

Staminate capitulum not seen. Pistillate capitulum 1-3 axillary, sessile; receptacle inconspicuous; subtended by marginal bracts, bracts c. 5, equal, membranous, semi-transparent, narrowly ovate or broadly linear, 0.8-1.0 × 0.15-0.4 mm, glabrous, with 2-3 black-brown longitudinal stripes; bracteoles numerous, dense, membranous, semitransparent, narrowly linear, 0.6-1.0 mm long, glabrous, with 1 black-brown longitudinal stripe. Pistillate flower short peduncled, tepal inconspicuous; pistil c. 0.7 mm long; ovary narrowly ellipsoid, 0.4 mm long; stigma 0.3 mm long. Infructescence not seen.

Habitat: Limestone cave, 20 individuals.

Etymology: Elatostema atrostriatum is named after pistillate bracts with 2-3 black-brown longitudinal stripes.

Vernacular name: Heiwen louticao (China).

Note: Elatostema atrostriatum is morphologically similar to *E. stewardii* Merr. but differs in leaves, pistillate receptacle, bracts and bracteoles (Table 1).

Table 1. Morphological comparison between *Elatostema atrostriatum* and *E. stewardii*

Characters	*E. atrostriatum*	*E. stewardii*
Leaves	28-105 × 10-34 mm, apex acuminate or blunt	70-125 × 28-45 mm, cuspidate
Pistillate receptacle	small, inconspicuous	subquadrate, c. 3 mm long
Pistillate bracts	narrowly ovate or broadly linear, glabrous, not corniculate with 2-3 black-brown longitudinal stripes	triangular, densely pubescent, corniculate with no stripes
Pistillate bracteoles	glabrous, with 1 black-brown longitudinal stripe	margin densely pubescent with no stripes

2. Elatostema jingxiense W.T. Wang & Y.G. Wei **sp. nov.** **(Fig. 2).**

Type: CHINA. Guangxi: Jingxi, Dizhou, limestone cave, 22 March 2009, Wei Yigang g067 (*Holotype*: PE, *Isotype*: IBK).

Diagnosis: *Species nova haec est similis E. malacotrico* W.T. Wang & Y.G. Wei, *quod plantis totis siccitate nigrescentibus, cauli staminati foliis normaliter evolutis eis cauli pistillati majoribus cystolithis densis praeditis, involucri staminati bracteis paucioribus ca. 8 depresse orbicularibus dorso puberulis distinguitur.*

Herbs perennial, terrestrial, dioecious. Male plant: stems 150–500 × 2–3 mm, erect, dark brown when fresh, simple, densely puberulent, the hairs c. 0.1 mm long, 3-4 densely degraded microphyll at the top of the stem, underneath staminate capitulum bearing at naked nodes. Stipules lanceolate-linear, 3-4 mm long, glabrous. Leaves distichous, alternate, sessile or short petioled; leaf blades 12-14 × 4-16 mm, chartaceous, oblique ovoid, nerves pinnate, lateral nerves 4 or 6 paired, 45-60° to the midrib, upper surface sparsely hispid, hairs c. 0.5 mm long, weakly curved, appressed, cystoliths absent, lower surface nerves densely hispid, hairs c. 0.5 mm long, weakly curved, cystoliths absent; base asymmetric, broader-half oblique, broadly cuneate, narrower-half cuneate; margin dentate, apex acuminate, acumen entire. Female plant: stems 450-550 × c. 5 mm, erect, green when fresh, simple, densely puberulous, hairs c. 0.1 mm long; leaves distichous, alternate, short petioled, petiole 4-20 mm long; leaf blades 50-160 × 35-65 mm, chartaceous, oblique oblong or elliptic, nerves pinnate, lateral nerves 5 or 6 paired, 45-60° to the midrib, upper surface sparsely hispid, hairs c. 0.5 mm long, weakly curved, appressed, cystoliths absent, lower surface nerves densely hispid, hairs c. 1 mm long, weakly curved, cystoliths absent; base asymmetric, broader-half rounded, narrower-half obliquely cuneate; margin dentate, broader-half

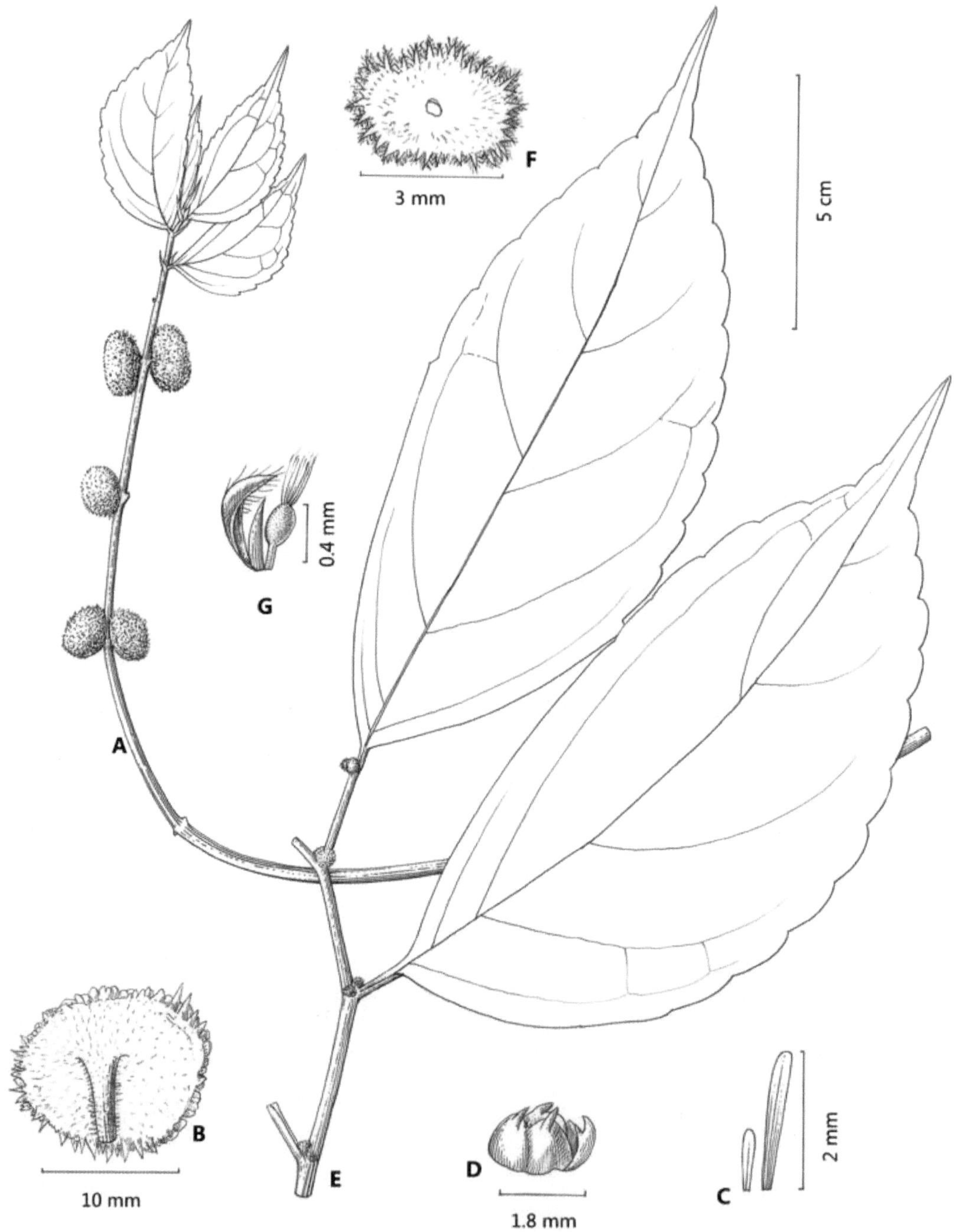

Fig. 2. *Elatostema jingxiense* W.T. Wang & Y.G. Wei: A. staminate stem; B. staminate capitulum; C. staminate bracteoles; D. staminate flower; E. upper part of pistillate stem; F. pistillate capitulum; G. two pistillate bracteoles and pistillate flower (Drawn by Y.B. Sun from holotype).

basal ≤1/3 entire, narrower-half basal ≤1/2 entire, apex acuminate, acumen entire. Staminate capitulum solitary or didymous, axillary, short pedunculate, c. 6 mm long, puberulent; receptacle subround, c. 10 mm in diam., puberulent, subtended by marginal bracts, the bracts c. 15, membranous, triangular, c. 1 mm long, glabrous; bracteoles dense, membranous, semi-transparent, linear, 0.8-2.0 mm long, glabrous, with 1-3 brown stripes. Staminate flowers short peduncled, tepals 5, broadly ovate, c. 1 mm long, subapical appendage c. 0.4 mm long, corniculate, glabrous. Pistillate capitulum didymous, axillary, sessile; receptacle elliptic, c. 3.0 × 2.5 mm, puberulent, subtended by marginal bracts, bracts c. 25, narrowly triangular or linear, 0.3-0.5 mm long, puberulent or sometimes glabrous; bracteoles numerous, dense, membranous, cymbiform or linear-lanceolate, 0.5-0.8 mm long, puberulent or glabrous. Pistillate flower short peduncled, tepals inconspicuous; pistil 0.65 mm long; ovary ellipsoidal, c. 0.25 mm long; stigma 0.4 mm long.

Habitat: Limestone cave, 20 individuals.

Etymology: Elatostema jingxiense is named after the type locality near Jingxi County.

Vernacular name: Jingxi louticao (China).

Note: Elatostema jingxiense is morphologically similar to *E. malacotrichum* W.T. Wang & Y.G. Wei but differs in colour of whole plant when dying, leaves of male stem and staminate bract (Table 2).

Table 2. Morphological comparison between *Elatostema jingxiense* and *E. malacotrichum*

Characters	*E. jingxiense*	*E. malacotrichum*
Whole plant turning black while dying?	no	yes
Leaves of male stem	much smaller than those of pistillate stem and lacking cystoliths	larger than those of pistillate stem and bearing numerous dense cystoliths
Staminate bracts	c. 15, triangular and glabrous	c. 8, depressed-orbicular and puberulous abaxially

3. Elatostema schizodiscum W.T. Wang, & Y.G. Wei sp. nov. (Fig. 3).

Type: CHINA. Guizhou Province, Anlong County, Dushan town, limestone cave, 29 March 2010, F. Wen 1036, (*Holotype*: PE, *Isotype*: IBK).

Diagnosis: *Species nova haec est similis E. angulaticauli* W.T. Wang & Y.G. Wei, *quod foliis eorum nervis secundariis 6-8-jugis et eorum cystolithis minoribus, 0.05–0.1 mm longis, capituli pistillati receptaculo indiviso, floribus pistillatis ad receptaculum totum dense crescentibus, stigmatibus subglobosis recedit.*

Herbs perennial, terrestrial. Stems c. 350 × 4 mm, decumbent, dark green, lower part drying 4-sulcate, distal part cystoliths dense, rod-shaped, 0.1-0.4 mm long, branched, glabrous. Stipules caducous. Leaves distichous, alternate, short petioled, petioles 2 mm long, glabrous; leaf blades 60-150 × 30-35 mm, oblique elliptic or oblique ovoid, chartaceous, nerves pinnate, lateral nerves 6-8 pairs, 45-60° to the midrib, above sunken, beneath bulged out, conspicuous, upper surface sparsely hispid, hairs c. 1 mm long, weakly curved, cystoliths small, dense, rod-shaped or punctiform, 0.05-0.1 mm long, lower surface glabrous, cystoliths absent; base asymmetrical, broader-half broadly cuneate or oblique rounded and narrower-half cuneate; margin serrate, the apex of serrate blunt or rounded; apex long acuminate or acuminate, acumen entire. Staminate inflorescence not seen. Pistillate capitulum didymous, axillary, c. 5 mm wide, sessile; receptacle white, 2-partite, glabrous, lobes oblong or subquadrate, c. 2 mm long, undivided or 3-partite, apex bearing flowers dense, glabrous, subtended by bracts, bracts 3, white, broadly triangular, c. 1 mm long, glabrous, bracteoles white, narrow, c. 0.7 mm long, glabrous. Pistillate peduncle white, c. 0.7 mm long, glabrous, tepals c. 2, equal, narrowly linear, c. 0.3 mm long, glabrous; ovary elliptic, c. 0.22 mm long, stigma subglobose. Achene brown, narrowly ovate, c. 0.7 ×0.3 mm, with 4 longitudinal ribbs; staminodium c. 3, white, narrowly ovate, c. 0.3 mm long, glabrous.

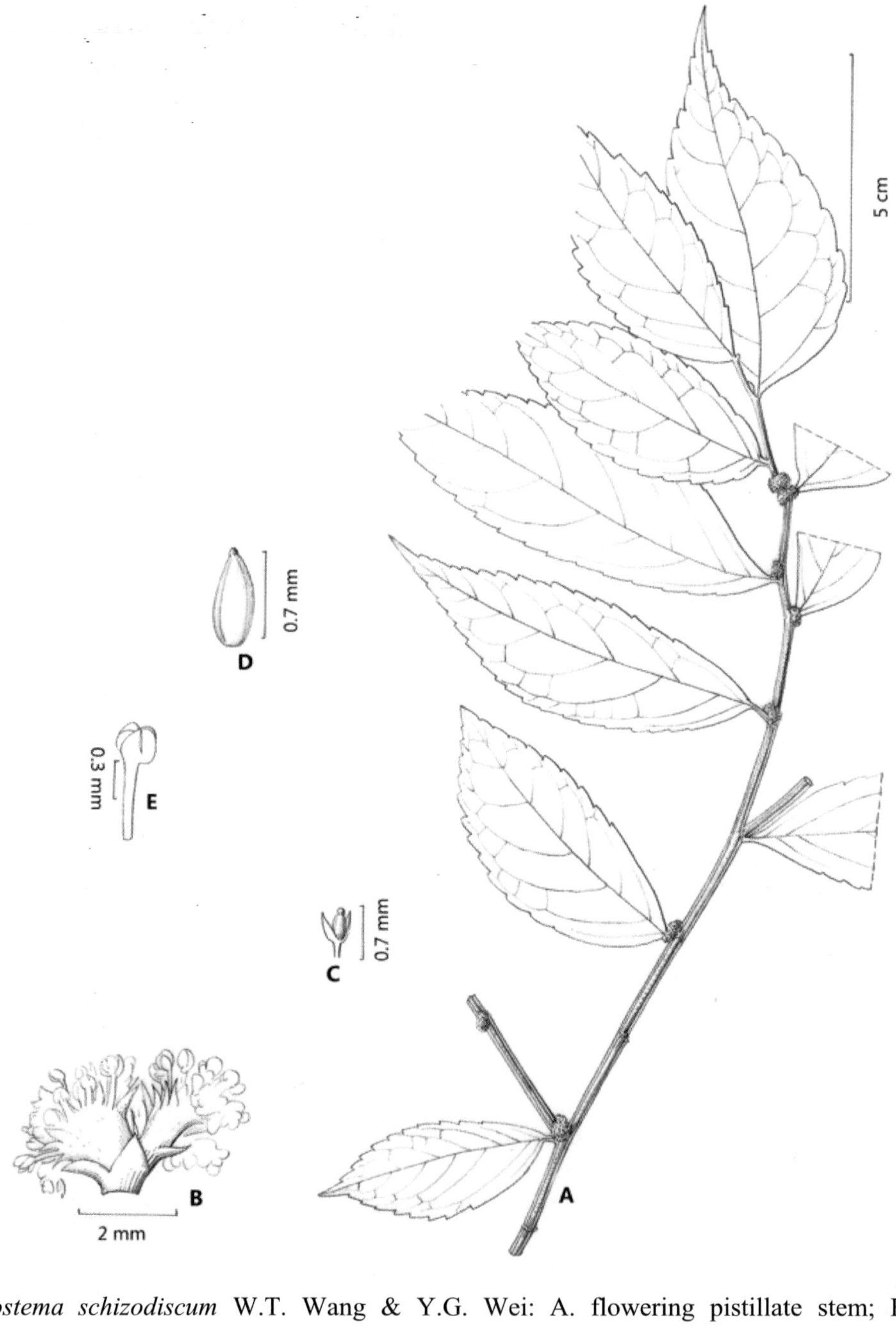

Fig. 3. *Elatostema schizodiscum* W.T. Wang & Y.G. Wei: A. flowering pistillate stem; B. pistillate capitulum; C. pistillate bracteoles and pistillate flower; D. achene; E. fruiting pedicel and staminodes (Drawn by Y.B. Sun from holotype).

Habitat: Limestone cave, 35 individuals.

Etymology: Elatostema schizodiscum is named after receptacle 2-parted.

Vernacular name: Lietuo louticao (China).

Note: Elatostema schizodiscum is morphologically similar to *E. angulaticaule* W.T. Wang & Y.G. Wei but differs in leaves, cystoliths, pistillate receptacle and stigma (Table 3).

Table 3. Morphological comparison between *Elatostema schizodiscum* and *E. angulaticaule*

Characters	*E. schizodiscum*	*E. angulaticaule*
Leaves	lateral nerves 6-8 pairs	lateral nerves 4-6 pairs
Cystoliths	0.05-0.1 mm long	0.1-0.3 mm long
Pistillate receptacle	2-parted, flowers densely growing on apex of receptacle	undivided, flowers growing on the whole receptacle
Stigma	subglobose	penicillate

Acknowledgements

We thank Mr. Ying-Bao Sun for the drawing, and Prof. Fa-Nan Wei for help. We also like to acknowledge the support of the National Natural Science Foundation of China (Grant No. 3116039), the Science Research Foundation of Guangxi Institute of Botany (Guizhiye11001), and the Science Research Foundation of Guangxi Institute of Botany (No. 10004) supporting the fieldwork.

References

Bi, H.Y., Yang, Z.R. and Lin, Q. 2011. New taxa of *Elatostema* (Urticaceae) from Thailand and India. Bangladesh J. Plant Taxon. **18**(2): 149-152.

Chen, C.J., Lin, Q., Friis, I., Wilmot-Dear, C.M. and Monro, A.K. 2003. Urticaceae. *In*: Wu, Z. and Raven, P.H. (Ed.), Flora of China, Vol. **5**. Science Press, Bejing, pp. 76-189.

Duan, L.D. 2010. *Elatostema cataractum* (Urticaceae), a new species from Guizhou Province, China. Ann. Bot. Fenn. **47**: 229-232.

Lin, Q and Duan, L.D. 2008. Two new species and a new series of *Elatostema* (Urticaceae) from China, Bot. J. Linn. Soc. **158**: 674-680.

Lin, Q., Shui, Y.M. and Duan, L.D. 2011. *Elatostema oppositum* (Urticaceae), a new species from Yunnan, China. Novon **21**(2): 212-215.

Wang, W.T. 2010a. New taxa of *Pellionia* and *Elatostema* (Urticaceae) from China. Guihaia **30**: 1-12.

Wang, W.T. 2010b. Notes on the genus *Elatostema* (Urticaceae). Guihaia **30**(6): 713-727.

Wang, W.T. 2011a. Two new species of *Elatostema* (Urticaceaa) from Southeastern Yunnan. Guihaia **31**(2): 143-147.

Wang, W.T. 2011b. Six new species of *Elatostema* (Urticaceae) from Gaoligong Shan, Yunnan. Plant Divers. Resour. **33**: 145-156.

Wang, W.T. 2012. Four new species and one variety of *Elatostema* (Urticaceae) from China. Plant Divers. Resour. **34**(2): 137-144.

Wei, Y.G., Monro, A.K. and Wang, W.T. 2011. Additions to the Flora of China: seven new species of *Elatostema* (Urticaceae) from the karst landscapes of Guangxi and Yunnan. Phytotaxa **29**: 1-27.

Wei, Y.G. 2009. Notes on *Elatostema* J. R. *et* G. Forst (Urticeceae) from Guangxi. Guihaia **29**(6): 711-718.

Wei, Y.G. and Wang, W.T. 2011a. *Elatostema xanthotrichum* and *E. bamaense* (Urticaceae), two new species from Guangxi, China. Ann. Bot. Fenn. **48**: 93-95.

Wei, Y.G. and Wang, W.T. 2011b. *Elatostema recurviramum* (Urticaceae), a new cave-dwelling species from Guangxi, China. Novon **21**(2): 281-284.

Wu, Z.Y., Wang, W.T., Wang, H. and Li, D.Z. 2011a. *Elatostema densistriolatum* sp. nov., *E. latistipulum* sp. nov. and *E. cyrtandrifolium* var. *hirsutum* var. nov. (Urticaceae) from southwest China. Nord. J. Bot. **29**: 227-232.

Wu, Z.Y., Wang, W.T., Wang, H. and Li, D.Z. 2011b. Two new species of *Elatostema* (Urticaceae) from southeast Yunnan, China. PhytoKeys **7**: 57-62.

Wu, Z.Y., Li, D.Z., Wang, H. and Wang, W.T. 2012. One new series with its only new species of *Elatostema* (Urticaceae) from Southeast Yunnan, China. Plant Divers. Resour. **34**(2): 151-154.

2

SONERILA GADGILIANA, A NEW SCAPIGEROUS SPECIES OF MELASTOMATACEAE FROM INDIA

M.K. Ratheesh Narayanan, M. Sivadasan[1,2], C.N. Sunil[3], M.K. Nandakumar[4], T. Shaju[5], A.H. Alfarhan[2] and A.S.M. Amal Tamimi[6]

Department of Botany, Payyanur College, Edat P.O., Kannur 670 327, Kerala, India

Keywords: Kerala; Melastomatoideae; New species; Sonerileae.

Abstract

Sonerila gadgiliana, a new species collected from high altitude moss-covered dripping rocks of grassland-shola margins in Wayanad district, Kerala, India is described and illustrated. The species resembles *S. raghaviana* Ratheesh *et al.*, *S. rotundifolia* Bedd. and *S. veldkampiana* Ratheesh *et al.*, but differs from leaf, inflorescence, peduncle, hypanthium, petal, anther, capsule and seed characteristics.

Introduction

The Melastomataceae Jussieu is a large family of about 188 genera with around 5,100 species, mainly distributed in tropics and also in subtropics, out of which about 1,550 species occur in the Old World (Stevens, 2012). The members of this family are easily recognized among dicots by having leaves with a characteristic acrodromous venation (Hickey, 1973) and numerous small, exalbuminous seeds. The family appears to be the largest clade of flowering plants characterized by this type of venation; only a few isolated taxa, e.g. *Heterocentron* Hook. & Arn., *Sonerila* Roxb., *Loreya nigricans* Triana and *Macairea rufescens* DC. have pinnate venation (Renner, 1993; Clausing and Renner, 2001).

The tribe Sonerileae (Melastomatoideae) occurs mainly in Southeast Asia and Madagascar, with a few species in the neotropics (Renner, 1993). Most of the Asiatic Sonerileae species belong to the taxonomically very poorly understood genus *Sonerila*, represented by caulescent and acaulescent herbaceous plants of shady habitats, often with basal rosettes of large, somewhat turgescent leaves, sometimes with tubers. Uniparous (scorpioid) cymes are particularly frequent in Sonerileae as commonly seen in *Sonerila*. Lundin and Nordenstam (2009) estimated the genus to have about 175 species distributed from Sri Lanka and India to the Indo-Pacific (Cellinese, 1997). Subsequent to later publications of additional new species, the genus is now with about 180 species and represents the largest and the only consistently trimerous genus in Sonerileae (except for the monotypic *Stussenia* C. Hansen and *Lithobium* Bongard) with the flowers having one or rarely two whorls of stamens, and as such easily differentiated.

Clarke (1879) in Hooker's Flora of British India recognized 43 species of *Sonerila* and provided names of three doubtful species. Out of the 43 species, eight species were treated as "stemless or almost stemless species" and only three, *viz.* *S. wallichii* Benn., *S. scapigera* Dalzell

---

[1]Corresponding author. E-mail: drmsivadasan@gmail.com
[2] Department of Botany & Microbiology, College of Science, King Saud University, P.O. Box 2455, Riyadh 11451, Kingdom of Saudi Arabia
[3]Department of Botany, S.N.M .College, Maliankara P.O., Ernakulam 683 516, Kerala, India
[4]M.S. Swaminathan Research Foundation, Puthoorvayal P.O., Wayanad 673 121, Kerala, India
[5]Jawaharlal Nehru Tropical Botanic Garden and Research Institute, Palode P.O., Thiruvananthapuram 695 562, Kerala, India
[6]Department of Biology, College of Science, Princess Nora bint Abdulrahman University, P.O. Box 87991, Riyadh 11652, Kingdom of Saudi Arabia

and *S. rotundifolia* Bedd. were from Peninsular India. Gamble (1919) also recognized the above three acaulescent species among the total 13 species of *Sonerila* in his Flora of the Presidency of Madras. Lundin (1998) made an extensive study of Melastomataceae with special emphasis on *Sonerila* of South India after his taxonomic study of the genus in Ceylon (Lundin, 1983). After Gamble's (1919) treatment, many new species were described and the genus is presently represented by about 50 species in India with high species diversity in the Western Ghats having 33 species and two varieties (Nayar, 1976; Giri and Nayar, 1985, 1986a, b, c, 1987a, b; Prakash and Mehrotra, 1988; Gopalan and Henry, 1989; Giri *et al.*, 1992; Ravikumar, 1999; Murugan and Manickam, 2002; Josephine *et al.*, 2003; Lundin and Nordenstam, 2009; Murugesan and Balasubramaniam, 2011; Ratheesh Narayanan *et al.*, 2013, 2014; Deepthikumary and Pandurangan, 2014; Sunil *et al.*, 2014).

During the field exploration for systematic studies in *Sonerila* of the Western Ghats, an interesting scapigerous herb was collected from Banasuramala, Wayanad District, Kerala, at altitude about 1700 m. Detailed observations and study revealed its novelty and distinctness from the hitherto known species, and is described and illustrated here as a new species.

Sonerila gadgiliana Ratheesh & Sivadasan, **sp. nov.** (**Figs 1 & 2**).

Diagnosis: *Sonerila gadgiliana* differs from *Sonerila rotundifolia* by having inflorescence with up to 12 flowers, an obscurely ridged hypanthium with sparsely glandular hairs, deeply cordate anthers, and tubercled seeds, and from *S. veldkampiana* Ratheesh *et al.* by having angular glandular hairy peduncle, sparsely glandular hairy hypanthium, petals with 2–4 glandular hairs along the midvein on abaxial side, shortly acuminate non-beaked anthers, and glandular hairy capsule. The new species differs from *S. raghaviana* Ratheesh *et al.* by having sparsely glandular hairy plants with non-overlapping basal-lobed green leaves, usually 2 inflorescences per plant, and seeds with prominent raphe.

Types: India, Kerala: Wayanad District, Banasuramala, moss-covered moist rocks in grassland-shola margins, about 1700 m, 16 Sep 2012, *M.K. Ratheesh Narayanan MSSH 2388* (*Holotype*: CAL; *Isotypes*: MH, TBGT).

Paratypes: India, Kerala: Wayanad District, Kurichiarmala, about 1600 m, 18 Aug 2014, *M.K. Ratheesh Narayanan & M.K. Nandakumar MSSH 1966* (TBGT).

Scapigerous herbs, attaining up to 15 cm high; rhizome small, up to 0.5 cm diameter, orbicular, white. Leaves radical, 3–5; petiole 3.5–5.5 cm long, adaxially grooved, glabrous; lamina ovate, 5–6 × 4–6 cm, base cordate, without overlapping margins, green with pink tinge below, leathery, margins distantly serrate, acute to shortly acuminate at apex, hairy above, glabrous below, pinnately veined, main nerves 3 pairs from base, prominent below, pinkish. Inflorescence unbranched scorpioid cyme, usually 2 per plant, 6–12-flowered; peduncle quadrangular, 8–15 cm long, sparsely hairy, hairs gland-tipped, light green; bract and bracteoles not prominent. Flowers 3-merous, light pink, pedicel angular, 0.4–0.5 cm long, longer than hypanthium, sparsely glandular hairy, light pink with green tinge; hypanthium c. 0.3 cm long, light pink with green tinge, glandular hairy, obscurely ridged. Calyx lobes 3, c. 0.1 cm long, broadly triangular, glabrous, greenish. Petals 3, elliptic to broadly obovate, pink, 0.5–0.6 × 0.4–0.5 cm, glabrous, shortly acute, mucronate at apex, midrib prominent with 2–4 glandular hairs on abaxial side. Stamens 3; filaments c. 0.3 mm long, filiform, glabrous, pinkish; anthers yellow, cordate at base, shortly acuminate at apex, non-beaked, glabrous. Style 0.7–0.8 cm long, deep pinkish at base; stigma capitate, glabrous. Capsules obcampanulate, pedicellate, 0.3–0.4 cm long, sparsely glandular hairy, green with pink tinge. Seeds many, minute, tubercled, greenish yellow, broadly oblong, raphe prominent, non-excurrent.

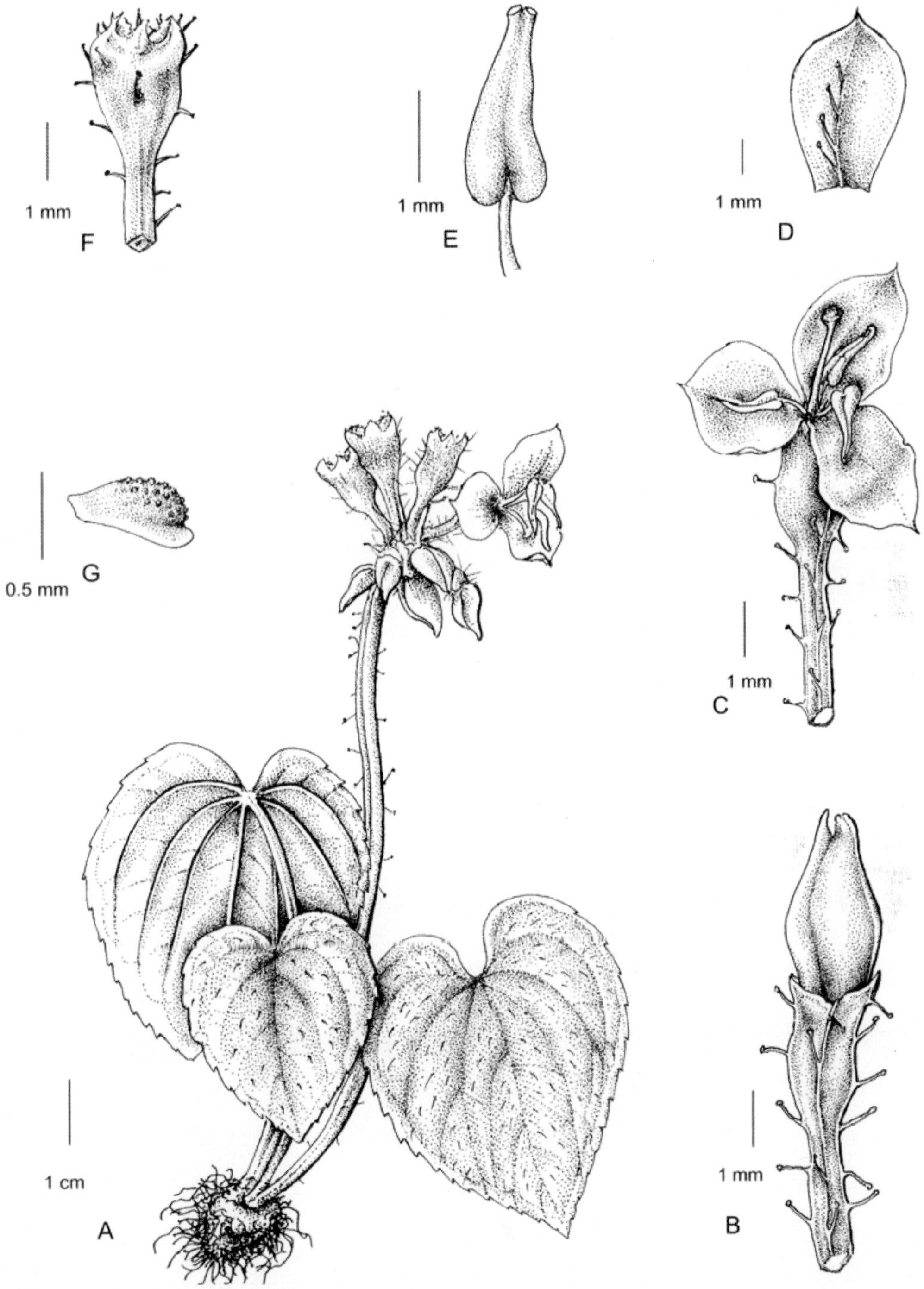

Fig. 1. *Sonerila gadgiliana* Ratheesh & Sivadasan, **sp. nov.** A. Habit; B. Flower bud; C. Mature flower; D. Single petal - abaxial side; E. Stamen; F. Young fruit; G. Seed. (Drawings by T. Shaju from live specimens)

Phenology: Flowering commences from early July with peak in August; fruiting during August–September.

Fig. 2. *Sonerila gadgiliana* Ratheesh & Sivadasan, **sp. nov**. A. Habit; B. Plant with inflorescence; C. Leaves; D. Single flower; E. Flowers, flower buds and young fruit; F. Single flower and fruits - mature and young.

Comparison of morphological characters of *Sonerila gadgiliana* and the related species are provided in Table 1.

Table 1. Comparison of characters of *Sonerila gadgiliana* sp. nov. and related species.

Characters	*Sonerila gadgiliana* sp. nov.	*S. raghaviana*	*S. rotundifolia*	*S. veldkampiana*
Habitat	Shady wet rocks along shola margins	Open grasslands	Shady moist rocky places	Exposed rocky grasslands
Leaf	4–6 cm wide, green, basal lobes not overlapping, 8–10-nerved, 3 pairs of nerves arising from base and 1 or 2 pairs from the midrib, sparsely glandular hairy above, glabrous below	3–6 cm wide, brownish, basal lobes overlapping; 14-nerved, 3 pairs of nerves arising from base of the leaf and 4 pairs from the midrib, densely hairy above, sparsely hairy below	1.5–3.0 cm wide, pinkish, basal lobes not overlapping, 3 or 4 pairs of nerves from near the base, glabrous on both sides	5–7 cm wide, pinkish, basal lobes overlapping, 12-nerved, 4 pairs of nerves arising from base and 2 pairs from the midrib, sparsely hairy on both sides
Inflorescence	Usually 2 per plant, unbranched scorpioid cymes with about 6–12 flowers	Usually 5 or 6 per plant, unbranched scorpioid cymes with more than 12 flowers	Usually 1, unbranched scorpioid cymes with 1–4 flowers	Usually 1–3 per plant, branched or unbranched scorpioid cymes with about 35 flowers
Peduncle	Quadrangular, up to 15 cm long, sparsely glandular hairy	Quadrangular, up to 15 cm long, densely glandular bristly	Angular, up to 8 cm long, sparsely non-glandular hairy	Terete, up to 30 cm long, glabrous
Bract	Not prominent	Not prominent	Not prominent	Prominent, persistent, linear lanceolate
Hypanthium	Obscurely ridged, sparsely glandular hairy	Obscurely ridged, densely glandular bristly	Not ridged, glabrous	Obscurely ridged, glabrous
Petals	5–6 mm long, elliptic to obovate-acuminate, 2–4 glandular hairs on the midrib below	7–9 mm long, broadly obovate, 3–5 glandular hairs on the midrib below	8–9 mm long, orbicular-obovate, glabrous	9–10 mm long, broadly obovate, glabrous
Anthers	Shortly acuminate, non-beaked, deeply cordate at base	Shortly acuminate, non-beaked, deeply cordate at base	Shortly acuminate, non-beaked, truncate at base	Long acuminate, beaked, deeply cordate at base
Capsule	Obcampanulate, crowned by a narrow margin, sparsely glandular hairy	Obcampanulate, crowned by a broad margin, densely bristly	Hemispheric, crowned by a narrow margin, glabrous	Obcampanulate, crowned by a broad margin, glabrous
Seeds	Minutely tubercled; raphae non-excurrent	Prominently tubercled; raphae not prominent	Not tubercled; raphae non-excurrent	Minutely tubercled; raphae sub-excurrent

Etymology: The epithet is named in honour of Prof. Madhav Dhananjaya Gadgil, a renowned Indian ecologist, in recognition of his valuable research on ecology of Western Ghats and contributions to conservation of biodiversity.

Distribution and habitat: *Sonerila gadgiliana* grows on high altitude moss-covered moist rocks in grassland-shola margins at altitudes of 1600–1700 m in contrast to *S. raghaviana* which grows in open grasslands. It is known only from the Banasuramala-Kurichiarmala hill ranges of the Wayanad district, Kerala. Small populations of this species are distributed at shady shola-grassland merging areas of the locality. The species appears during the onset of the South-West monsoon (June–July). Some of the associated species are *Argostemma courtallense* Arn., *Eria nana* A. Rich., *Impatiens lawsonii* Hook. f., *I. scapiflora* B. Heyne *ex* Roxb., *Peristylus spiralis* A. Rich. and *Strobilanthes* sp.

Conservation status: Only two well-separated populations of the species were noticed, one in Kurichiarmala and the other in Banasuramala at altitudes of 1600−1700 m and each population was with less than 50 individuals. Both the habitats were interspersed with patches of grasslands rendering the populations prone to forest fire. Effective efforts are essential to protect the existing populations. In the absence of any detailed observations and data on populations, the species is categorized as 'Data Deficient' (DD) (IUCN, 2012).

Acknowledgements

The authors are grateful to the President and Principal of the Payyanur College, Payyanur; Manager, S.N.M. College, Maliankara; Director, M.S. Swaminathan Research Foundation (MSSRF), Chennai; and the Director, Jawaharlal Nehru Tropical Botanic Garden and Research Institute, Thiruvananthapuram. Sincere thanks are expressed to Dr. J.F. Veldkamp, for critical remarks on the manuscript, and to Ms. V. Mini, Mr. Prajeesh Parameshwaran, Mr. Jayesh P. Joseph and Mr. K.T. Satheesh of the Community Agrobiodiversity Centre, MSSRF, Wayanad for their help. The second and sixth authors thankfully acknowledge the encouragements and support extended by the Deanship of Scientific Research, King Saud University, through the research group project no. RGP-VPP-135.

References

Cellinese, N. 1997. A new *Sonerila* (Melastomataceae) from Central Kalimantan, Borneo. Novon **7**: 103–105.

Clarke, C.B. 1879. *Sonerila*. *In*: Hooker, J.D. (Ed.), The Flora of British India. Vol. **2**. L. Reeve & Co., London. pp. 529–540.

Clausing, G. and Renner, S.S. 2001. Molecular phylogenetics of Melastomataceae and Memecylaceae: Implications for character evolution. Amer. J. Bot. **88**: 486–498.

Deepthikumary, K.P. and Pandurangan, A.G. 2014. A new species of *Sonerila* (Melastomataceae) from the Western Ghats of Kerala, India. Taprobanica **6**: 72–75 + pls. 1–2.

Gamble, J.S. 1919. *Sonerila*. *In*: The Flora of the Presidency of Madras. Vol. **1**, Part 3. Adlard & Son Limited, London, pp. 497–500.

Giri, G.S. and Nayar, M.P. 1985. A new species of *Sonerila* Roxb. (Melastomataceae) from India. Blumea **31**: 235–237.

Giri, G.S. and Nayar, M.P. 1986a. A new species of *Sonerila* (Melastomataceae) from India. J. Jap. Bot. **61**: 344–348.

Giri, G.S. and Nayar, M.P. 1986b (1984, published in 1986). A new species and notes on two species of *Sonerila* Roxb. (Melastomataceae) from South India. Bull. Bot. Surv. India **26**: 174–180.

Giri, G.S. and Nayar, M.P. 1986c (1984, published in 1986). A new variety of *Sonerila sadasivanii* Nayar (Melastomataceae) from southern India. Bull. Bot. Surv. India **26**: 189–190.

Giri, G.S. and Nayar, M.P. 1987a (1985, published in 1987). A new variety of *Sonerila elegans* Wight (Melastomataceae) from Kerala, India. Bull. Bot. Surv. India **27**: 72–74.

Giri, G.S. and Nayar, M.P. 1987b (1985, published in 1987). A new species of *Sonerila* Roxb. (Melastomataceae) from southern India. Bull. Bot. Surv. India **27**: 86–89.

Giri, G.S., Pramanik, A. and Chowdhery, H.J. 1992. A new species of *Sonerila* Roxb. (Melastomataceae) from Arunachal Pradesh, India. Indian J. Forest. **15**: 95–96.

Gopalan, R. and Henry, A.N. 1989. A new *Sonerila* Roxb. (Melastomataceae) from southern India. J. Bombay Nat. Hist. Soc. **86**: 82–83.

Hickey, L.J. 1973. Classification of the architecture of dicotyledonous leaves. Amer. J. Bot. **60**: 1–33.

IUCN 2012. IUCN Red List Categories and Criteria: Version 3.1. 2nd edition. IUCN Species Survival Commission, Gland, Switzerland and Cambridge, pp. iv + 32.

Josephine, M.M., Manickam, V.S., Murugan, C., Sundaresan, V. and Jothi, G.J. 2003. *Sonerila longipetiolata* Manickam *et al.*, a new species of Melastomaceae from Tamil Nadu, India. J. Bombay Nat. Hist. Soc. **100**: 81–83.

Lundin, R. 1983. Taxonomy of *Sonerila* (Melastomataceae) in Ceylon. Nordic J. Bot. **3**: 633–656.

Lundin, R. 1998. Studies in Melastomataceae with special emphasis on the Ceylonese and South Indian *Sonerila* species. Licentiatavhandling, Stockholmsuniversitet, pp. 1–94.

Lundin, R. and Nordenstam, B. 2009. Two new species of *Sonerila* (Melastomataceae) from South India. Novon **19**: 76–79.

Murugan, C. and Manickam, V.S. 2002. New species of *Memecylon* and *Sonerila* (Melastomataceae) from southern Western Ghats of India. J. Econ. Taxon. Bot. **25**: 509–513.

Murugesan, M. and Balasubramaniam, V. 2011. A new species of *Sonerila* (Melastomataceae) from the Western Ghats, India. Taprobanica **3**: 93–95.

Nayar, M.P. 1976. A new species of *Sonerila* (Melastomataceae) from Kerala, S. India. J. Bombay Nat. Hist. Soc. **71**: 632–633.

Prakash, V. and Mehrotra, B.N. 1988. A new species of *Sonerila* Roxb. (Melastomataceae) from Meghalaya, India. Candollea **43**: 143–145.

Ratheesh Narayanan, M.K., Mini, V., Shaju, T., Sujana, K.A., Joseph, J.P., Sivadasan, M. and Alfarhan, A.H. 2013. A new species of *Sonerila* Roxb. (Melastomataceae) from Kerala, India. Phytotaxa **110**: 55–60.

Ratheesh Narayanan, M.K., Sunil, C.N., Nandakumar, M.K., Shaju, T., Mini, V., Satheesh, K.T. and Balakrishnan, V. 2014. *Sonerila raghaviana* sp. nov. (Melastomataceae), a new species from southern Western Ghats of Karnataka, India. Int. J. Advanced Res. **2**: 772–777.

Ravikumar, K. 1999. Novelties from High Wavy Mountains, Southern Western Ghats, Theni District, Tamil Nadu, India. Rheedea **9**: 55–75.

Renner, S.S. 1993. Phylogeny and classification of the Melastomataceae and Memecylaceae. Nordic J. Bot. **13**: 519–540.

Stevens, P.F. 2012. Angiosperm Phylogeny Website. Version 12, July 2012. <http://www.mobot.org/MOBOT/research/APweb>. Retrieved on 13 November 2014.

Sunil, C.N., Naveen Kumar, V.V. and Rajeev, T.S. 2014. *Sonerila sreenarayaniana* (Melastomataceae), a new species from Kerala, India. Int. J. Advanced Res. **2**: 701–704.

HABENARIA NICOBARICA (ORCHIDACEAE), A NEW SPECIES FROM ANDAMAN AND NICOBAR ISLANDS, INDIA

C. MURUGAN[1], JOJU P. ALAPPATT[2], S. PRABHU[3] AND W. ARISDASON[4]

Botanical Survey of India (BSI), Andaman and Nicobar Regional Centre, Port Blair 744 102, Andaman and Nicobar Islands, India

Keywords: Habenaria nicobarica; India; Andaman and Nicobar Islands; Orchidaceae.

Abstract

A new species of terrestrial orchid *Habenaria nicobarica* Murugan, Alappatt, Prabhu & Arisdason sp. nov. is described from South Nicobar Islands. Key to species of *Habenaria* in Andaman and Nicobar Islands, illustration and photograph of the new species are also provided.

Introduction

Orchidaceae include c. 788 genera (Mabberley, 2008) and c. 25,971 species (Joppa *et al.*, 2011) and distributed widely in the world. In India, orchids are represented by 184 genera and 1129 species (Jalal *et al.*, 2008), of which 150 species belonging to 59 genera are found in Andaman and Nicobar Islands.

The Andaman and Nicobar Islands in India is a repository of diverse species, including 25 endemic orchid species (Misra *et al.*, 2012). The genus *Habenaria* Willd. is estimated to have approximately 750 species and distributed in every continent of the world (Comber, 1990). Among these, 100 species are recorded from India (Santapau and Henry, 1973) and only 1 species, namely *H. andamanica* Hook. f., is reported from Andaman and Nicobar Islands (Choudhury *et al.*, 2011).

During the course of botanical explorations in Great Nicobar Biosphere Reserve and Little Nicobar Tribal Reserve, Andaman and Nicobar Islands, from 2008 to 2011, the authors have collected few interesting specimens of *Habenaria*. On critical study of the specimens and perusal of literature (Hooker, 1890; Jayaweera, 1981; Comber, 1990; Chowdhery, 1998; Mathew, 1998; Sinha, 1999; Pandey and Diwakar, 2008), the authors found it as an undescribed species, which is very closely allied to *Habenaria koodersii* J. J. Sm., but differs in leaf, flower and lip characters (Table 1). Hence, it is described here as a new species, *viz., Habenaria nicobarica* along with key to species, illustration and photograph, for further collection and identification in the field.

Habenaria nicobarica Murugan, Alappatt, Prabhu & Arisdason, **sp. nov.** **(Figs 1, 2 b-e).**

Type: INDIA, Andaman and Nicobar Islands: South Nicobar, Little Nicobar Tribal Reserve, Pulopaha (E), 25.11.2008, *C. Murugan* 26630 (*Holotype:* CAL; *Isotype*: PBL).

Diagnosis: *Habenaria nicobarica* is closely related to *H. koodersii* J. J. Sm. but differs in bract size, flower colour and lip characters.

[1]Corresponding author. E-mail: sivanthimurugan@rediffmail.com

[2]Forest Training Institute, Wimberligunj, Port Blair-744 206, South Andaman, Andaman and Nicobar Islands, India

[3]Botanical Survey of India (BSI), Andaman and Nicobar Regional Centre, Port Blair-744 102, Andaman and Nicobar Islands, India

[4]Botanical Survey of India, CGO Complex, Salt Lake City, Kolkata-711 064, West Bengal, India

Tuberous, terrestrial herbs, up to 1 m high. Stems terete. Leaves spiral, 5-8, confined to the middle of stem, oblong-lanceolate, 7-12 × 2-3 cm, rounded with auricled at base, entire at margin, acuminate at apex, 3-veined. Racemes terminal, 10-20 cm long, lax; peduncles terete, 10-25 cm long. Flowers lax, light brown; bracts linear, acuminate at apex, 1.5-2.0 × 0.2-0.3 cm. Dorsal sepal 10-17 mm long with 7 mm long filiform extension, 3-nerved; lateral sepals spreading, c. 10×4 mm with 9 mm long tail, 3-nerved. Petals 2, bilobed; posterior lobe shorter than sepals, erect, linear, c. 10.0 × 0.1 mm; anterior lobe short, acuminate at apex. Lip 3-lobed to the base, mid lobe c. 12 × 2 mm, flat; lateral lobe linear, c. 10 × 1 mm; spur cylindric, narrowed towards apex, up to 27 mm long. Gynostemium c. 4 mm long. Anther canal c. 2 mm long; lateral rostellar arms as long as anther canals, projecting forward; central lobe indistinct; auricle 2, prominent. Stigma processes large, c. 2.0 × 1.5 mm, rounded at apex. Ovary obconic, c. 2.3 × 0.2 cm, trigonous. Capsules oblong, c. 2.0 × 0.5 cm.

Flowering and fruiting: October - April.

Distribution: India (Nicobar Islands).

Paratypes: INDIA, Andaman and Nicobar Islands: S. Nicobar, Little Nicobar Tribal Reserve, Pulo ulon (N), 26.11.2008, *C. Murugan* 26680 (PBL); Pulo Panja (N), 31.10.2009, *C. Murugan* 27828 (PBL). Pulopaha, 8.4.2011, *C. Murugan* 28322 (PBL); Great Nicobar Biosphere Reserve, 10 km on East-west Road, 31.10.2011, *Joju P. Alappatt* 337 (PBL).

Etymology: The species is named after the Nicobar Islands, the type locality and one of the plant diversity hotspots in India.

Table 1. Comparison of exomorphic characters among *Habenaria koodersii*, *H. nicobarica* sp. nov. and *H. andamanica*.

Characters	*Habenaria koodersii*	*H. nicobarica* **sp. nov.**	*H. andamanica*
Plant height	40-60 cm	60-100 cm	70-80 cm
Number of leaves	7-9	5-8	7
Leaf size	13.5-17.0 cm long	7-12 cm long	5-15 cm long
Inflorescence	21-31 cm long	10-20 cm long	25-40 cm long
Peduncle	12-19 cm long	10-25 cm long	5-7 cm long
Bract	c. 2.8 cm long	1.5-2.0 cm long	2-3 cm long
Flower colour	Greenish white	Light brown	White with brown
Dorsal sepal	1.1 cm with c. 4 mm long thread	1.0-1.7 cm with c. 7 mm long thread	1.2 cm without thread
Petals	1.0-1.1 cm long, unlobed at base.	1 cm long, lobed at base	1.0-1.4 cm long, unlobed at base
Lip	Mid lobe shorter than side lobe	Mid lobe (c. 1.2 cm long) longer than side lobe	Mid lobe linear, 3-fid (c. 1.5 cm long) longer than side lobe
Spur	Tubular, narrow towards apex	Tubular, c. 2.7 cm long narrow towards apex	clavate

Key to species of *Habenaria* Willd. in Andaman and Nicobar Islands.

1 Flowers white; sepals without filiform attachment near apex; lip laciniate. ... *H. andamanica*

- Flowers light brown; sepals with filiform attachment near apex; lip 3-fid. ... *H. nicobarica* **sp. nov.**

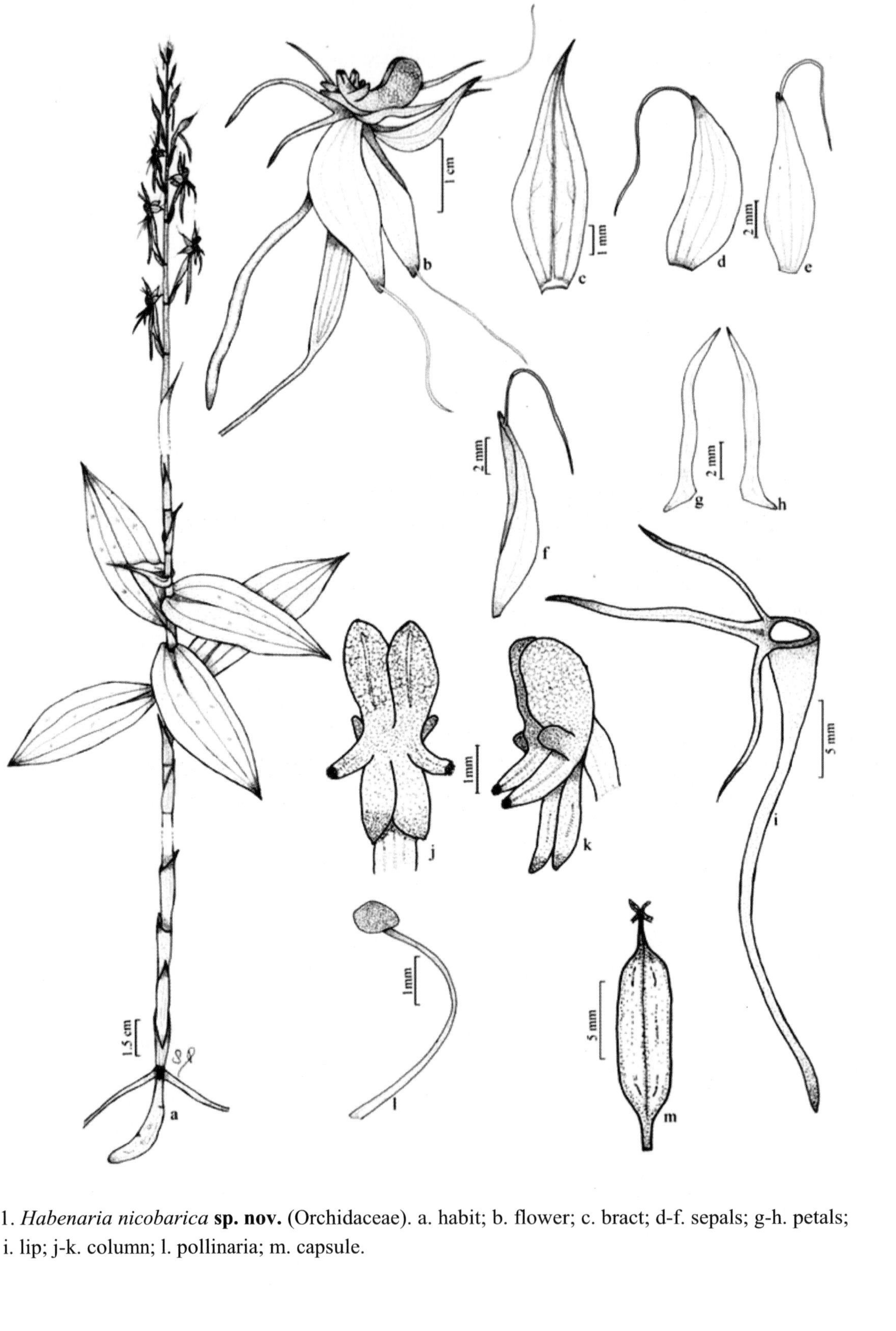

Fig. 1. *Habenaria nicobarica* **sp. nov.** (Orchidaceae). a. habit; b. flower; c. bract; d-f. sepals; g-h. petals; i. lip; j-k. column; l. pollinaria; m. capsule.

Fig. 2a. *Habenaria andamanica* Hook. f. (Habit with inflorescence); b-e. *Habenaria nicobarica* **sp. nov.** b. habit; c. inflorescence; d. flower; e. capsule.

Acknowledgements

The authors are thankful to Dr. M. Sanjappa, former Director and Dr. P. Singh, Director, Botanical Survey of India, Kolkata, for providing facilities and constant support. They are also grateful to Shri H.J. Chowdhery, former Secretary & PCCF, Shri D.V. Negi, former PCCF (WL), Dr. Shashi Kumar, Secretary & PCCF, the former Divisional Forest Officers, Dr. P. Viswakannan, and Dr. S. Dinesh Kannan, Department of Environment and Forests, Andaman and Nicobar Islands for their permission and encouragement during the field studies and Dr. Huber Kurzweil, Singapore Herbarium, Singapore, for confirmation of novelty.

References

Choudhury, S., Mukheerjee, S.K. and Chowdhery, H.J. 2011. Distribution and diversity of the genus *Habenaria* Willdenow in India. *In*: Ghosh, C. and Das, A.P. (Eds), Recent Studies in Biodiversity and Traditional Knowledge in India. Sarat Book House, Kolkata, pp. 81-90.

Chowdhery, H.J. 1998. Orchid Flora of Arunachal Pradesh. Bishen Singh Mahendra Pal Singh, Dehra Dun, India, pp. 440-452.

Comber, J.B. 1990. Orchids of Java. Royal Botanic Gardens, Kew, pp. 60-66.

Hooker, J.D. 1890. Orchidaceae. *In*: Hooker, J.D. (Ed.), The Flora of British India, Vol. **6**. L. Reeve & Co. Ltd., London, pp. 1-198.

Jalal, J.S., Kumar, P., Rawat, G.S. and Pangtey, Y.P.S. 2008. List of species. Orchidaceae, Uttarakhand, Western Himalaya, India. Check List **4**: 304-320.

Jayaweera, D.M.A. 1981. Orchidaceae. *In*: Dassanayake, M.D. (Ed.), Revised Handbook to the Flora of Ceylon, Vol. **2**. Oxford & IBH Publishing Co. Ltd., New Delhi, pp. 4-388.

Joppa, L.N., Roberts, D.L. and Pimm, S.L. 2011. How many species of flowering plants are there? Proc. R. Soc. B **278**: 554-559.

Mabberley, D.J. 2008. Mabberley's Plant Book - A portable dictionary of Plants, their Classification and Uses (3rd Edition.). Cambridge University Press, Cambridge.

Mathew, S.P. 1998. A supplementary report on the flora and vegetation of the Bay Islands, India. J. Econ. Taxon. Bot. **22**: 249-272.

Misra, S., Nayakm P.H. and Panda, S.P. 2012. *Aerides rosea* Lodd. *ex* Lindl. & Paxton (Orchidaceae) – A new record from the Andaman and Nicobar Islands, India. J. Bombay Nat. Hist. Soc. **109**: 226-229.

Pandey, R.P. and Diwakar, P.G. 2008. An integrated checklist of Andaman and Nicobar Islands, India. J. Econ. Taxon. Bot. **32**: 403-500.

Santapau, H. and Henry, A.N. 1973. A Dictionary of the Flowering Plants in India. Council of Scientific & Industrial Research, New Delhi.

Sinha, B.K. 1999. Orchidaceae. *In*: Hajra, P.K. and Rao, P.S.N. (Eds), Flora of Great Nicobar Islands. Botanical Survey of India, Calcutta, pp. 420-447.

4

A NEW SPECIES OF *HYDROCOTYLE* L. (ARALIACEAE) FROM INDIA

S. Karuppusamy, M. Ajmal Ali[1], K. M. Rajasekaran, Joongku Lee[2], Soo-Yong Kim[2], Arun K. Pandey[3] and Fahad M. A. Al-Hemaid[1]

Department of Botany, The Madura College, Madurai-625011, Tamil Nadu, India

Keywords: Hydrocotyle kollimalayensis; India; ITS; New species; nrDNA.

Abstract

Hydrocotyle kollimalayensis, a new species is described and illustrated from Kolli hills of South Eastern Ghats, Tamil Nadu, India. The new species is morphologically closer to *H. sibthorpioides*, but differs from the latter by its filiform stem, glabrous peduncle, shorter petiole and narrower flowers. The most prominent features of the new species *Hydrocotyle kollimalayensis* are: stem filiform with silky setaceous hairs; leaves sparsely hirsute above and dense below, 5-lobed; flowers 6-12 per inflorescence with glabrous peduncle; and fruits ellipsoidal with 3 obscure ribs. A key to distinguish the new species from other *Hydrocotyle* species of Tamil Nadu, India is provided. We also inferred the relationship of *H. kollimalayensis* with allied species using molecular phylogenetic analyses based on nrDNA ITS sequence data.

Introduction

Hydrocotyle L. (Araliaceae) comprises some 130 species worldwide (Du and Ren, 2010). Most of them inhabit marshy, moist shady and understory environments in tropical and subtropical forests. The genus *Hydrocotyle* possess small, bisexual regular flowers on pedunculate, axillary or terminal umbel, and the plant body is covered with hispid hairs (Hiroe, 1979; Pimenov and Leonov, 1993). Karthikeyan *et al.* (2009) recognized eight taxa of *Hydrocotyle* from India i.e. *Hydrocotyle conferta* Wight, *H. hookeri* (C.B. Clarke) Craib, *H. javanica* var. *hookeri* C.B. Clarke, *H. nepalensis* Hook., *H. podantha* Molk., *H. ramiflora* Maxim., *H. siamica* Craib and *H. sibthorpioides* Lam.

While the first author was carrying out a survey of plant diversity across Eastern Ghats region of India in 2009, he came across some populations (about 90 patches in about 2 sq. km) of *Hydrocotyle* in Solakkadu area of Kolli hills with novel characters in the leaves, inflorescence, flowers and fruits, which were different from the other known species of *Hydrocotyle*. This led us to carry out detailed morphological and molecular phylogenetic studies for its proper identity and comparison with its allied species. The results revealed that these specimens belong to a hitherto undescribed species that is morphologically allied to *H. sibthorpioides*, *H. conferta* and *H. javanica*. Based on extensive morphological and molecular studies, we herein describe and illustrate the species as a new entity, *Hydrocotyle kollimalayensis* S. Karup. & A. Ali.

Materials and Methods

The morphological characters of *Hydrocotyle kollimalayensis* were critically studied, and further compared with related taxonomic literature (Clarke, 1879; Gamble, 1935; Mathias, 1936; Hiroe, 1979; Matthew, 1983, 1996; Mukherjee and Constance, 1993; Pullaiah *et al.*, 2007), and herbarium specimens housed at Botanical Survey of India, Central National Herbarium, Howrah

[1]Department of Botany and Microbiology, College of Science, King Saud University, Riyadh-11451, Saudi Arabia. Corresponding author. Email: majmalali@rediffmail.com

[2]International Biological Material Research Center, Korea Research Institute of Bioscience and Biotechnology, Daejeon-305806, South Korea

[3]Department of Botany, University of Delhi, Delhi-110007, India

(CAL), Madras Herbarium, Southern Regional Center, Coimbatore (MH), Sikkim Himalayan Regional Center, Gangtok (BSHC), Arunachal Pradesh Regional Center, Itanagar (ARUN), Eastern Regional Center, Shillong (ASSAM), and Northern Regional Center, Dehradun (DD).

For molecular study, the taxa along with sources and GenBank accession numbers are provided in Appendix 1. Leaf sample of *H. kollimalayensis*, *H. sibthorpioides, H. javanica* and *H. conferta* were collected during various plant exploration trips in Tamil Nadu, India. All the voucher specimens including the holotype of *H. kollimalayensis* have been deposited at the Madras Herbarium (MH). The leaf materials were fixed in silica gel. Total DNA was extracted using the DNeasy Plant Mini Kit (Qiagen, Valencia, CA, USA). The nuclear ribosomal DNA (nrDNA) Internal Transcribed Spacer (ITS) region was amplified using the primers ITS1 and ITS4 (White *et al.*, 1990). PCR products were purified using SolGent PCR Purification Kit-Ultra (Solgent, Daejeon, South Korea) and sequenced employing the primers ITS1 and ITS4 in 10 μl reactions including 2 μl BigDye, 1 μl primers (20 pM) and template DNA, and deionized water to reach the final reaction volume. Cycle sequencing used 25 cycles of 96°C for 10 s, 50°C for 5 s, and 60°C for 4 min. Sequencing products were visualized on an ABI Prism 377 automated DNA sequencer. Each sample was sequenced in both the sense and anti-sense direction. The sequences were analyzed by the ABI Sequence Navigator software (Perkin-Elmer/Applied Biosystems). Nucleotide sequences of both the DNA strands were obtained and compared with the forward and reverse sequence to ensure the accuracy.

The nrDNA ITS sequences of *Hydrocotyle* species available in the GenBank were retrieved (Appendix 1) for comparison and phylogenetic analysis. The GenBank retrieved sequence of *Centella asiatica* (L.) Urban was selected as outgroup in the analysis as a follow up of Choi and Park (2012). Sequences were aligned using Clustal X (Thompson *et al.*, 1997). The aligned sequence was subsequently adjusted manually using BioEdit (Hall, 1999). All sequences generated in the present study were deposited in GenBank. We constructed phylogenies using the Maximum Parsimony (MP) method implemented in MEGA v. 4.0 (Tamura *et al.*, 2007). In the analyses, gaps were treated as missing data. Support for internal nodes was assessed using bootstrap analysis (Felsenstein, 1985) of 1000 replicates with 100 random additions per replicate and holding 10 trees at each step.

Results and Discussion

Hydrocotyle kollimalayensis S. Karup. & A. Ali, **sp. nov.** **(Fig. 1**).

Diagnosis: *Planta H. sibthorpioides simulans; caulibus filiformis, sparsim pilosis, folis angulatis, supra sparsim pilosis secus nervos villosis, subtus dense villosis vel molliter hirsutis, profunde angulate 5-lobatis, lobis triangulates, crenatis acutis, acuminatisque duplo dentatibus obtusis, pedunculis quam foliis multo brevioribus, umbellis simplicibus globosis 6-12-floris, pedicillis glabris, petalis albis, stylis divaricatis, stylopodio depresso, fructibus ellipsoidus glabratis, costis obscures.*

Type: India. Tamil Nadu, Kolli hills, Solakkadu, 78°17'-78°27' E longitude, 11°55'-11°21' N latitude, 1250 m, 14 Jul 2009, *S. Karuppusamy 24375* (*Holotype:* MH).

Perennial herbs. Stem weak, slender, filiform, silky setaceous, creeping, rooting at nodes. Leaves simple, pale green, sparsely hirsute above and dense below, membranous, base cordate, 1.5-2.0 cm wide, 1.0-1.5 cm long, angular, margin shallowly 5-lobed, lobes acute or obscurely triangular, middle lobe larger than the others, crenate, subequal. Petioles slender, 2-4 cm long, with white reflexed hirsute hairs above. Umbels simple, 6-12 flowered, globose; peduncle 1.0-2.5 cm long, glabrous, axillary, slender. Flowers sub-sessile or very shortly pedicillate. Petals ovate, white, entire, minute, c. 0.2 mm long, triangular, acute. Stamens 5, minute, slightly incurved in bud; filaments short, c. 0.1 mm long; anthers bilobed. Ovary minute, ellipsoidal, greenish; style 1, persistent; stylopodium slightly elevated. Fruits ellipsoid, compressed, with obscure dorsal edges, glabrous, 3–ribbed, ribs obscure, subequal, c. 1.2 mm long and c. 1 mm wide, glabrous above, minutely granulate at base, pale brown.

Phenology: Flowering and fruiting occurs almost throughout the year.

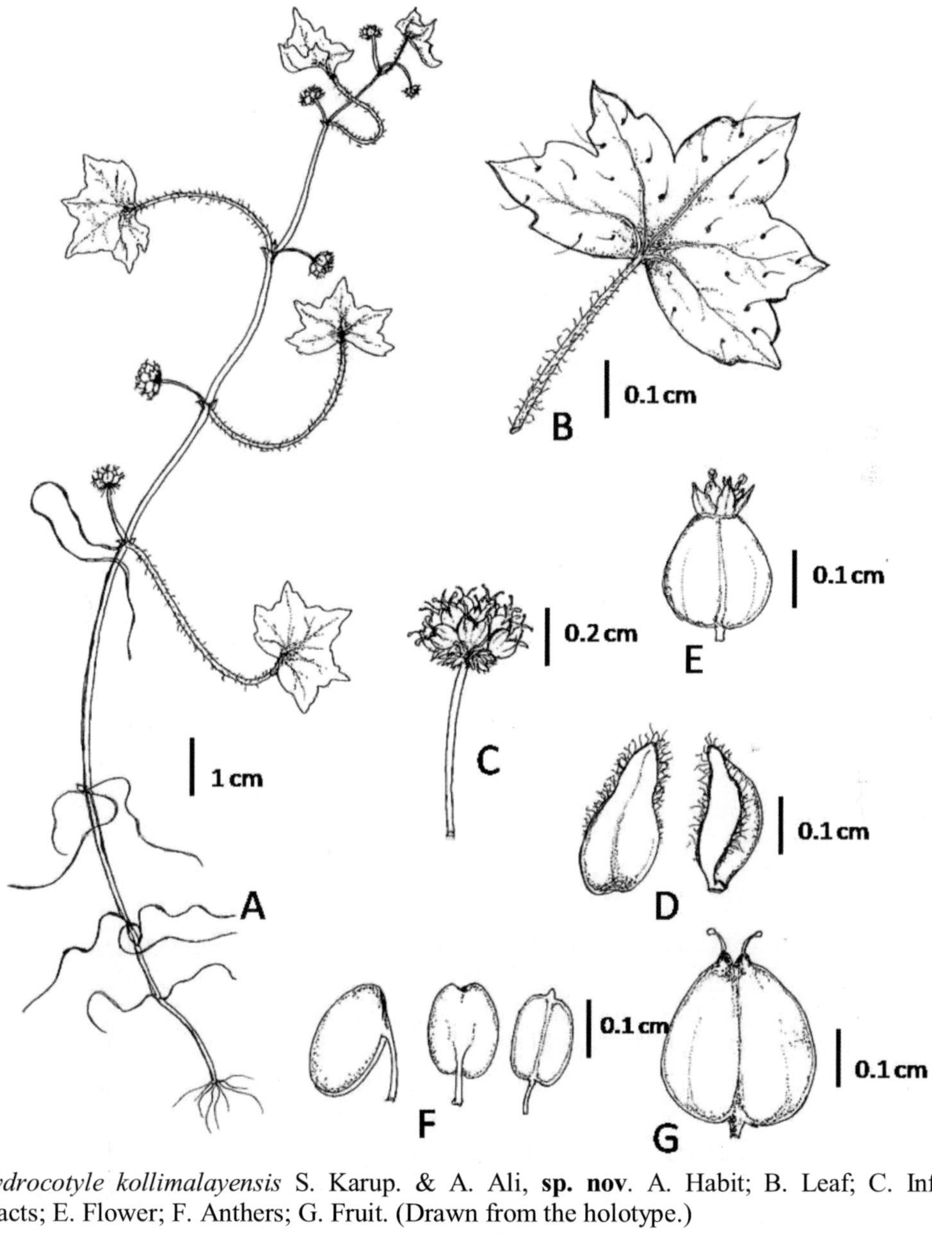

Fig. 1. *Hydrocotyle kollimalayensis* S. Karup. & A. Ali, **sp. nov**. A. Habit; B. Leaf; C. Inflorescence; D. Bracts; E. Flower; F. Anthers; G. Fruit. (Drawn from the holotype.)

Distribution: Hydrocotyle kollimalayensis is known from Solakkadu, near observatory, Kolli hills of Southern Eastern Ghats in Tamil Nadu, India (Fig. 2). It grows in between 1200 to 1350 m elevation.

Etymology: The new species is named after the type locality Kollimalai, from where we collected the plant populations.

Conservation status: The new species, according to IUCN Red list category (IUCN, 2012), is considered under 'Endangered' category (EN, criteria B).

Notes: Hydrocotyle kollimalayensis is morphologically similar to *H. sibthorpioides*, however, it differs from the latter by its filiform stem with silky setaceous hairs, more number of flowers per inflorescence, glabrous peduncle and obscure ribs on fruits. A comparative account of *H. kollimalayensis* with its closely related species *H. sibthorpioides, H. conferta* and *H. javanica* is presented in Table 1. The morphological characters which are used to delimit the species of *Hydrocotyle* include habit, stem, leaves, flowers and fruits. However, the genus *Hydrocotyle* shows much variation in morphological characters, which often creates difficulties in demarcating the taxonomic ranks at or below the species level (Hiroe, 1979). Earlier floristic reports (Gamble, 1935; Matthew, 1983) from Tamil Nadu (India) represented only three species of *Hydrocotyle* L. i.e. *H. conferta, H. javanica* and *H. sibthorpioides. Hydrocotyle kollimalayensis* is allied to *H. sibthorpioides* which can be evidenced from the morphological comparison (Table 1).

Table 1. Comparison of diagnostic morphological characters of *Hydrocotyle kollimalayensis* sp. nov. with its allied species.

Characters	*H. kollimalayensis*	*H. sibthorpioides*	*H. conferta*	*H. javanica*
Stem	Filiform	Wiry	Wiry	Wiry
Stem hairs	Silky setaceous	Hirsute	Hirsute	Hirsute
Leaves	5-lobed	5-7-lobed	7-9-lobed	7-11-lobed
Leaf indumentum	Sparse hirsute above and dense below	Hirsute on both surface	Hirsute on both surface	Hirsute on both surface
Petiole	2-4 cm long	2-3 cm long	2-6 cm long	5-15 cm long
No. of flowers per inflorescence	6-12	3-10	10-20	20-30
Peduncle	Glabrous	Hirsute	Hirsute	Hirsute
Flowers	0.1 mm wide	0.2 mm wide	0.3 mm wide	0.5 mm wide
Fruits	Ellipsoidal	Ellipsoidal	Suborbicular	Suborbicular
Ribs on fruits	3, obscure	3, distinct	3, obscure	5, distinct

Phylogenetic relationship

The phylogenetic relationship of *Hydrocotyle kollimalayensis* with its allied species is presented in Figure 3. The combined length of the entire ITS region (ITS1, 5.8S and ITS2) in the species included ranged from 604–625 bp. The length of ITS1 region ranged from 208-230 bp, the 5.8S gene was 161 bp, and the length of ITS2 region varied from 227-237 bp. In *H. kollimalayensis*, the combined length of the ITS region was 606 bp (the length of ITS1 and ITS2 region was 210 bp and 235 bp, respectively). Aligned data matrix has a total number of 656 characters of which 450 characters were constant, 113 characters were variable but parsimony-uninformative, and 99 were parsimony-informative. Insertions and deletions (indels) were necessary to align the sequences. Indels ranged from 1-17 bp. The parsimony analysis of the entire ITS region resulted into 14 Maximum Parsimony Trees (MPTs) with a length of 88 steps, a consistency index (CI) of 0.711, a homoplasy index (HI) of 0.201, rescaled consistency index (RC) of 0.503, and a retention index (RI) of 0.727.

In MPTs, *H. kollimalayensis* shows proximity (bootstrap support 86%) with *H. conferta* and *H. javanica*, and is clearly distinct from *H. sibthorpioides* (Fig. 3). A comparison of nrDNA ITS sequence of *H. kollimalayensis* with the allied species *H. javanica, H. sibthorpioides* and *H. conferta* reveals the differences of 25, 33 and 42 bp, respectively.

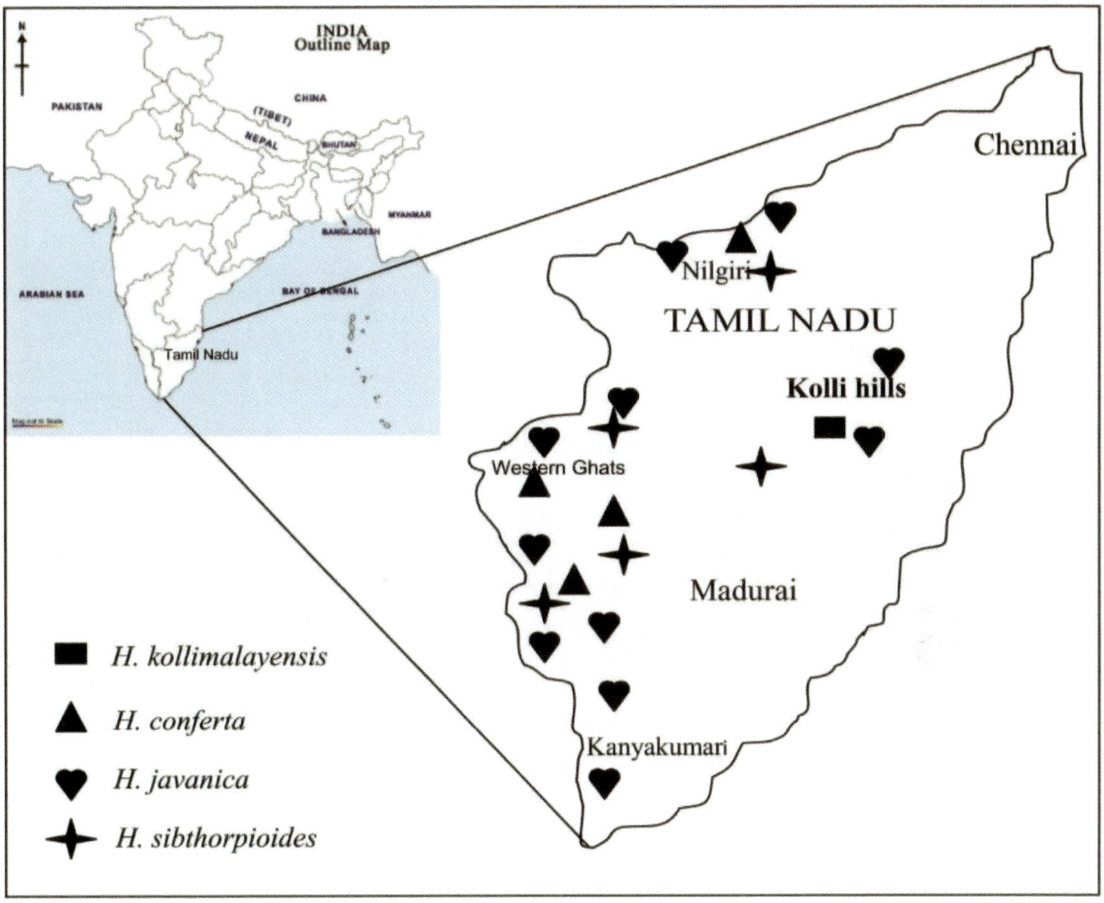

Fig. 2. Distribution map of *Hydrocotyle kollimalayensis* and its related species in Tamil Nadu, India.

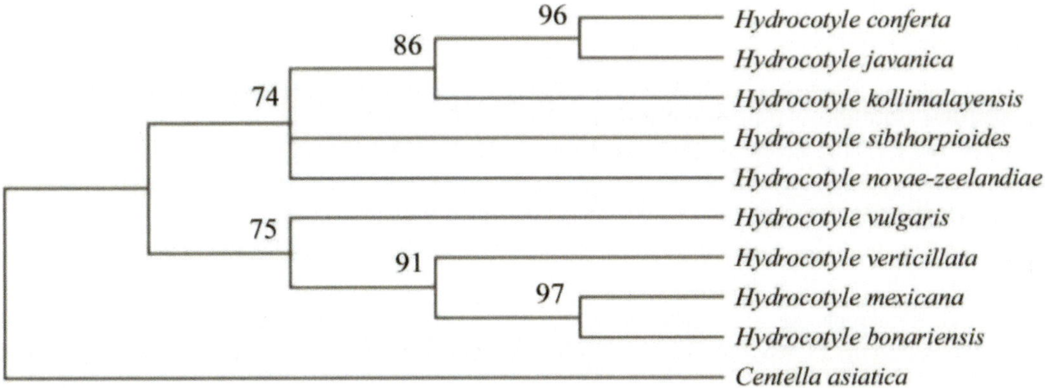

Fig. 3. The bootstrap strict consensus tree of 14 maximally parsimonious trees of *Hydrocotyle* inferred from ITS sequences of nrDNA data. Numbers above the line indicate bootstrap values in 1000 bootstrap replicates.

Key to the species of *Hydrocotyle* L. in Tamil Nadu, India

Subsequent to the discovery of the new species *Hydrocotyle kollimalayensis*, the total number of species of *Hydrocotyle* in Tamil Nadu, India has been increased to four. A key is provided herewith to facilitate easy identification of the taxa.

1. Stem wiry; Leaves orbicular, reniform; peduncle hirsute; flowers 0.2-0.5 mm wide.
    2. Umbel solitary on axillary nodes; mericarps compressed; ribs 3 on fruits.
        3. Leaves > 1.5 cm in diameter; peduncle very short or sessile. — *H. conferta*
        3. Leaves < 1.5 cm in diameter; peduncle c. 1.2 cm long. — *H. sibthorpioides*
    2. Umbel clustered on terminal and subterminal nodes; mericarps slightly compressed; ribs 5 on fruits. — *H. javanica*
1. Stem filiform; leaves angular, not reniform; peduncle glabrous; flowers c. 0.1 mm wide. — *H. kollimalayensis*

Acknowledgement

We are thankful to the curators and staffs of MH, CAL, BSHC, ARUN, ASSAM and DD for providing facility for herbarium and library consultation. Research supported by the King Saud University, Deanship of Scientific Research, College of Science, Research Center.

References

Choi, K.S. and Park, S. 2012. Molecular phylogenetic studies of Korean *Hydrocotyle* L. Korean J Plant Res. **25**(4): 490-497.

Clarke, C.B. 1879. *Hydrocotyle. In*: Hooker, J.D., Flora of British India, **2**: 667-669. Reeve & Co., England.

Du, X.C. and Ren, Y. 2010. *Hydrocotyle changanensi* (Araliaceae), a new species from Shaanxi, China. Ann. Bot. Fenn. **47**: 403-407.

Felsenstein, J. 1985. Confidence limits on phylogenies: An approach using the bootstrap. Evolution **39**: 783-791.

Gamble, J.S. 1935. *Hydrocotyle. In:* Flora of the Presidency of Madras. Adlard & Son, London, pp. 555-556.

Hall, T.A. 1999. BioEdit: a user-friendly biological sequence alignment editor and analysis program for Windows 95/98/NT. Nucleic Acids Symp. Ser. **41**: 95-98.

Hiroe, M. 1979. *Hydrocotyle. In:* Umbelliferae of World. Ariake Book Company, Matsuo Biru, Tokyo, Japan, pp. 103-168.

IUCN 2012. IUCN Red List Categories and Criteria: Version 3.1. Second edition. Gland, Switzerland and Cambridge, UK: IUCN. iv + 32 pp.

Karthikeyan, S., Sanjappa, M. and Moorthy, S. 2009. Apiaceae [Umbelliferae]. Flowering plants of India, Dicotyledons. Vol. **I** (Acanthaceae–Avicenniaceae). Botanical Survey of India, Howrah, India, pp. 98-127.

Mathias, M.E. 1936. The genus *Hydrocotyle* in northern South America. Brittonia **2**: 201-237.

Matthew, K.M. 1983. Apiaceae. *In:* Flora of Tamil Nadu Carnatic, Vol. **1**. Rapinet Herbarium, Tiruchirapalli, Tamil Nadu, India, pp. 815-827.

Matthew, K.M. 1996. Apiaceae. *In:* Flora of Palni hills, Vol. **1**. Rapinet Herbarium, Tiruchirapalli, Tamil Nadu, India., pp. 757-765.

Mukherjee, P.K. and Constance, L. 1993. Umbelliferae (Apiaceae) of India. Oxford & IBH Pub. Co. Ltd., New Delhi, pp. 279-286.

Pimenov, M.G. and Leonov, M.V. 1993. The Genera of the Umbelliferae: A nomenclature. Royal Botanic Gardens, Kew, pp. 5-161.

Pullaiah, T., Ramamurthy, K.S. and Karuppusamy, S. 2007. Flora of Eastern Ghats, Hill ranges in South India, Vol. **3**. Regency Publications, New Delhi, India, pp. 125-127.

Tamura, K., Dudley, J. Nei, M. and Kumar, S. 2007. MEGA 4: Molecular evolutionary genetics analysis (MEGA) software, version 4.0. Mol. Biol. Evol. **24**: 1596-1599.

Thompson, J.D., Gibson, T.J., Plewniak, F., Jeanmougin, F. and Higgins, D.G. 1997. The Clustal X windows interface: flexible strategies for multiple sequence alignment aided by quality analysis tools. Nucleic Acids Research **24**: 4876-4882.

White, T.J., Bruns, T., Lee, S. and Taylor, J. 1990. Amplification and direct sequencing of fungal ribosomal RNA genes for phylogenetics. *In*: Innis, M., Gelfand, D., Sninksky, J. and White, T. (Eds), PCR protocols: a guide to method and application. Academic Press, San Diego, California, USA, pp. 315-322.

5

NITELLA ZAMANII SP. NOV. (CHAROPHYTA) FROM BANGLADESH

NASRIN JAHAN DIBA, SABRINA NAZ[1] AND HENDRIK SCHUBERT[2]

Department of Botany, University of Rajshahi, Rajshahi-6205, Bangladesh

Keywords: New species; *Nitella zamanii*; Characeae; Bangladesh.

Abstract

Nitella zamanii has been described and illustrated as a new species from a rice field at Charghat upazila in Bangladesh. The new species resembles *Nitella furcata* and *N. polycarpa* but differs by presence of accessory branchlets, spiky dactyls and size of oogonium, oospore and antheridium.

Introduction

Nitella Ag. had been described by different researchers (Islam and Sarma, 1976; Aziz and Tanbir, 2003; Islam and Irfanullah, 2005; Aziz, 2009; Naz *et al.*, 2009; Diba and Naz, 2011) from different parts of Bangladesh and morphologically described with their habitats. A new species of *Nitella* is presented in this paper with illustration. This *Nitella* species was collected from a rice field at Rajshahi district in Bangladesh. A total of seventeen taxa of *Nitella* have been so far reported from Bangladesh (Naz *et al.*, 2011).

Specimens were collected from only one location and 10 cm depth of water by hand and preserved in Transeau`s solution (Transeau, 1916).

Class: Charophyceae; Order: Charales; Family: Characeae; Genus: *Nitella* Ag.

Nitella zamanii Naz, Diba & Schubert, **sp. nov.** **(Figs 1-14; Pl. 1)**.

Diagnosis: *Nitella zamnii* is closely related to *N. furcata* and *N. polycarpa* but differs from them by presence of accessory branchlets, spiky dactyls, size and position of gametangia, size of oogonium and oospore, and rare occurrence of antheridium.

Type: Bangladesh, Rajshahi district, Charghat, Chamta beel, rice field, growing on soft mud, 10 cm of water depth, 4 January 2004, Nasrin Jahan Diba N16 (*Holotype*: PLLBDRB).

Plant monoecious, height 5-6 cm, spiky in appearance, mucus absent, slightly encrusted; lower whorls slightly spreading, stem 250 μm in diameter; internodes 1-2 cm, branchlets 5-8, 172 μm in diameter, height 1-2 cm; furcation irregular; branchlet two types: short accessory branchlet and long branchlets; accessory branchlet few, sometimes curved, primaries 3, of which one is not furcate, two rays again forked into 1 or 2, secondaries two, unequal; dactyls 2-celled, acute; a long unbranched branchlet arise from the base of a whorl and primaries 2-celled, dactyls 2 or 3, unequal, short, acute; long branchlet 1-3 times furcate, primaries 1/3 of the total branchlet, 4-6 in number, forked into 3, 4, 6 secondaries, few are again forked into 3 or 4 tertiaries; dactyls 1-4, different types, abbreviated, spine like, spreading, tapering, 1 or 2 celled, penultimate cell tapering distally to base of end cell, sometimes acuminate, ultimate cell acute; gametangia conjoined;

[1]Corresponding author: Email: drsabrina_naz@yahoo.com

[2]Institute of Bioscience, University of Rostock, Albert-Einstein-Str. 3, D-18059, Rostock, Germany

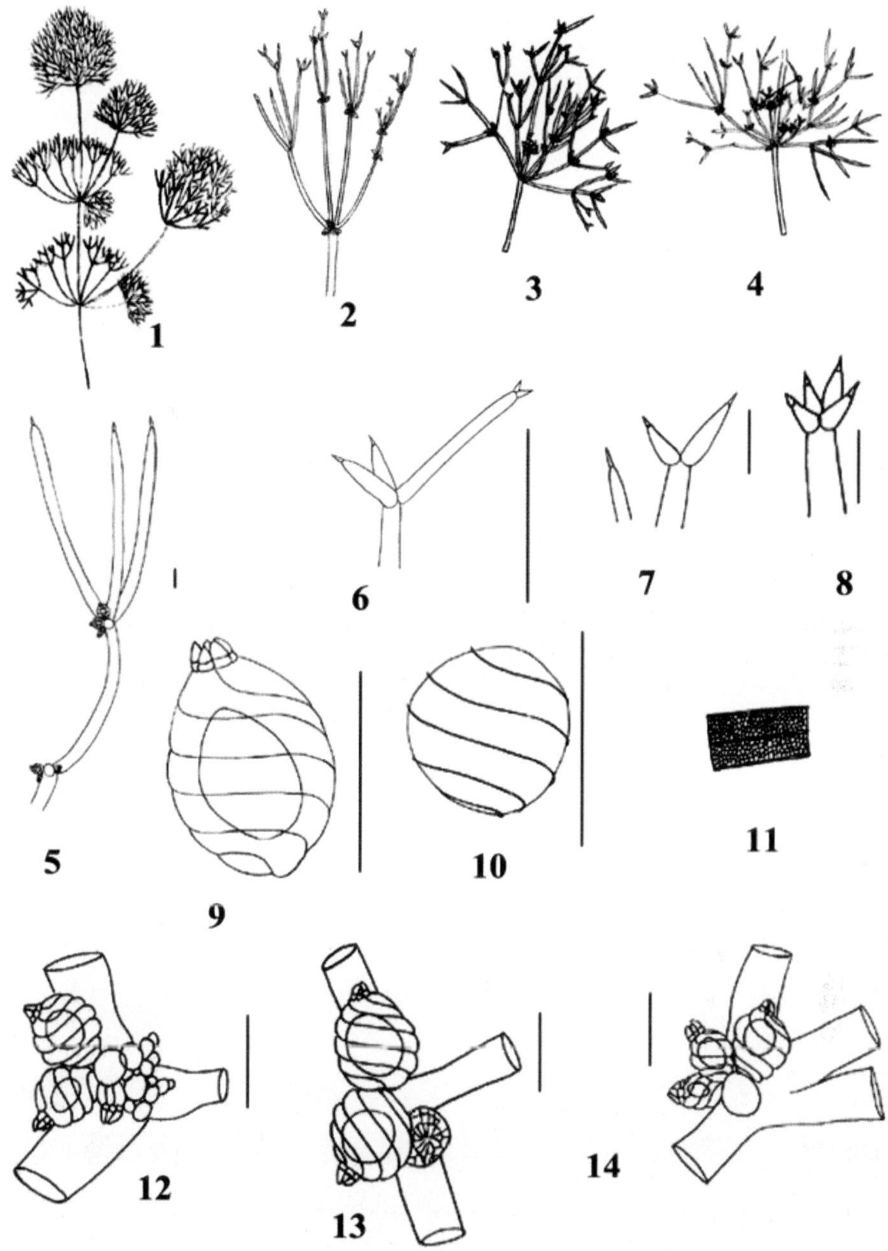

Figs 1-14. *Nitella zamanii* Naz, Diba & Schubert, **sp. nov.** 1. Habit; 2. Upper portion of a plant; 3-4. A whorl; 5. Upper portion of a branchlet; 6–8. Dactyls; 9. Oogonium; 10. Oospore; 11. Oospore membrane; 12–14. Branchlet node with aggregated gametangia. (Scales: 1–4 & 11 = freehand drawing; rest = 0.2 mm)

oogonium usually aggregated in all branchlet nodes and at the base of whorl; lateral, short whorl arise from all stem nodes, oogonium 63-287 µm long, 35-208 µm wide, convolutions 6-9; corona two tiers, 39-40 µm long, 40-63 µm wide at the base, upper cell elongated, acute, sometimes unequal; oospore globose, 29-198 µm long, 22-165 µm wide, with 5 prominent ridges; membrane reticulate, antheridia rare, 185 µm in diameter.

Habitat: *Nitella zamanii* was collected from a shallow water habitat. This low land was cultivated with different crops, e.g. rice, wheat and sugarcane, in different seasons. After rainy season, charophytes densely grow in this location. Water depth ranged 5-15 cm. *Nitella zamanii* was found associated with *Nitella furcata* subsp. *furcata* f. *sieberi* (A. Br.) R.D.W.

Distribution: Only known from Chamta beel, Charghat Upazila, Rajshahi, Bangladesh

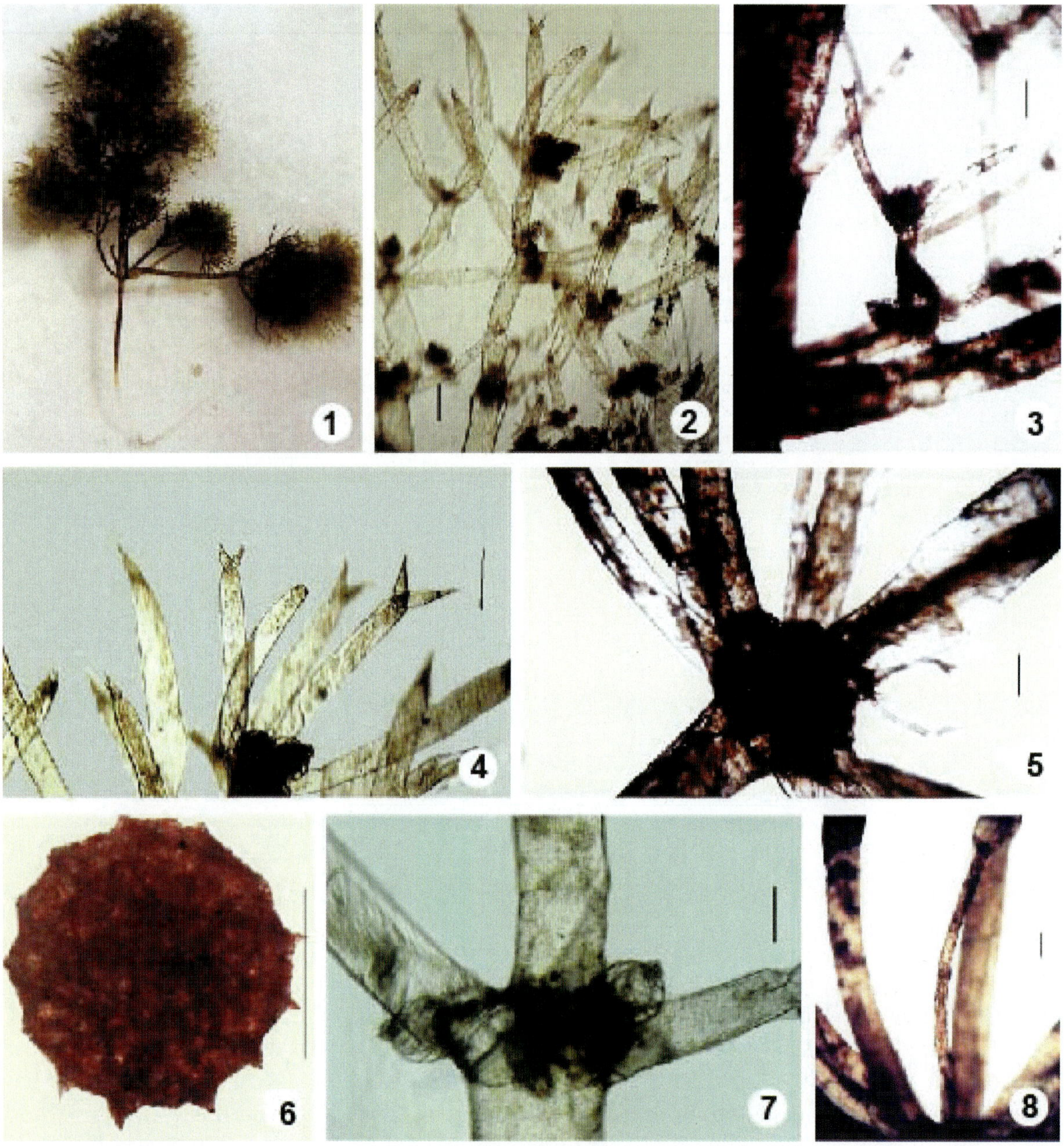

Plate 1. *Nitella zamanii* Naz, Diba & Schubert **sp. nov.** 1. Habit; 2. Fertile branchlet; 3. Tiny, short accessory branchlet; 4. Dactyls; 5. Short curved accessory branchlet; 6. Oospore; 7. Stem node and oogonium; 8. A long and unbranched branchlet. (Scales = 0.2 mm).

Etymology: Nitella zamanii has been named in honour of Professor M. Zaman, Department of Botany, University of Rajshahi, Bangladesh for his lifelong contribution to Phycology.

Notes: The new species *Nitella zamanii* has unique characteristics (Table 1). The whole plant is spike like. Accessory branchlets arise from the base of whorl and 1st furcation at a branchlet; accessory branchlets are of different types, elongated, short and sometimes curved. Dactyls are spine like. Fertile specimens of *N. zamanii* can easily be distinguished from *Nitella furcata*. The gametangia of *Nitella furcata* are restricted to the branchlet nodes (Zaneveld, 1940; Wood and Imahori, 1965), whereas oogonia of *Nitella zamanii* are aggregated at the base of whorl and each branchlet nodes. *Nitella zamanii* is similar to *Nitella polycarpa* which has oogonia at the base of whorls. But it differs from *Nitella polycarpa* (Pal *et al.*, 1962; Naz *et al.*, 2009) by the presence of accessory branchlets and the whorls of *N. polycarpa* are diffuse but the whorls of *N. zamanii* are compact. It is morphologically closely related to another *Nitella*, i.e. *Nitella furcata* subsp. *megacarpa* f. *megacarpa* (Wood and Imahori, 1965). But it differs by the position of gametangia and oospore membrane, the gametangia of *Nitella furcata* subsp. *megacarpa* f. *megacarpa* are restricted to the branchlet nodes and oospore membrane is finely to coarsely papillate, appearing incompletely reticulate at lower focus (Wood and Imahori, 1965). Oospore membrane of *Nitella zamanii* is finely reticulate.

Table 1. Comparison of morphological features of *Nitella zamanii* with closely related taxa of *Nitella*.

Morphological features	*Nitella zamanii*	*N. furcata* (Zaneveld,1940)	*N. polycarpa* (Pal *et al.*, 1962)	*N. furcata* subsp. *megacarpa* f. *megacarpa* (Wood and Imahori, 1965)
Accessory branchlets	Present	Absent	Absent	Absent
Oogonium	63-287 μm long, 35-208 μm wide	240-450 μm long, 220-320 μm wide	480 μm long, 370 μm wide	300-700 μm long, 260-510 μm wide
Antheridium	185 μm	240-350 μm	275 μm	220-300 (-445) μm
Oospore	29-198 μm long, 22-165 μm wide	190-300 μm long, 180-270 μm wide	260 μm long, 250 μm wide	280-420 μm long, 245-380 μm wide

References

Aziz, A. 2009. Genus *Nitella* (C. Agardh) Hooker. *In*: Ahmed, Z.U., Khondker, M., Begum, Z.N.T., Hasan, M.A., Kabir, S.M.H., Ahmad, M., Ahmed, A.T.A. and Rahman, A.K.A. (Eds), Encyclopedia of Flora and Fauna of Bangladesh, Vol. **4**. Algae: Charophyta-Rhodophyta (Achnanthaceae-Vaucheriaceae). Asiatic Society of Bangladesh, Dhaka, pp. 15-23.

Aziz, A. and Tanbir, M. 2003. Algal flora of some northern districts of Bangladesh. Bangladesh J. Plant Taxon. **10**(1): 68-78.

Diba, N.J. and Naz, S. 2011. New record for Bangladesh: two species and two forms of *Nitella* Agardh. J. Plant Studies **1**(1): 47-59.

Islam, A.K.M. Nurul and Irfanullah, H.M. 2005. Hydrobiological studies within the tea gardens at Srimangal, Bangladesh. II. Algal flora (excluding Chlorophyceae). Bangladesh J. Plant Taxon. **12**(1): 33-52.

Islam, A.K.M. Nurul and Sarma, D. 1976. The Characeae of Bangladesh II. Genus *Nitella*. J. Asiat. Soc. Bangladesh (Sci.) **2**(1): 43-61.

Naz, S., Diba N.J. and Azam, S.M.G.G. 2009. A new record for Bangladesh: *Nitella polycarpa* Pal. J. Life & Earth Science **3-4**: 47-50.

Naz, S., Diba, N.J. and Schubert, H. 2011. Monograph on Charophytes of Bangladesh. VDM Verlag Dr. Müller GmbH & Co. KG, pp. 88-127.

Pal, B.P., Kundu, B.C., Sundaralingam, V.S. and Venkataraman, G.S. 1962. Charophyta - Monographs on algae, Vol. **5**. Indian Council of Agricultural Research, New Delhi, pp. 43-76.

Transeau, E.N. 1916. The periodicity of fresh water algae. Amer. J. Bot. **3**: 121-133.

Wood, R.D. and Imahori, K. 1965. A revision of the Characeae. Part - I. Monograph of the Characeae. Verlag Von, J. Cramer, Weinheim, pp. 357-716 .

Zaneveld, J.S. 1940. The Charophyta of Malaysia and adjacent countries. Blumea **4**(1): 49-109.

6

THE GENUS *SORASTRUM* KÜTZING (HYDRODICTYACEAE, SPHAEROPLEALES, CHLOROPHYTA) FROM INDIA, WITH A NEW SPECIES *S. PHILIPOSIANUM*

JAI PRAKASH KESHRI[1] AND PRASANT MALLICK

Centre for Advanced Studies in Botany, University of Burdwan, Golapbag, Burdwan 713 104, West Bengal, India.

Keywords: Sorastrum philiposianum; Chlorophyta; New species; New records; India.

Abstract

In the present paper five species of *Sorastrum* Kützing including a new species, *Sorastrum philiposianum* have been described from Bankura and Purulia districts of West Bengal, India. Amongst these species *S. indicum* Bernard and *S. hathoris* (Cohn) Schmidle are being reported for the first time from India. Moreover, *S. americanum* (Bohlin) Schmidle is an addition to West Bengal algae. All the specimens have been collected from desmids habitats of West Bengal having a low pH (5.0-6.5) and those were growing lodged on submerged portions of the aquatic weeds like *Ceratophyllum demersum* L., *Hydrilla verticillata* (L.f.) C. Presl and *Ipomoea aquatica* Forssk.

Introduction

Sorastrum Kützing (1845) is a rare planktonic coenobial member of the Family Hydrodictyaceae under the Order Chlorococcales occurring mostly in a mixed assemblage with other planktonic members and lodged on submerged aquatic weeds. Although the genus is treated under Chlorococcales, the recent taxonomic revisions (Deason *et al.*, 1991; Buchheim *et al.*, 2005; McManas and Lewis, 2005; Graham *et al.*, 2009) suggest its inclusion under the members of Hydrodictyaceae belonging to Sphaeropleales. Due to its small size and sparse occurrence the genus is easily overlooked. Komárek and Fott (1983) recognized 7 species of this genus world over. Patel and George (1984) added one more species *S. sphericum* Patel *et* George to this list. Earlier following taxa of the this genus have been recorded from India, *viz.*, *S. bengalicum* Philipose, *S. americanum* (Bohlin) Schmidle, *S. americanum* var. *undulatum* G. M. Smith, *S. sphericum* Patel *et* George and *S. spinulosum* Nägeli (Carter, 1869; Turner, 1892; Carter, 1926; Subba Raju, 1963; Philipose, 1967; Anand, 1975, 1987; Sarma and Khan, 1980, 1991; Patel, 1970; Patel and George, 1984; Patel and Isabella, 1977; Kamat,1974; Freitas, 1980; Ashtekar and Kamat, 1980; Compère, 1983; Habib and Chaturvedi, 2001; Jaiswal and Tiwari, 2003; Gupta 2012). During the systematic investigations on the planktonic algae of West Bengal the authors recorded five species of the genus from the desmid habitats of Bankura and Purulia districts of West Bengal.

Materials and Methods

The algal samples were collected from different desmid habitats of Bankura and Purulia districts of West Bengal, India. The pH, temperature and detailed ecological notes were recorded at collection spots. The specimens were preserved in 5% formalin. Camera Lucida drawings were made both from live and preserved specimens using G.W.F. solution (Bando, 1988).

---

[1]Corresponding author. Email: keshrijp@gmail.com

Results and Discussion

Taxonomic treatment

1. **Sorastrum americanum** (Bohlin) Schmidle 1900 **(Pl. 1; Figs 1, 2)**.

(Komárek and Fott, 1983, p. 312, pl. 94, f. 1a-g; Comas, 1996, p. 47, f. 5A-C).

Colony of 16 heart shaped or pyramidal cells, 42-47 µm in diameter (26-32 µm without spines), with distinct pith; cells 6.0-9.5 µm broad, 10-14 µm long, narrowed towards the base, attached to the centre by a short cylindrical stalk and with two long stout outwardly directed spines at each end; spines 12-14 µm long, 3.5-6.0 µm thick.

Collection No. PM 664, dated 1.12.2001, Saheb-bandh, Purulia (Dist. Purulia) growing lodged on submerged leaves of *Ceratophyllum demersum* L. in a creamy colour assemblage (pH 5, temp. 24°C).

Distribution in India: Maharashtra (Kamat, 1974), Kashmir (Subba Raju, 1963), U.P (Mathur and Pathak, 1990; Jaiswal and Tiwari, 2003).

Note: This is the first report of the species from West Bengal, India.

2. **Sorastrum indicum** Bernard 1908 **(Pl. 1, Fig. 3)**.

(Komárek and Fott, 1983, p. 312, pl. 94, f. 6).

Colony of 8 cells, 32-35 µm in diameter (19-22 µm without spines), cells reniform, 9-12 µm broad, 3-4 µm long and 3 µm thick, both ends blunt with two sharp 5.5-7.5 µm long spines at each end; cells attached to a short central stalk.

Collection No. PM 1140, dated 17.2.2003, Lal-bandh, Bishnupur (Dist. Bankura), growing lodged on rotting submerged *Hydrilla verticillata* (L.f.) C. Presl leaves in a form of brownish colour mass in a lake (pH 5.5, temp. 23.5°C).

Note: This is the first report of the species from India.

3. **Sorastrum hathoris** (Cohn) Schmidle 1900 **(Pl. 1, Fig. 4)**.

(Komárek and Fott, 1983, p. 312, pl. 93, f. 8).

Colony of 16 cells, 38 µm in diameter (27 µm without spines); cells 9-14 µm broad, 6-7 µm long and 8-9 µm thick, crescent shaped to cuneate with long stalk and having two short pointed spines from each angle; spines 7-8 µm long.

Collection No. 344, dated 17.10.2001, Lal-bandh, Bishnupur (Dist. Bankura) growing lodged on the *Ipomoea aquatica* Forssk. stems in a form of a yellowish mass with some filamentous green algae (pH 6, temp. 33°C).

Note: This is the first record of the species from India.

4. **Sorastrum spinulosum** Nägeli 1849 **(Pl. 2, Figs 8, 9)**.

(Philipose, 1967, p. 132, f. 47; Hindak, 1980, p. 183, pl. 69, f. 16; Komárek and Fott, 1983, p. 310, pl. 93, f. 5a-e; Comas, 1996, p. 48, f. 4B).

Colony of 4-8 cells, 27-30 µm in diameter (19-23 µm without spines); cells 10-12 µm broad, 5.5-7.0 µm long, 4-6 µm thick, reniform to cuneate, three-angled with a short stalk, having two sharp pointed spines from each angle; spines 6.0-7.5 µm long.

Collection No. PM 1116, dated 31.12.2002, Lal-bandh, Bishnupur (Dist. Bankura) growing on submerged and rotting leaves of *Hydrilla verticillata* (L.f.) C. Presl and *Ceratophyllum demersum* L. in a form of a brown colour mass (pH 6, temp. 19°C).

Distribution in India: Andhra Pradesh (Philipose, 1967), Assam (Carter, 1926), Bihar (Singh and Saha, 1982), Jammu (Anand, 1975, 1987), Kashmir (Compère, 1983), Maharashtra (Carter, 1869; Gonzalves and Joshi, 1946; Kamat, 1963; Freitas, 1980), Meghalaya (Turner, 1892), M.P. (Mathur and Pathak, 1990), Orissa (Philipose, 1967), Tamilnadu (Rani *et al.,* 2007), Uttarakhand (Gupta, 2005), U.P. (Pandey *et al.,* 1981; Pandey *et al.,* 1983; Habib *et al.,* 1988; Habib and Charaurvedi, 2001; Jaiswal and Tiwari, 2003) and West Bengal (Turner, 1892; Mallick and Keshri, 2008).

Note: This is a fairly common and cosmopolitan species of the genus.

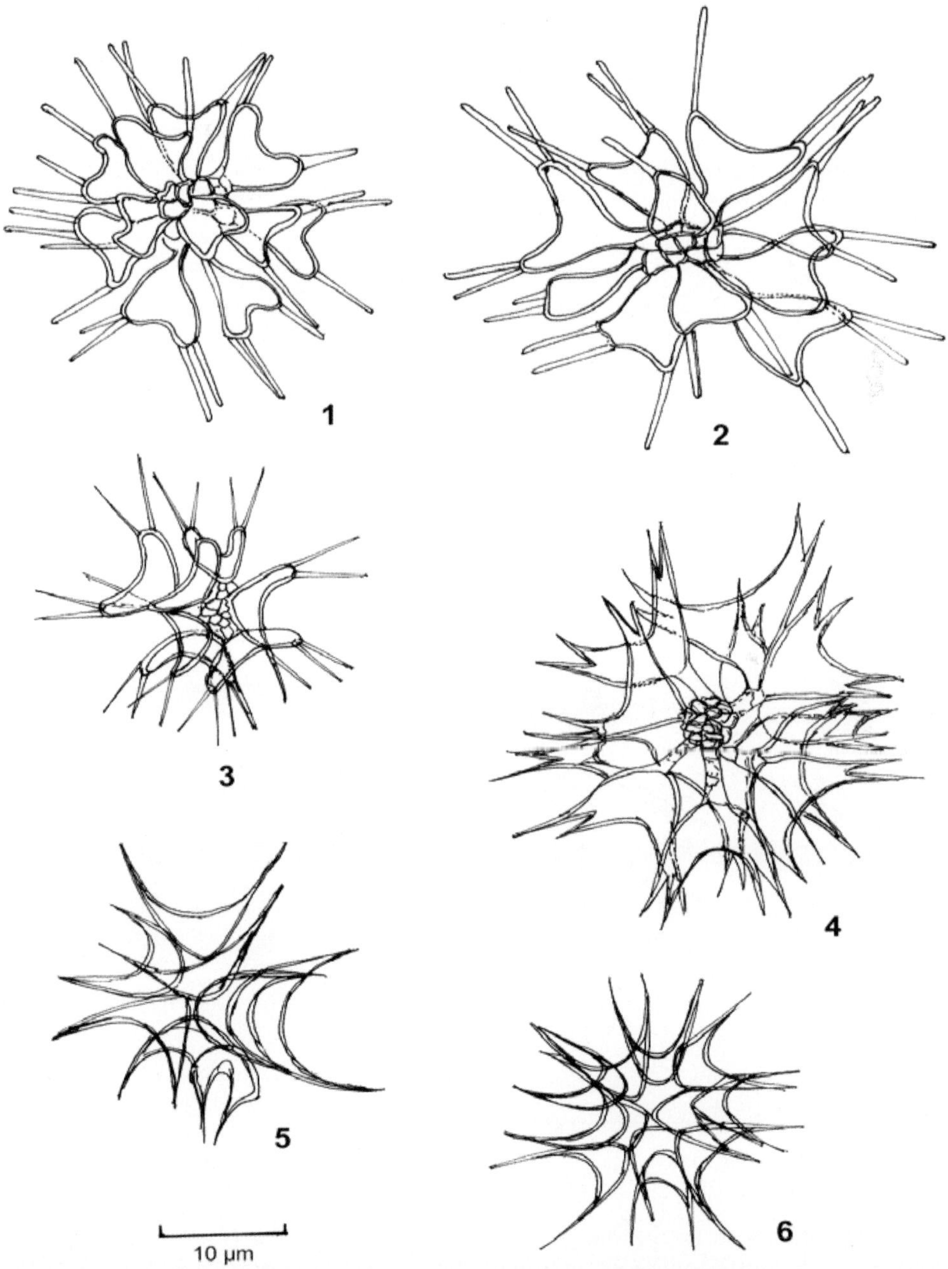

Plate 1. Figs 1 & 2. *Sorastrum americanum* (Bohlin) Schmidle; Fig. 3. *S. indicum* Bernard; Fig. 4. *S. hathoris* (Cohn) Schmidle; Figs 5 & 6. *S. philiposianum* Keshri *et* Mallick, **sp. nov**.

5. Sorastrum philiposianum Keshri *et* Mallick, **sp. nov.** **(Pl. 1, Figs 5, 6; Pl. 2, Fig. 7)**.

Diagnosis: *Colonia 8-16 cellularum, medulla praetermissa, 36-42 μm in diametro (28-30 μm sine spinis); Cellulae pulchre curvatae, semi lunatae vel crescentiformes, 9-12 μm longae, cum extremis mutatis in spinis acutis dispositis subparallis ad axem verticallem; spinae 6-9 μm longae et 5.0-6.5 μm crassae ad basin.*

Holotypus: lectus die 1.12.2001 sub numero PM 664, ad locum Saheb-bandh, Purulia, crescens affixus in *Ceratophyllum demersum* L. foliis submerses in lacu (pH 5, temp. 24°C).

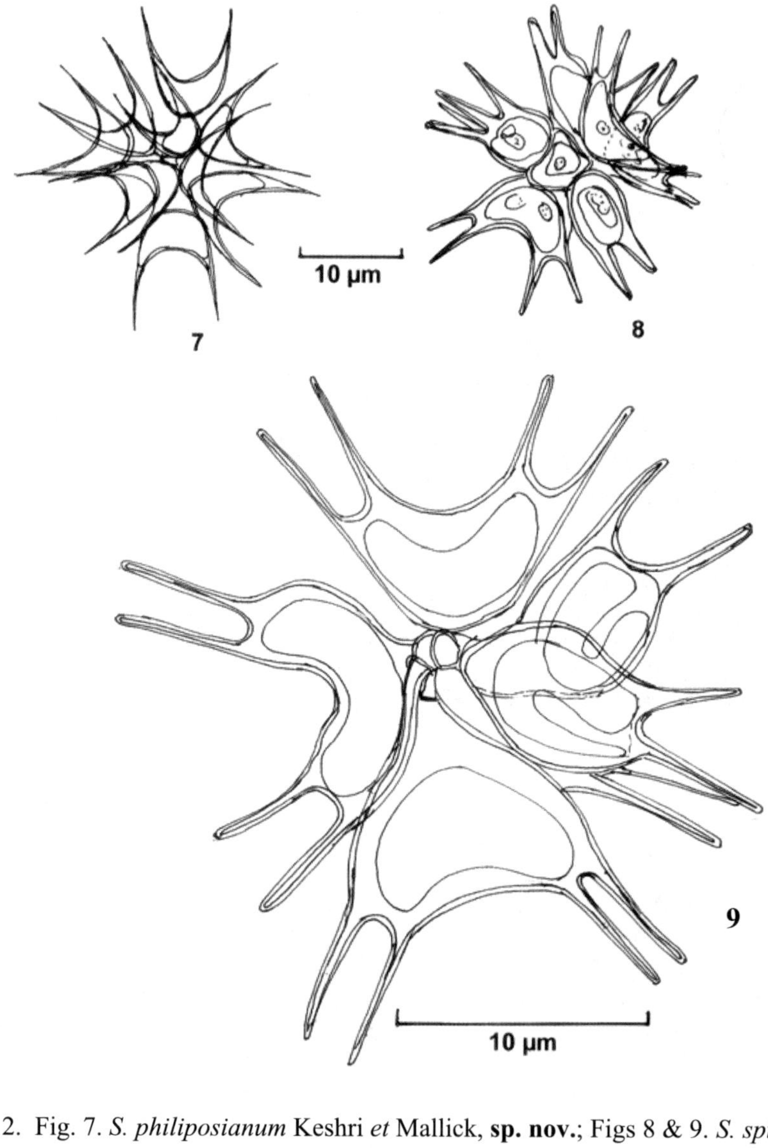

Plate 2. Fig. 7. *S. philiposianum* Keshri *et* Mallick, **sp. nov.**; Figs 8 & 9. *S. spinulosum* Nägeli (Scale Bar = 10 μm).

Colony of 8-16 cells with negligible pith, 36-42 μm in diameter (28-30 μm without spines); cells gracefully curved, semi-lunate to crescent shaped, 9-12 μm broad, 6.5-8 μm long with ends into a sharp spine disposed almost parallel to the vertical axis; spines 6-9 μm long and 5.0-6.5 μm thick at base.

Table 1. A comparative account of the *Sorastrum* species.

Species	Cell Shape	Stalk of the cell	Pith	Spine	Colony Size	Cell size
S. americanum	Heart shaped	Distinct, short & broad	Distinct, multicellular	4, long, blunt, apart	22-61 µm	4.8-20.0 µm broad, 4-8 µm thick
S. bengalicum	Spherical to ovoid	Short, not discernible	Short, not discernible	2, small, apart	25 µm	3-5 µm diameter
S. bidentatum	More or less rectangular with extended arms	Short, not discernible	Short, not discernible	Absent, mere elevation only	28-32 µm	*
S. hathoris	Crescent shaped	Distinct, long	Distinct, multicellular	4, sharp, narrow	27-38 µm	16-22 µm broad, 10 µm thick
S. indicum	Reniform	Indistinct, short	Negligible	4, sharp, not broad at base	15-20 µm	6 x 10 µm
S. minimum	Triangular	Distinct	Short, distinct	2, blunt	Not recorded	4-6 µm broad
S. philiposianum **sp. nov.**	Semilunate to crescent shaped	Negligible	Negligible	2, sharp, long	36-42 µm	9-12 µm broad (thick), 6.5-8.0 µm long
S. sphericum	Spherical	Not discernible	Short, distinct	2, stout, closely placed	60-80 µm	22-30 µm
S. spinulosum	Reniform or cuneate	Distinct, short	Distinct, multicellular	4, pointed or blunt, short, broad at base	27-35 µm	6-8 µm long, 8-20 µm broad, 5-8 µm thick

*Not mentioned in the description of type species.

Holotype: No. PM 664, dated 1.12.2001, Saheb-bandh, Purulia growing lodged on submerged *Ceratophyllum demersum* L. leaves in a form of creamy colour assemblage in a lake (pH 5, temp. 24°C).

Repository: Algae Herbarium, Department of Botany, The University of Burdwan, West Bengal, India (BURD).

Etymology: The species has been named in honour of Late Professor M. T. Philipose, who has significantly contributed to the taxonomy of this group.

Notes: This new species differs from all the existing species of *Sorastrum* in the distinctive nature of colony, its cell shape and nature and disposition of spines. A comparative account of the species is appended in Table 1.

Conclusion

Our knowledge about the planktonic green algae of India is still insufficient (Sarma and Khan, 1980, 1991; Gupta, 2012). Since they play distinctive roles in aquatic ecosystem this knowledge may be utilized for various purposes. *Sorastrum* Kützing is a rare genus, not well explored. In this work attention has been given to taxonomic and ecological perspective. It has been observed that some aquatic angiosperms like *Ceratophyllum demersum* L., *Hydrilla verticillata* (L.f.) C. Presl and *Ipomoea aquatica* Forssk. provide suitable substance for the growth and survival of the coenobial members of Sphaeropleales as well as to the desmids. Little acidic pH is also a determining factor.

Acknowledgements

Thanks are due to the Head of the Department of Botany, the University of Burdwan for laboratory facilities; University Grants Commission for financial assistance; to Late Dr. N.C. Majumdar for Latin diagnosis of the new species and Prof. G.L. Tiwari for providing valuable literatures. The constant encouragement of our revered teacher Prof. Pranjit Sarma throughout this work is gratefully acknowledged.

References

Anand, V.K. 1975. A checklist of planktonic algae from Mansar Lake, Jammu. Phykos **14**(1&2): 77-79.

Anand V.K. 1987. Studies on the algal flora of Jammu (Jammu & Kashmir) [India] 1. Chlorococcales. J. Econ. Taxon. Bot. **11**(2): 487-492.

Ashtekar, P.V. and Kamat, N.D. 1980. Chlorococcales of Aurangabad, Maharashtra. Phykos **19**(1): 115-119.

Bando, T. 1988. A revision of the genera *Docidium, Haplotaenium* and *Pleurotaenium* (Desmidiaceae, Chlorophyta) of Japan. J. Sci., Hiroshima Univ., Ser. B, **22**: 1-63.

Buccheim, M., Buccheim, J., Carlson, T., Braband, T., Hepperle, D., Krienitz, L., Wolf, M. and Hegewald, E. 2005. Phylogeny of the Hydrodictyaceae (Chlorophyceae) inferences from rDNA data. J. Phycol. **41**: 1039-1054.

Carter, H.J. 1869. A description with illustrations of the development of *Sorastrum spinulosum* to which is added that of a new form of *Protococcus.* Ann. Mag. Nat. Hist. Ser. IV, **4**: 420-435.

Carter, N. 1926. Freshwater algae from India. Rec. Bot. Surv. India **9**(4): 263-302.

Comas, A. 1996. Das Chlorococcales dulciacuicolas de Cuba. Bibliotheca Phycologica Band **99**, pp. 141.

Compère, P. 1983. Some algae from Kashmir and Ladakh, W. Himalayas. Bulletin de la Société Royale de Botanique de Belgique **116**(2): 141-160.

Deason, T.R., Silva, P.C., Watanabe, S. and Floyd, G.L., 1991. Taxonomic status of the species of the green algal genus *Neochloris.* Plant Syst. Evol. **177**: 213-219.

Freitas, J.F. 1980. A checklist of Chlorococcales of Nagpur. Phykos **19**(1): 111-114.

Gonzalves, E.A. and Joshi, D.B. 1946. Freshwater algae near Bombay. J. Bombay Nat. Hist. Soc. **46**(1): 154-176.

Graham, L.E., Graham, M. and Wilcox, L.W. 2009. *Algae*. 2nd Edition. Benjamin Cummings, USA, 616 pp.

Gupta, R.K. 2005. Algal flora of Dehradun District, Uttaranchal. Botanical Survey of India, 298 pp.

Gupta, R.K. 2012. Algae of India, Vol. **2**. A check list of Chlorophyceae, Xanthophyceae, Chrysophyceae & Euglenophyceae. Botanical Survey of India, 428 pp.

Habib, I. and Chaturvedi, U.K. 2001. A systematic account of Chlorococcales from Mahoba, India. Phykos **40**(1&2): 107-113.

Habib, I., Pandey, U.C. and Shukla, H.M. 1988. Chlorococcales of Shahjahanpur (India). Mendel **5**(4): 293-304.

Hindak, F. 1980. Studies on the Chlorococcal algae (Chlorophyceae). II. Veda, Bratislava, 195 pp.

Jaiswal, K.K. and Tiwari, G.L. 2003. Chlorococcales (Green Algae). Bioved Research Society, Allahabad, 116 pp.

Kamat, N.D. 1963. The algae of Kolhapur, India. Hydrobiologia **22**(3/4): 209-305.

Kamat, N.D. 1974. Algae of Marathawada, Maharashtra. Phykos **13**(1): 22-32.

Komárek, J. and Fott, B. 1983. Chlorophyceae (Grünalgen) Ordnung: Chlorococcales. *In*: Huber–Pestallozi, G. (Ed.) Das Phytoplankton des süsswasers. Systematik und Biologie. 7. Teil, 1. Hälfte [= Thienemann, A. (Ed.), Die Binnengewässer Bd. XVI , 7. Teil, 1. Hälfte]. Stuttgart: Schweizerbart. 1044 pp.

McManus, H.A. and Lewis, L.A. 2005. Molecular phylogenetics, morphological variation and colony evolution in the family Hydrodictyaceae (Sphaeropleales, Chlorophyta). Phycologia **44**(6): 582-596.

Mallick, P. and Keshri, J.P. 2008. A study of chlorococcalean algae and their associated plants in Bankura district, West Bengal. Int. J. Plant Sci. **4**(1): 285-288.

Mathur, M. and Pathak, N. 1990. Chlorococcales from the rock shelters of Hosangabad. Phykos **29**(1&2): 111-113.

Patel, R.J. 1970. An enumeration of Chlorococcales of Gujarat. J. Bombay Nat. Hist. Soc. **66**(3): 665-669.

Patel, R.J. and George, I. 1984. New species of *Sorastrum* Kuetz. from Gujarat. Phykos **23**(1&2): 88-89.

Patel, R.J. and Isabella, P.K. 1977. Chlorococcales of Gujarat, India - *Pediastrum* Meyen, *Sorastrum* Kützing and *Hydrodictyon* Roth. J. Indian Bot. Soc. **56**(2): 172-178.

Pandey, U.C., Tiwari, G.L. and Pandey, D.C. 1981. Additions to the algal flora of Allahabad, India: 7. Chlorophyta, Chlorococcales. Proc. Indian Nat. Acad. Sci., Part B, Biol. Sci. **47**(2): 255-259.

Pandey, U.C., Tiwari, R.K. and Pandey, D.C. 1983. An enumeration of Chlorococcales from Allahabad, U.P. India. Bibliotheca Phycologica Band **66**: 115-126.

Philipose, M.T. 1967. *Chlorococcales*. Indian Council of Agricultural Research, New Delhi, 365 pp.

Rani, V., Laila Banu, N.R. and Prakash, J.W. 2007. Algal diversity of a rural pond in Kalkulum Taluk, Kanyakumari district, Tamilnadu. J. Basic Appl. Biol. **1**: 35-37.

Sarma, Y.S.R.K. and Khan, M. 1980. Algal Taxonomy in India. Today & Tomorrow's Printers and Publishers, New Delhi, 153 pp.

Sarma, Y.S.R.K. and Khan, M. 1991. Fresh water algae. *In:* Khan, M. (Ed.), Indian Phycological Review, Vol. **I**. Bishen Singh Mahendra Pal Singh, Dehradun, pp. 1-56.

Singh, N.K. and Saha, L.C. 1982. Chlorococcales of Bhagalpur, India: 1. Bihar. J. Econ. Taxon. Bot. **3**(1): 197-200.

Subba Raju, N. 1963. The algal flora of Kashmir-I. J. Osmania Univ., Sci. **1**: 9-17.

Turner, W.B. 1892. The freshwater algae (principally Desmidieae) of East India. Kongl. Svensk. Vet. Akad. Handl. **25**(5): 1-187.

7

RHODODENDRON LEIGONGSHANENSE (ERICACEAE), A NEW SPECIES FROM CHINA

CHENG-HUA YANG, ZHENG-GUO XIE[1], YONG-FU YU[1] AND ZHI-RONG YANG[2]

Forestry Academy of Guizhou Province, Guiyang 550005, Guizhou, People's Republic of China

Keywords: China; Ericaceae; New species; *Rhododendron*.

Abstract

Rhododendron leigongshanense, a new species from Leishan County, Guizhou Province, China, is described and illustrated. The new species is close to *R. magniflorun* W.K. Hu and *R. glanduliferum* Franchet, but differs by having stipitatus glands on leaf abaxial surface, corolla trumpet-shaped with 7 lobes, 7.5–8.0 cm long, 8–10 cm in diameter and calyx 7- lobed.

Introduction

The genus *Rhododendron* L. is one of the largest genera of the family Ericaceae with many ornamental species with high horticultural value (Fang, 1999). The genus consists of about 1000 species and is distributed in Asia, Europe and North America. So far, 574 species have been recorded in China, of which 409 are endemic (Fang *et al.*, 2005). The southwest-central region of China, including Guizhou Province is possibly the region of the geographic origin of the genus *Rhododendron* (Fang and Ming, 1995).

Leigongshan National Nature Reserve is located at southeast Guizhou Province, China, which has rich natural plant resources, including approximately 24 species of *Rhododendron*. In 2011 and 2012, the authors conducted an intensive taxonomic survey of *Rhododendron* in this area, when a previously unknown specimen belonging to *Rhododendron* was found. The plant grew in evergreen broad-leaved forests on limestone hills (latitude 26°21′50″N, longitude 108°08′29″E) at an altitude of 1,400 m. After critical examination of herbarium specimens from GF, HGAS, GZAC and PE herbaria and carefully consulting relevant literature (Zhou and Yao, 1989; Zhang and Chen, 1990; Hu and Fang, 1994; Fang *et al.*, 1999; Fang *et al.*, 2005; Zhang and Zhang, 2007; Chen *et al.*, 2010a, b; Chen *et al.*, 2012; Yang *et al.*, 2012), this species was very distinct when compared with its morphologically closest matches. Thus it is described and illustrated herein as *Rhododendron leigongshanense* sp. nov.

Rhododendron leigongshanense C.H. Yang, Z.G. Xie, Y.F. Yu & Z.R. Yang, **sp. nov.** (**Figs 1& 2**).

Diagnosis: The new species is morphologically similar to *R. magniflorum* and *R. glanduliferum*, but differs by its leaf abaxial surface with stipitatus glands, corolla trumpet-shaped with 7 lobes, 7.5–8.0 cm long, 8–10 cm in diameter, and calyx 7-lobed. More detailed morphological differences among these three species are given in Table 1.

Types: China. Guizhou: Leishan County, Leigongshan National Nature Reserve, Huxiongpo, 26°21′50″N, 108°08′29″E, altitude 1400 m, 25 June 2012, *Chenghua Yang* 7590 (fl) (*Holotype*: GF; *Isotypes*: HGAS, PE).

---

[1]Administrative Bureau of Leigongshan National Nature Reserve, Leishan 557100, Guizhou, People's Republic of China.

[2]Corresponding author. State Key Laboratory of Systematic and Evolutionary Botany, Institute of Botany, Chinese Academy of Sciences, Beijing 100093, People's Republic of China. Email: zry@ibcas.ac.cn

Paratypes: China. Guizhou: Leishan County, Leigongshan National Nature Reserve, Huxiongpo, 3 July 2011, *Chenghua Yang & Zhengguo Xie* 7667 (fl) (GF); *ibid.* 8 July 2011, *Chenghua Yang & Zhengguo Xie* 7751 (GF); *ibid.* 8 September 2011, *Chenghua Yang & Yongfu Yu* 9232 (fr.) (GF).

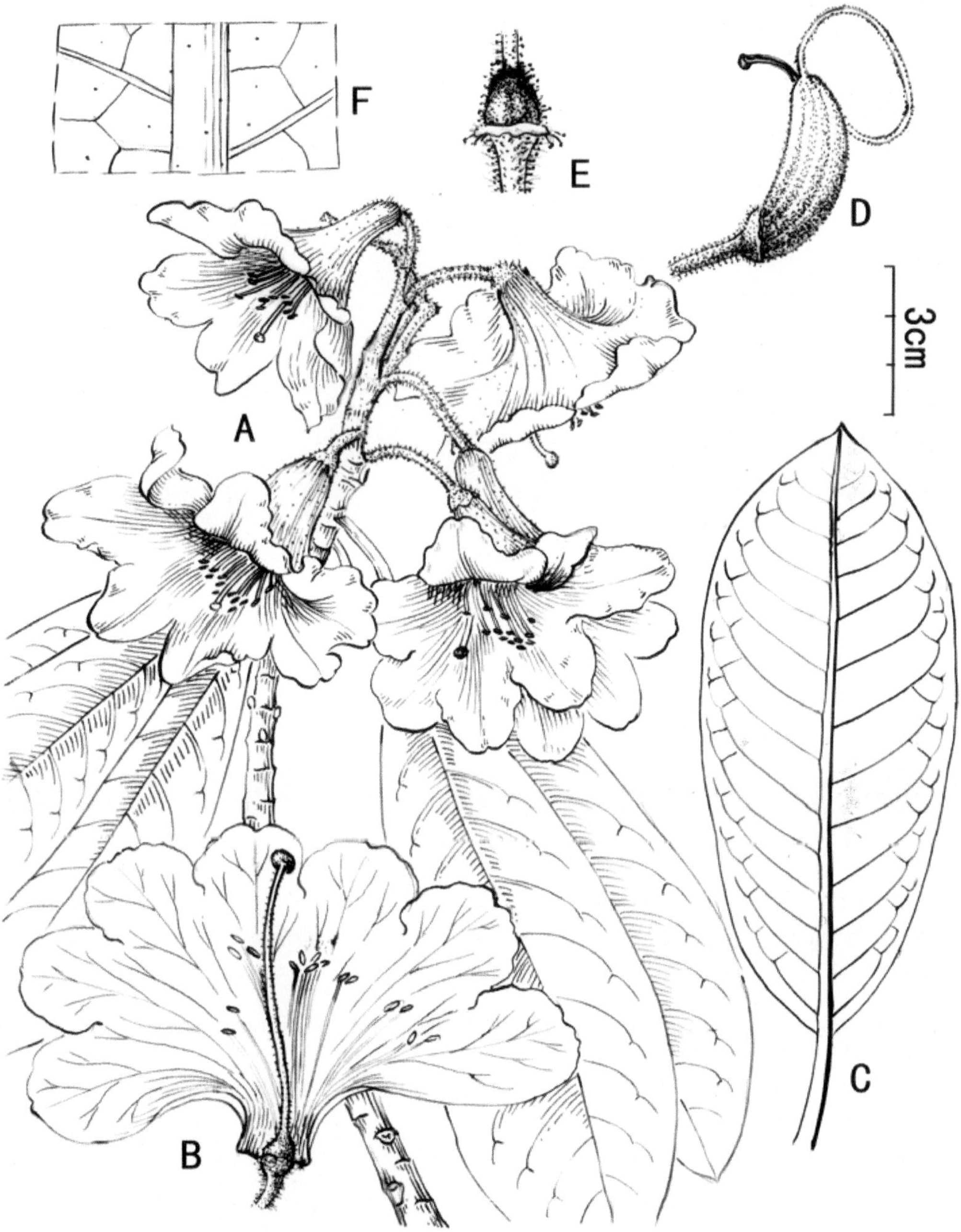

Fig. 1. *Rhododendron leigongshanense* C.H. Yang, Z.G. Xie, Y.F. Yu & Z.R. Yang, **sp. nov.** A. Flowering branch; B. Corolla (displaying); C. Leaf blade (adaxial surface); D. Capsule; E. Ovary (ampliate); F. Leaf blade abaxial surface (ampliate).

Fig. 2. *Rhododendron leigongshanense* C.H. Yang, Z.G. Xie, Y.F. Yu & Z.R. Yang, **sp. nov.** A. Habitat and habit; B. Leaf blade showing stipitatus glands on leaf abaxial surface; C. Flower, showing trumpet-shaped corolla with 7 petals; D. Capsule.

Table 1. Comparison of morphological characteristics among *Rhododendron leigongshanense* sp. nov., *R. magniflorum* and *R. glanduliferum*.

Characters	*R. leigongshanense* **sp. nov**.	*R. magniflorum*	*R. glanduliferum*
Leaf blade	Oblong-elliptic	Oblong	Oblong-lanceolate or oblanceolate
Leaf surface	Abaxial surface stipitate, glandular-hairy	Abaxial surface smooth	Abaxial surface smooth
Lobes of calyx	7	5	7 or 8
Lobes of corolla	7	5	7 or 8
Corolla size	7.5–8.0 cm long	9.5–10.0 cm long	5–6 cm long
Corolla shape	Trumpet shaped	Tabulate - funneliform	Funneliform - campanulate

Evergreen small tree, 5 m tall, c. 16 cm in diameter; branchlets thick, cylindrical with stipitate glands. Leaf blade leathery or thick-leathery, 6–9 leaves on branch-top, pendulous; young leaves violet, densely stipitate glands and floccose; leaf blade oblong-elliptic, apex acute, base broadly cuneate, slightly asymmetry, 15–21 × 5.0–7.5 cm, margin revolute, abaxial surface stipitate glands; midrib, veins and lateral veins slightly grooved adaxially, midrib and lateral veins prominent abaxially, lateral veins 15–18 pairs. Petiole 2–4 cm long, glandular. Inflorescence racemose, 10 or 11-flowered, rachis light brown, 7.0–9.5 cm long, densely stipitate glands; bract lanceolate, abaxial thinly stipitate or non-stipitate glands, adaxial densily glands and villous; pedicels 3.5–4.5 cm long with densely long glands. Calyx 7-lobed, lobes semi-circle, disparate, 2–4 mm long, abaxial glandular, adaxial smooth. Corolla trumpet-shaped, 7-lobed, white, 7.5–8.0 × 8.0–10.0 cm, succulent, aromatic, rotund, abaxial stipitate or non-stipitate glands. Stamens 16, unequal, 4–6 cm long, filaments bald. Ovary conoid, 5–8 mm long, densely glands; style thick, 5–8 cm long, faint green, densely stipitate or non-stipitate glands to the tip; stigma pale, disk-like, 2–3 mm in diameter. Capsule cylindric, 3.0–3.5 × 1.0–1.2 cm, slightly curved, densely stipitate or non-stipitate glands.

Phenology: Flowering from June to July; fruiting from October to November.

Etymology: The specific epithet is named after the type locality Leigongshan.

Vernacular name: Leigongshan Dujuan.

Distribution: So far, this species is only known from the type locality Leigongshan Nature Reserve, Leishan County, Guizhou Province, southwest China.

Habitat: This species grows in evergreen broad-leaved forests on limestone hills at 1,400 m.

IUCN Red List category: Since only one population and a total of 50 mature individuals of this species were found in the area from where the type collections were made, the species seems to be very rare and restricted in distribution. A preliminary conservation assessment for the species gave it Critically Endangered (CR) status, based on criterion D (population less than 50 mature individuals) of the IUCN Red List criteria (IUCN, 2011). Therefore, careful protection of the species is warranted.

Acknowledgements

This work was supported by the National Natural Science Foundation of China (grant no. 31370657), through the projects 'Study on Magnoliaceae Plant in Guizhou' and 'Study on Wetland Plant in Guizhou'. The authors are grateful to curators of the herbaria GF, HGAS, GZAC and PE for allowing them to access their collections. They also thank Prof. Rui-Zheng Fang for revising the morphological description, Prof. Zu-Pei Liu for revising the English and Mr. Hua Xie for preparing the original line drawings. The authors also appreciate the support given by Administrative Bureau of Leigongshan National Nature Reserve and the assistance rendered by Xiao-Yong Dai and Yun-Li Jiang during fieldwork.

References

Chen, X., Laurie, C., Huang, J.Y. and Chen, X. 2010a. New taxa of *Rhododendron* (Ericaceae) from China. Ann. Bot. Fennici. **47**: 397–402.

Chen, X., Huang, J.Y., Laurie, C. and Chen, X. 2010b. Two new species of *Rhododendron* (Ericaceae) from Guizhou, China. Novon **20**(4): 386–391.

Chen, X., Yang, C.H., Xie, H. and Chen, X. 2012. A new species of *Rhododendron* (Ericaceae) from China. J. Trop. Subtrop. Bot. **20**(5): 513–516. [In Chinese].

Fang, M.Y., Fang, R.C., He, M.Y., Hu, L.Z., Yang, H.B. and Chamberlain, D.F. 2005. *Rhododendron*. *In*: Wu, Z.Y. and Raven, P.H. (Eds), Flora of China, Vol. **14**. Science Press, Beijing & Missouri Botanical Garden Press, St. Louis, pp. 260–331.

Fang, R.C. 1999. Flora Republicae Popularis Sinicae. Tomus **57**, No. 1. Science Press, Beijing, pp. 13–15. [In Chinese].

Fang, R.Z. and Ming, T.L. 1995. The floristic study on the genus *Rhododendron*. Acta Botanica Yunnanica **17**: 359–379. [In Chinese, with English abstract]

Hu, L.Ch. and Fang, M.Y. 1994. Flora Republicae Popularis Sinicae. Tomus **57**, No. 2. Science Press, Beijing, pp. 7–40. [In Chinese].

IUCN 2011. IUCN Red List Categories and Criteria, Version 9.0. Prepared by the IUCN Species Survival Commission. IUCN, Gland, Switzerland, and Cambridge, United Kingdom.

Yang, C.H., Yang, C.D. and Mu, J. 2012. One new species of *Rhododendron* (Ericaceae) from Guizhou, China. Guizhou. Sci. **30**(1): 95–96. [In Chinese]

Zhang, H.H. and Zhang, X. 2007. Study on the biological diversity of the Leigongshan Nature Reserve. Science Press of Guizhou, Guiyang, pp. 49–50. [In Chinese]

Zhang, X.S. and Chen, X. 1990. *Rhododendron*. *In*: Flora of Guizhou Editorial Committee (Eds), Flora of Guizhou. Vol. **3**. The People's Press of Guizhou, Guiyang, pp. 195–233. [In Chinese]

Zhou, Z.X. and Yao, M.S. 1989. Scientific Survey of the Leigong Mountain Nature Reserve. The People's Press of Guizhou, Guiyang, pp. 147–174. [In Chinese]

8

SYSTEMATICS OF *SISYRINCHIUM ANGUSTIFOLIUM* MILL. (IRIDACEAE), A NEWLY RECORDED SPECIES FROM TURKEY

ÖZGÜR EMINAGAOGLU[1] AND MELAHAT ÖZCAN[2]

Department of Forest Engineering, Faculty of Forestry, Artvin Coruh University, 08000 Artvin-Turkey

Keywords: New record; *Sisyrinchium*; Anatomy; Chromosome number; Turkey.

Abstract

Sisyrinchium angustifolium Mill. (Iridoideae, Sisyrinchieae) has been reported as a new record for the flora of Turkey from NE Anatolia, and taxonomic and conservation status of this species are evaluated. Detailed morphological description, leaf and stem anatomical properties, and chromosome numbers are provided. The present chromosome count is the first tetraploid level for this species reported from Eastern Anatolia, Turkey.

Introduction

Sisyrinchium L. (Iridoideae, Sisyrinchieae) includes approximately 200 species (Rudall *et al.*, 1986) and occurs in wet to dry grasslands, rocky or sandy soils, and also in highly transformed areas, such as roadsides or grazed and burned fields at elevations of up to 800 m (Bicknell, 1899). During fieldwork in Artvin Province (Turkey) some interesting specimens belonging to Iridaceae were collected. After critical study and consultation with relevant literature (Fedchenko, 1968; Davis, 1984, 1988; Güner *et al.*, 2000; Özhatay *et al.*, 2011), the specimens have been identified as *Sisyrinchium angustifolium* Mill. The species was not reported earlier from Turkey (Özhatay *et al.*, 2011; Eminağaoğlu *et al.*, 2012; Eminağaoğlu and Özcan, 2013) and hence reported as a new record for the flora of Turkey. *S. angustifolium* is the first species of the genus *Sisyrinchium* to have been reported for Turkey. The number of genera of Iridaceae in the flora of Turkey is increased to 6 with addition of *Sisyrinchium*. In this study, we aim to describe detailed morphological characters of *Sisyrinchium angustifolium*, to determine its conservation status, and distribution along with exploring the anatomical and cytological properties of the newly recorded species.

Materials and Methods

Morphological analysis: Plant materials were collected from distinct parts of Artvin, Turkey, at different altitudes in May and June, 2013. The collected materials were critically studied. The voucher specimens have been deposited at the Herbarium of Artvin Coruh University (ARTH), Artvin, Turkey.

Anatomical preparation: Anatomical observations were performed in stem and leaf from living specimens. Plant samples were stored in 70% alcohol for anatomical studies. Transverse sections of stem and leaf, and peripheral sections of upper and lower epidermis of leaves were taken by hand using commercial razor blades and stained in Haematoxylin for about 15 min. To remove the excess stain, sections were washed in water several times (Algan, 1981). Semi-permanent slides were mounted in glycerin or permanent slides were covered with glycerin-gelatin

[1]Corresponding author. Email: oeminagaoglu@artvin.edu.tr

[2]Department of Biology, Faculty of Science and Arts, Artvin Coruh University, 08000 Artvin-Turkey

(Vardar, 1987). Well stained sections were examined under a light microscope and photographed using an Olympus BX-53 microscope with digital camera attachment DP 73.

Chromosome count: For mitotic chromosome observation root tips growing in pots were cut off and pretreated with 0.05% colchicine at room temperature for 4.0-4.5 h (Ozcan *et al.*, 2008), then fixed in fresh Carnoy absolute alcohol-glacial acetic acid (3:1) for 24 h at 4ºC. For chromosome counts, root tips were hydrolyzed in 5N HCl for 10-12 min at room temperature and then rinsed with distilled water for 2-3 min. Staining was carried out in lacto-propionic orcein at least for 2 h at room temperature. Permanent slides were prepared from at least ten well-spread cells. The best metaphase plates were photographed with Olympus BX-53 microscope with digital camera attachment DP 73.

Results and Discussion

Sisyrinchium angustifolium Mill. Gard. Dict., ed. 8. n. 2 (1768). *Sisyrinchium angustifolium* f. *album* J.K. Sim & Y.S. Kim, Korean J. Pl. Taxon. 22: 3 (1992); *Sisyrinchium hibernicum* Á. Löve & D. Löve, Bot. Not. 114: 37 (1961); *Sisyrinchium membranaceum* E.P. Bicknell, Bull. Torrey Bot. Club 26: 612 (1899); *Bermudiana homomalla* (Klatt) Kuntze, Revis. Gen. Pl. 2: 700 (1891); *Sisyrinchium homomallum* Klatt, Abh. Naturf. Ges. Halle 15: 378 (1882); *Bermudiana graminea* Gaertn., Fruct. Sem. Pl. 1: 32 (1788); *Sisyrinchium gramineum* Lam., Encycl. 1: 408 (1785). (**Fig. 1**).

Caespitose, perennial herb, up to 40 cm tall. Stem erect or ascending, branched, conspicuously winged towards base, the wing 3-5 mm broad; with 1-2 nodes, 2.3-5.0 mm wide, glabrous, margin often minutely denticulate especially basally. Leaves radical, linear-ensiform, 3-5 mm broad, leaf blades glabrous. Flowers 1-4 in inflorescence, c. 15 mm in diameter; spathes usually green, wider than supporting branch, glabrous, keels denticulate to entire; outer 18-38 mm long, 2.0-9.5 mm longer than inner, usually tapering evenly towards apex, margin basally connate 4-6 mm; inner with keel evenly curved or straight, hyaline, margin 0.1-0.3 mm wide, apex acuminate to acute, ending 0.2-0.7 mm proximal to green apex. Flowers pale blue to violet, occasionally white, bases yellow; outer tepals 7.7-12.5 mm long, apex rounded or emarginate, aristate. Stamens inserted at the base of perianth, the filaments usually more or less adnate to the tube, filaments connate ± entirely, stipitate-glandular basally; anthers erect or ascending. Ovary spherical or shortly oblong, 3-celled, with numerous ovules in each cell; style filiform, short or exceeding the stamens, the tips of style branches acting as receptive stigmatic surfaces; ovary similar in colour to foliage. Capsules dark brown or black, sometimes with purplish tinge, ± globose, 4-7 mm long; fruiting pedicel spreading or ascending. Seeds globose to obconic, lacking obvious depression, 0.5-1.2 mm long, rugulose.

Flowering period: March - July; *fruiting period*: June - August.

Specimens examined: **Turkey**: A8 Artvin, Hopa, Kemalpaşa, damp roadside, grassland, 20 m, 8 June 2013, 41°30′30″N, 41°32′16″E, *Ö. Emin.* 15855 (ARTH 5248); Artvin, Hopa, Hopa to Borçka roadside, gravelly soil, 270 m, 21 June 2013, 41°23′42″N, 41°29′54″E, *Ö. Emin.* 16948 (ARTH 5249); Artvin, Hopa, Kemalpaşa, damp roadside, grassland, 3 m, 8 May 2014, 41°30′16″N, 41°32′07″E, *Ö. Emin.* 19361 (ARTH 5266).

Distribution: Austria, Brazil, Canada, Colombia, Czechoslovakia, Germany, Great Britain, Mexico, United States, Russia and Georgia. New to Turkey.

Ecology: *Sisyrinchium angustifolium* grows on damp, humid and gravelly soil including roadsides and forest sides at 20-300 m with *Plantago major* L., *Polygonum aviculare* L, *Oxalis corniculata* L., *Equisetum fluviatile* L., and *Juncus effusus* L.

Conservation status: Sisyrinchium angustifolium is only known from two localities in Artvin district. The natural habitat of the species is under threat due to animal grazing and road construction. In this state, according to the IUCN (2013), the species is considered to be vulnerable (VU B2b(ii) - the estimated area of occupancy is less than 2.000 km^2, and the field observations indicate a continuing decline of habitat quality). Therefore, *S. angustifolium* should be included in the List of Threatened Species of Turkey's Flora.

Fig. 1. *Sisyrinchium angustifolium;* A. Habitat; B. Front and lateral view of the flowers; C. Capsules. Scale bars: A, B = 1 cm; C = 0.5 cm

Anatomical characteristics:

Anatomical studies reveal that stem is ancipital in two directions. The epidermis contains a single layer of cells. Cuticle is very thick and cell walls of the epidermis are prominently thickened. There is no papilla above the epidermis. Several sunken stomata are observed in this layer. Mechanical tissue does not develop. Collateral vascular bundles in two circles and surrounded by sclerenchymatic ring. The larger bundles are present near to the centre (Fig. 2).

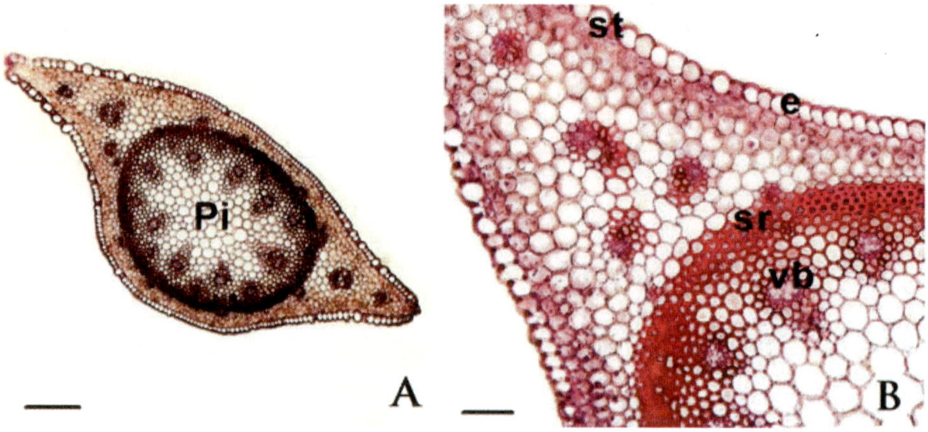

Fig. 2. Stem anatomy of *Sisyrinchium angustifolium;* A. Cross section of stem; B. Vascular bundles in stem. e: epidermis, pi: pith, sr: sclerenchymatic ring, st: stomata, vb: vascular bundle. Scale bars: A = 200 μm; B = 50 μm.

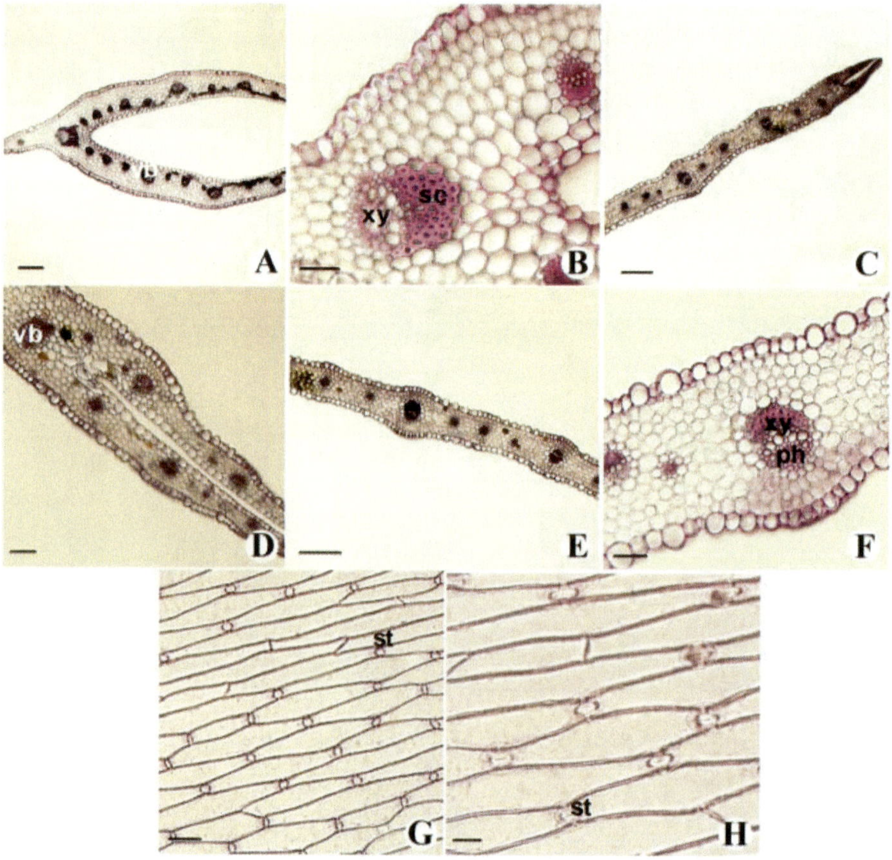

Fig. 3. Leaf anatomy of *Sisyrinchium angustifolium*. A-F. Cross section of leaf; G,H. Peripheral section. ph: phloem, sc: sclerenchymatic cap, st: stomata, xy: xylem, Scale bars: A,C,E = 200 μm; D = 100 μm; B,F,G = 50 μm; H = 20 μm.

The innermost part of the cylinder is occupied by a pith with thick-walled parenchyma. In some cases, hollow areas were observed instead of these cells in the center. The sheathing base is unifacial with both abaxial and adaxial epidermis, and leaf blade is unifacial, with an abaxial epidermis only. Surfaces do not contain any trichomes. The cuticle is smooth, epidermal cells of both surfaces are arranged in a single layer. Adaxial epidermal cells are thin-walled and stomata are absent in this surface, whereas abaxial epidermal cells are thick-walled, longer and narrower over veins than in intercostals areas (Fig. 3A-F). In the leaf blade, mesophyll is almost entirely chlorenchymatic. Vascular bundles with different sizes are collateral and surrounded by parenchymatous sheath cells, and they are arranged in a single row. Sclerenchymatic caps are observed in xylem pole. In peripheral sections, cells next to the stomata sometimes with slightly protruding over guard cells. Stomata are anomocytic in abaxial parts and sunken (Fig. 3G, H).

Mechanical tissue does not exist in the stems of examined samples. This feature has been previously reported in *Sisyrinchium* by Holm (1908). Leaf blade are longitudinally furrowed and vascular bundles with sclerenchymatic cap at the xylem poles. The study also reveals that sunken stomata are only found in abaxial surface. Rudall *et al.* (1986) investigated anatomical and chromosomal characteristics of 11 taxa and reported in these characteristics in the section of *Bermudiana* in *Sisyrichium* which is supported by the present study.

Cytological investigation:

In the present study the somatic chromosome number in *Sisyrinchium angustifolium* has been determined as $2n = 4x = 32$ (Fig. 4).

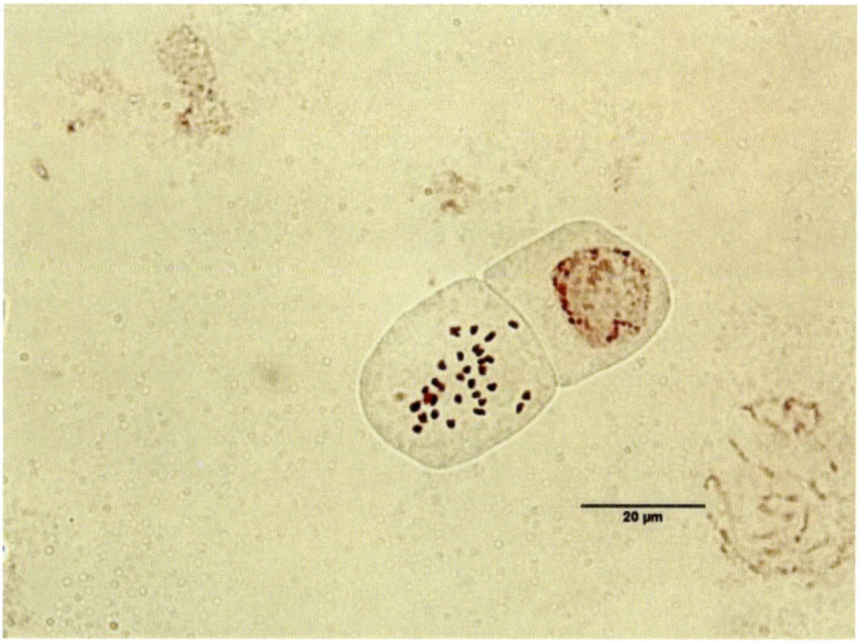

Fig. 4. Somatic metaphase of *Sisyrinchium angustifolium*. $2n = 32$

Basic chromosome number of the genus *Sisyrinchium* has been reported by Oliver and Lewis (1962) as $x = 8$, 9 and 17. In addition, Rudal *et al.* (1986) reported that basic chromosome number of the section of *Bermudiana* as x = 8 in the genus. This species belongs to the section *Bermudiana*. Our chromosome count from Eastern Anatolia is the first tetraploid level in *S.*

angustifolium. Till now, only one previous record in a dodecaploid level ($2n = 12x = 96$) for this species has been reported by Murin and Majovsky (1976).

Acknowledgements

The authors are thankful to the Ziraat Bank in Turkey and the Research Fund of Artvin Coruh University (project number: 2012.F15.02.21) for their financial supports.

References

Algan, G. 1981. Bitkisel Dokular İçin Mikroteknik, İstanbul. Fırat University Science Faculty Press (in Turkish).

Bicknell, E.P. 1899. Studies in *Sisyrinchium* IV: *S. angustifolium* and related species of the West and Northwest. Bull. Torrey Bot. Club **26**: 445-457.

Davis, P.H. (Ed.) 1984. Flora of Turkey and the East Aegean Islands. Vol. **8**. Edinburgh University Press, Edinburgh, 632 pp.

Davis, P.H. (Ed.) 1988. Flora of Turkey and the East Aegean Islands.Vol. **10**. Edinburgh University Press, Edinburgh, 590 pp.

Eminağaoğlu, Ö. and Özcan, M. 2013. *Euonymus leiophloeus* (Celastraceae) - A new record for the flora of Turkey. Bangladesh J. Plant Taxon. **20**(2): 263-266.

Eminağaoğlu, Ö., Özcan, M. and Kültür, Ş. 2012. Contributions to the leaf and stem anatomy of *Tradescantia fluminensis*: an alien species new to the flora of Turkey. ACU J. For. Fac. **13**(2): 270-277.

Fedchenko, B.A. 1968. *Sisyrinchium angustifolium* Mill. *In*: Komarov, V.L. (Ed.), Flora of the U.S.S.R. Vol.**4**, Jerusalem: Israel Program for Scientific Translation, pp. 577-578.

Güner, A., Özhatay, N., Ekim, T. and Başer, K.H.C. (Eds). 2000. Flora of Turkey and the East Aegean Islands. Vol. **11**. Edinburgh University Press, Edinburgh, 680 pp.

Holm, T. 1908. *Sisyrinchium*: Anatomical studies of North American species. Bot. Gaz. **46**(3): 179-192.

IUCN 2013. The IUCN red list of threatened species, version 2013.1. IUCN Red List Unit, Cambridge, U.K. http://www.iucnredlist.org [accessed on 31 October 2013].

Murin, A. and Majovsky, J. 1976. In IOPB chromosome number reports LIII. Taxon **25**: 483-500.

Oliver, R.L. and Lewis, W.H. 1962. Chromosome numbers of *Sisyrinchium* (Iridaceae) is Eastern North America. Sida **1**: 43-48.

Ozcan, M., Hayırlıoğlu-Ayaz, S. and Inceer, H. 2008. Chromosome counts of some *Cirsium* (Asteraceae, Cardueae) taxa from Turkey. Caryologia **61**(4): 375-382.

Özhatay, F.N., Kültür, Ş. and Gürdal, M.B. 2011. Checklist of additional taxa to the supplement Flora of Turkey V. Turk. J. Bot. **35**: 589-624.

Rudall, P., Kenton, A.Y. and Lawrence, T.J. 1986. An anatomical and chromosomal investigation of *Sisyrinchium* and allied genera. Bot. Gaz. **147**(4): 466-477.

Vardar, Y. 1987. Botanikte Preparasyon Teknigi. Izmir, Ege University Science Faculty Press (in Turkish).

9

STENOCHLAENA RIAUENSIS (BLECHNACEAE), A NEW FERN SPECIES FROM RIAU, INDONESIA

NERY SOFIYANTI[1], DYAH IRIANI, FITMAWATI AND ANDESBA A. ROZA

Laboratory of Botany, Department of Biology, Faculty of Mathematics and Natural Sciences, University of Riau, Kampus Bina Widya Panam, Pekanbaru, Riau, Indonesia

Keywords: Blechnaceae; *Stenochlaena*; New species; Indonesia.

Abstract

Stenochlaena riauensis, a new fern species from Riau, Indonesia is described and illustrated. This species can be distinguished from its most closely related species *S. palustris* by the position of fertile and sterile pinnae. In the new species, both fertile and sterile pinnae are located on the same frond, the lower part of lamina consisted of sterile pinnae that gradually narrower towards apex forming linear fertile pinnae.

Introduction

Stenochlaena J. Smith belonging to the family Blechnaceae is a small genus consisting of only six species, and widely distributed in tropical and sub-tropical regions (Holttum, 1971; Chamber, 2013; Wang *et al.*, 2013). Most of the species of the genus have both sterile and fertile pinnae. Chamber (2013) reported that this genus is strongly dimorphic, in which fertile and sterile pinnae are located at the different fronds, with imparipinnate pinnae. Furthermore, *Stenochlaena* is charcterized by possessing radially symmetrical rhizome (Chamber, 2013); purplish-red to red fronds when young; sterile pinnae entire (Piggot, 1996); lanceolate to oblong-lanceolate, lateral veins arising from a closely parallel series of areolae; acrotischoid sori, that densely covering the dorsal surface or the lower surface of the fertile pinnae and not arrranged in discrete lines or dots; spores with a very thin and closely appressed perispore (Chamber, 2013).

During our expedition in Riau Province (Sumatera Island, Indonesia) from 2012 to 2014, we discovered that *S. palustris* (Burm.f.) Bedd (1876) is the the most widely distributed and abundant fern species in this province (Fig. 1). Local people consume the jouvenile leaves of this species as vegetables. Surprisingly, some specimens collected from conservation forest (Taman Hutan Raya Sultan Syarif Hasyim, Siak) and Pekanbaru, Riau showed a different type of fronds. Both fertile and sterile pinnae are located in one frond, the lower part of lamina are consisted of sterile pinnae that gradually narrower forming fertile pinnae on the tip of leaves. The other morphological characters are similar to *S. palustris.* Our detail observation of morphological characters of these specimens, brings us to a conclusion to a new *Stenochlaena* species, *Stenochlaena riauensis* sp. nov.

Stenochlaena riauensis Sofiyanti, Iriani, Fitmawati & Roza, **sp. nov.** (**Fig. 2**).

Diagnosis: *Stenochlaena riauensis* morphogically differs from its closest related species *S. palustris* (Burm. F.) Bedd by being monomorphic, and having both fertile and sterile pinnae located on one frond, the lower part of lamina consisted of sterile pinnae that gradually narrower

[1]Corresponding author. Department of Biology, Faculty of Mathematics and Natural Sciences, University of Riau, Kampus Bina Widya Jl. Bangkinang-Pekanbaru Km 12.5 Panam, Pekanbaru, Riau, Indonesia. Email: nery_yusuf@yahoo.com

towards apex, forming fertile pinnae on the tip, and the acrotischoid sori with monolete and reniform (bean-shaped) spores.

Types: Indonesia. Riau Province: Taman Hutan Raya Sultan Syarif Hasyim, 11 June 2012, *N. Sofiyanti, Fitmawati & A.A. Roza,* STR2 (*Holotype*: ANDA); Pekanbaru, 3 December 2014, *N. Sofiyanti & Fitmawati,* STRPKU1 (*Paratype*: ANDA).

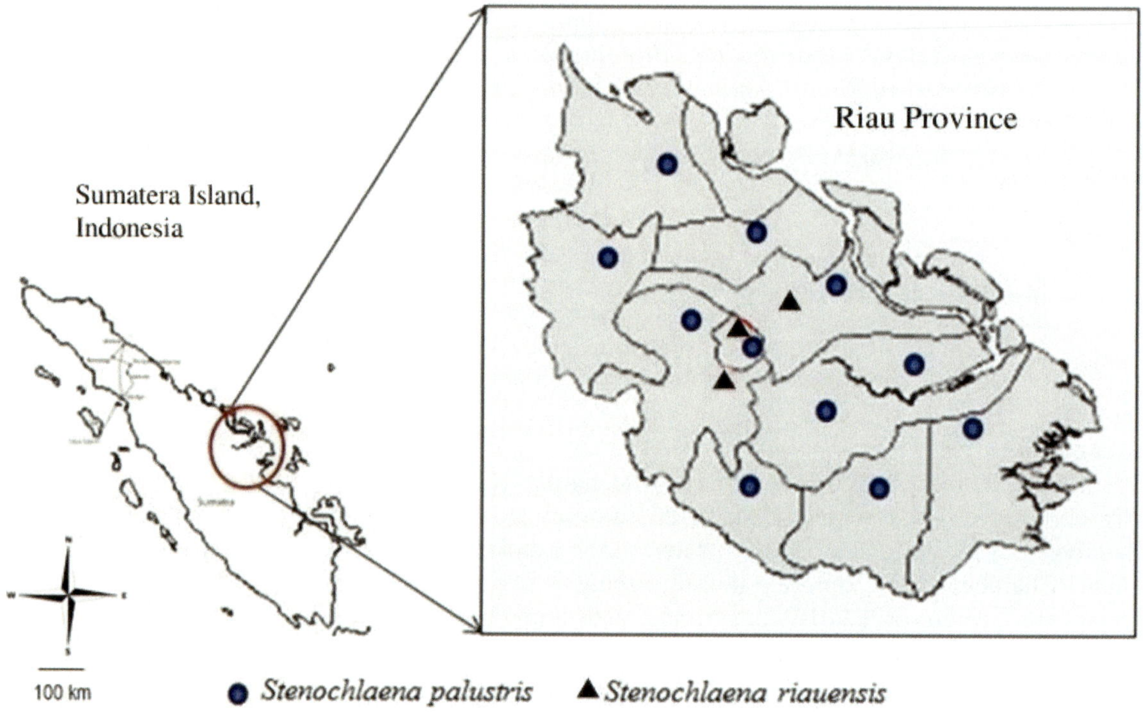

Fig. 1. Distribution map of *Stenochlaena* species in Riau, Indonesia.

Terrestrial, scrambling. Rhizome creeping or climbing, stout, 5–8 mm in diameter, light brown, covered by brown scales, especially towards the apex. Frond up to 90 cm long, pinnate. Stypes up to 20 cm long, glabrous, greenish brown at the base, green towards laminae. Rachis glabrous, dark green, c. 4 mm in diameter, having grove along the rachis. Laminae purplish red when young, dark green at maturity, c. 60–70 cm long, 1-pinnate, consisted of 17 pinnae. Pinnae sterile at the base of laminae, gradually narrower towards the apex and form fertile pinnae. Sterile pinnae 8 per frond, lanceolate to oblong-lanceolate, mostly cuneate, papery, glabrous on both surfaces, c. 11 cm long, 3 cm wide, rounded at the base, shortly stalked, blackish green, stalk c. 3 mm long, tip acute, serrate margin, veins anastomosing and forming single row of areoles along each side of costa. Fertile pinnae 9 per frond, started from the middle of laminae toward the apex, pinnae gradually reduce towards the apex of lamina, shortly stalked, form elongated fertile pinnae on the tip, usually asymetric at the base, the reduced sterile pinnae glabrous on both surfaces, c. 7 cm long, 1.5 cm wide, blackish green stalk c. 2.0–2.5 mm long, margin serrate, veins anastomosing and forming single row of areoles along each side of costa, at the fertile parts become yellowish green and narrower, linear, the top pinnae is the longest (c. 8 cm long, 2 mm wide), margin curve forming a linear groove at ventral surface. Sori cover the dorsal surface forming acrotischoid sori, brownish yellow. Spores monolete, reniform or bean-shaped.

Etymology: The specific epithet refers to the type locality, Riau Province of Indonesia.

Fig. 2. *Stenochlaena riauensis* Sofiyanti, Iriani, Fitmawati & Roza, **sp. nov.** A. Mature frond; B. Alternate pinnae arrangement; C. Sterile pinna; D. Fertile Pinna; E. Dorsal surface of fertile pinnae; F. Ventral surface of fertile pinna; G. Leaf base of fertile pinnae showing symetric pinna; H. Leaf base of sterile pinnae, showing asymetric pinna; I. Serrate leaf margin; J. Spore (scale bar = 10 μm).

Distribution: *Stenochlaena riauensis* was first found between 75 m and 150 m in conservation forest, Taman Hutan Raya Sultan Syarif Hasyim, Riau Province, Sumatera Island, Indonesia. This forest is a low land tropical forest that dominated by Dipterocarpaceae species, such as *Shorea acuminata* Dyer (Meranti Rambai), *Shorea leprosula* Miq. (Meranti pirang), *Shorea parvifolia*

Dyer (Meranti Bunga), *Parashorea aptera* Slooten (Meranti Batu), and *Dipterocarpus* sp. (Kruing). This species is also recorded from Pekanbaru District, Riau.

Habitat: This species commonly grows in open area or partly shaded area and mostly in streetside, at Riau Province, Indonesia.

Note: In Riau Province, *S. riauensis* shares the same distribution area with *S. palustris* (Fig. 1). However, *S. palustris* is more abundant than the new species, and sometime is found as epiphyte on the oil palm tree (Sofiyanti, 2013) or other woody species, as well as in the open space area or partly shaded area (Sofiyanti *et al.*, 2014). Morphologically, these two species have close similarity, especially when young. Both species have purplish red fronds when young, usually covered by scales, 1-pinnate lamina, short stalked pinnae, with broadly rounded base of fertile pinnae, usually cuneate, serrate margin of sterile pinnae, with acuminate tip. The veins of sterile pinnae are anastomosing and forming single row of areoles along each side of costa. The sori type and spore characteristic are also the same. A comparative account of *S. riauensis* sp. nov. with its closely related species, *S. palustris*, is presented in Table 1.

Table 1. Comparison of diagnostic morphological characters of *Stenochlaena riauensis* sp. nov. with *S. palustris*.

Characters	*S. riauensis*	*S. palustris*
Type of frond	Monomorphic	Dimorphic
Position of sterile pinnae	At the base of laminae	At the sterile frond
Number of sterile pinnae	8	17–29
Position of fertile pinnae	At the upper part of laminae	At the fertile frond
Number of fertile pinnae	9	16–28
Lower part of fertile pinnae	Broader at the base, green	Linear, brown
Upper part of fertile pinnae	Gradually narrower toward the apex forming linear structure	Linear

Most of *Stenochlaena* species usually have acrosticoid sori and lack true indusium (Chamber, 2013) with monolete spores (Holttum, 1932; Chamber, 2013). The spore investigation of six *Stenochlaena* species conducted by Chamber (2013), presented the similar spore shape and ornamentation. Their spores are monolete spores with a single line indicating the splitting axis of mother spore with thin perispore, as observed in other Blechnaceae genus, i.e. *Blechnum* L. (Mendoza-Ruiz and Perez-Garcia, 2009). These characteristics are also found in newly described *S. riauensis*. The genus *Stenochlaena* was previously reported as strongly dimorphic fern, with distant fronds (Holttum, 1932, 1971; Chamber, 2013; Wang *et al.*, 2013). Usually, the sterile pinnae are shorter and wider than fertile pinnae. However, *S. riauensis* is not dimorphic fern, because the sterile and fertile pinnae are located in the same frond. Sterile pinnae are located at the base of laminae, and gradually narrower and reduced toward apex. The narrow tips are elongated forming fertile pinnae. The reduced sterile pinnae is located of the base of pinna, usually with asymetric base. This is the strong characteristic of *S. riauensis* to be treated as a distinct *Stenochlaena* species.

Acknowledgements

This study, led by the first author, was supported by Penelitian Unggulan Perguruan Tinggi (PUPT) Research Grant from Directorate General of Higher Education 2015, Indonesia. The authors thank the Forestry Department of Riau Province for giving permit to conduct the research.

They are also grateful for the support of many people during the field work, especially forestry staff from Taman Hutan Raya Sultan Syarif Hasyim, Riau.

References

Chamber, T.C. 2013. A review of the genus *Stenochlaena* (Blechnaceae, subfamily Stenochlaenoideae). Telopea **15**: 13–36.

Holttum, R.E. 1932. Further notes on the *Stenochlaena*, *Lomariopsis* and *Tetophyllum*. Gard. Bull. Str. Sett. **9**: 139–144.

Holttum, R.E. 1971. The genus *Stenochlaena* J. Smith with the description of new taxa. Amer. Fern J. **61**: 119–123.

Mendoza-Ruiz, A. and Perez-Garcia, B. 2009. Morphogenesis of the gametophytes of eight Mexican species of *Blechnum* (Blechnaceae). Acta Bot. Mex. **88**: 59–72.

Piggot, A.G. 1996. Fern of Malaysia in Colour. Tropical Press SDN. BH. Kuala Lumpur, 486 pp.

Sofiyanti, N. 2013. The diversity of epiphytic fern on the oil palm tree (*Elaeis guineensis* Jacq.) in Pekanbaru, Riau. Jurnal Biologi **XVII** (2): 51–55.

Sofiyanti, N., Iriani, D. and Roza, A.A. 2014. Morfologi Tumbuhan Paku di Taman Hutan Raya Sultan Syarif Hasyim, Riau. UNRI Press. Pekanbaru, Riau, 96 pp. (in Indonesian).

Wang, F.G., Xing, F.W., Dong, S.Y. and Kato, M. 2013. Blechnaceae. *In*: Wu, Z.Y., Raven, P.H. and Hong, D.Y. (Eds), Flora of China, Vols. **2 & 3** (Pteridophytes). Science Press, Beijing; Missouri Botanical Garden Press, St. Louis, pp. 411–417.

PERSICARIA LANKESHANENSIS (POLYGONACEAE: PERSICARIEAE), A NEW SPECIES FROM GUANGDONG, CHINA

TONG-JUN LIANG AND BO LI[1,2]

Lushan Botanical Garden, Jiangxi Province and the Chinese Academy of Sciences, Jiujiang, 332900, China

Keywords: New species; *Persicaria lankeshanensis*; China; Micromorphology.

Abstract

Persicaria lankeshanensis, a new species of Polygonaceae from the Lankeshan Nature Reserve, Guangdong province of China, is described and illustrated. *Persicaria lankeshanensis* is similar to *P. taquetii* but differs from the latter by having stems, leaves, and ocreae densely pubescent, pedicels shorter than bracts, smaller flowers and achenes, and achenes opaque and minutely granular-striate. A comparison of achenes, pollen and abaxial leaf epidermis between the two species is provided.

Introduction

Persicaria [Tourn.] *ex* Mill. contains c. 150 species of prostrate or twining annual or perennial herbs, and are distributed mainly in northern temperate and tropical regions (Brandbyge, 1993). It is characterized by many-flowered, spike-like or capitate inflorescence; usually entire, ciliate, or pectinate ocreae; 4–5-parted perianths; tepals with trifid venation; the presence of nectaries; 4–8 stamens (Haraldson, 1978; Ronse Decraene and Akeroyd, 1988); and mostly rectangular to elongated tepal epidermis cells with straight or slightly undulating anticlinal walls (Hong *et al.,* 1998). Within *Persicaria*, Haraldson (1978) recognized four sections, viz. *Cephalophilon* (c. 16 species), *Echinocaulon* (c. 21 species), *Persicaria* (c. 60 species), and *Tovara* (c. 3 species). Qaiser (2001) accepted this treatment, while Freeman (2005) included *Rubrineva* (c. 2 species) and Hou (2006) excluded *Tovara*.

During a field trip in September 2009 in the Lankeshan Provincial Nature Reserve in Guangdong Province of China, a slender *Persicaria* with completely interrupted spike-like inflorescence, densely pubescent stems, leaves, and ocreae, and tiny flowers and fruits was collected. The plants superficially resemble *P. taquetii* (H. Lév.) Koidz., but can be distinguished by a suite of morphological and micromorphological characters. After critical study we recognized it as a new species, *Persicaria lankeshanensis* **sp. nov.**

Materials and Methods

Morphological comparisons of the new species and its related taxa were carried out based on both herbarium specimens and live plants. Characters were measured using a micrometer and a stereomicroscope. Field studies were carried out from 2009 to 2011.

For scanning electron microscopy (SEM) observations, samples of achenes and pollen were removed from mature fruits and unopened flowers, respectively, from the herbarium specimens. After cleaned in 95% ethanol, mounted onto cupreous stubs, and coated by JFC-1100E sputter coater (JEOL Led., Janpan), samples were examined under JSM-6360LV SEM (JEOL Led., Japan) at a voltage of 15 KV. Terminology for achenes and pollen follows Hou (2006) and Zhang and Zhou (1998), respectively.

---

[1]Laboratory of Subtropical Biodiversity, Jiangxi Agricultural University, Nanchang, 330045, China
[2]Corresponding author. Email: hanbolijx@163.com

For light microscopy (LM) investigation of leaf epidermis, samples were taken from mature leaves of live plants, fixed in FAA solution (18 formalin:1 acetic acid:1 ethanol), dissected under a Zeiss stereoscope, and stained in a solution of 1% safranin before being mounted in glycerine jelly. Measurements and counts were conducted randomly from 5 areas in the prepared slide per species under LM. Stomatal density was calculated using the formula S/M × 100%, where S is the number of stomata in a given area (M). Terminology follows Zhu *et al.* (2007).

Persicaria lankeshanensis T. J. Liang & B. Li, **sp. nov.** **(Figs 1 & 2)**.

Diagnosis: *Species P. taquetii (H. Lév.) Kodiz. affinis, sed caules, folia, ochreae dense pubescens, perianthia 0.7-1.0 mm, achenia striato–granulatis, 0.7–1.0 × 0.5–0.7 mm differt.*

Type: CHINA. Guangdong: Zhaoqinq city, Lankeshan Provincial Nature Reserve, growing in wet grassy slopes along river side, altitude 200 m, 23°08′38″N, 112°35′20″E, 24 Sep 2009, *Bo Li* 0075 (*Holotype*: IBSC; *Isotype*: JXAU).

Annual herbs. Stems erect, prostrate, or ascending at base, 15-35 cm tall, slender, densely pubescent, much-branched from base, rooting at proximal nodes. Leaves: ocreae tubular, 4-6 mm long, membranous, densely pubescent, apex truncate, cilia 3-5 mm long; petioles nearly absent; leaf

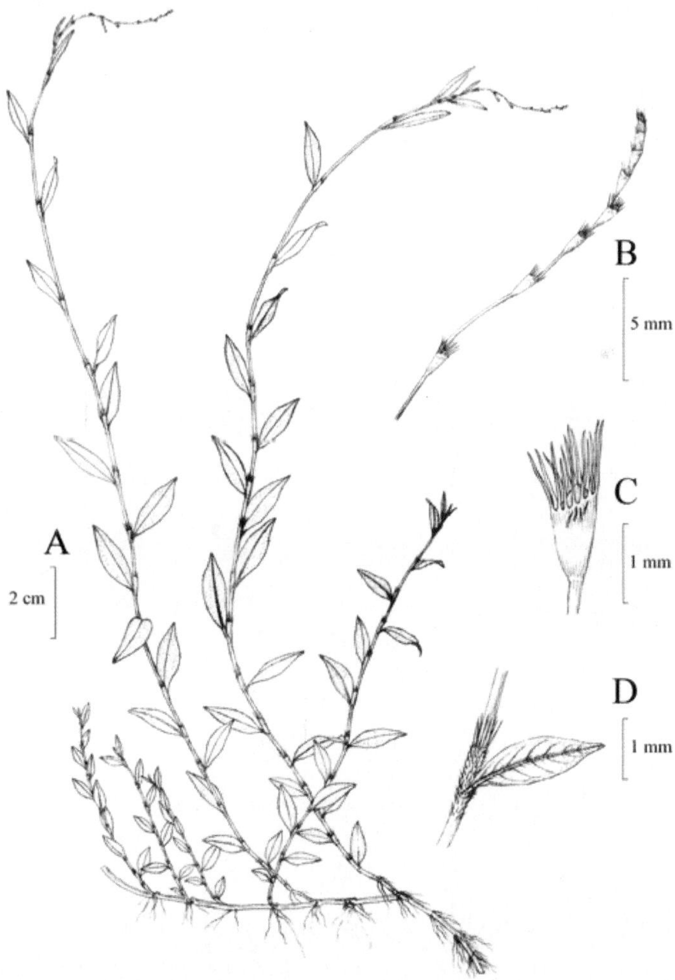

Fig. 1. *Persicaria lankeshanensis* T. J. Liang & B. Li, **sp. nov.** A. Habit; B. Inflorescence; C. Bract; D. Ocrea and leaf (all from the type, *Bo Li* 0075, IBSC).

blades narrowly lanceolate, 15-30 × 4-8 mm, both surfaces densely pubescent, base narrowly cuneate, or slightly rounded, margin entire, ciliate, apex acute. Inflorescence terminal or rarely axillary, spike-like, erect, 1.5-2.5 cm long, slender, interrupted. Bracts green, funnelform, 1.5-2.0 mm long, densely pubescent, margin submembranous, ciliate, each 2-3-flowered. Pedicels 0.8-1.3 mm long, much shorter than subtending bracts. Perianth white, 5-parted, lobes (or tepals) elliptic, 0.7-1.0 mm long. Stamens 7, included. Styles 3, included, connate to below middle; stigmas capitate. Achenes included or slightly exserted from persistent perianth, brown-black, minutely granular-striate, opaque, ovoid, trigonous, 0.7-1.0 x 0.5-0.7 mm.

Phenology: Flowering: August - October; fruiting; September - November.

Etymology: The specific epithet is derived from the type locality.

Distribution and habitat: *Persicaria lankeshanensis* is known only from two nearby localities in Lankeshan Provincial Nature Reserve, Guangdong province of China, from where it was first collected in 2009. It grows in wet, grassy slopes along river sides at an altitude of 100-200 m.

Notes: *Persicaria lankeshanensis* bears a superficial resemblance to *P. taquetii* in having very slender and completely interrupted inflorescence, tiny flowers and fruits, 7 stamens, small and narrowly lanceolate leaves, and slender stems. However, *P. lankeshanensis* differs from *P. taquetii* in plant indumentum, pedicel length, tepal length, and achene size and surface texture (Table 1; Fig. 3A vs. E).

Fig. 2 *Persicaria lankeshanensis* T.J. Liang & B. Li, **sp. nov.** A. Habit; B. A part of stem with ocrea showing pubescence; C. Pubescence of adaxial surface of leaf blade; D. Inflorescence.

In *Persicaria*, achenes (Yang *et al.*, 1991; Ronse Decraene *et al.*, 2000; Qu *et al.*, 2006), abaxial leaf epidermis (Meng *et al.*, 1997; Zhu *et al.*, 2007; Yasmin *et al.*, 2010a), and pollen (Hedberg, 1946; Wang and Feng, 1994; Zhang and Zhou, 1998; Yasmin *et al.*, 2010b) have been proved to be useful characters for the species delimitation. In this study, we found that *P. lankeshanensis* and *P. taquetii* also can be distinguished from each other by the epicarp sculpturing of the achenes, the shape of the epidermal cells on the abaxial leaf surface, the type of stomata, and pollen size and ornamentation (Table 1, Fig. 3).

Table 1. Comparison of *Persicaria lankeshanensis* sp.nov. with *P. taquetii*.

Characters	*P. lankeshanensis* sp. nov.	*P. taquetii*
Plant indumentum	Stems, leaves, and ocreae densely pubescent	Leaves and ocreae sparsely pubescent or rarely glabrous, stems glabrous
Pedicel	0.8-1.3 mm long, much shorter than bracts	1.8-2.7 mm long, obviously longer than bracts
Tepal	0.7-1.1 mm long	1.4-1.6 mm long
Achene		
Size (L × W)	0.8 (0.7-1.0) × 0.6 (0.5-0.7) mm	1.5 (1.3-1.6) × 0.8 (0.7-1.0) mm
Surface	Opaque, minutely granular striate	Shiny, smooth
Epicarp sculpturing	With irregular ridges formed by oblate tubercles following a longitudinal pattern	With irregular hollows
Leaf abaxial epidermis		
Epidermal cell shape and anticlinal wall	Polygonal, straight or curved	Irregular, sinuolate
Stomatal type	Anisocytic	Paracytic, rarely anisocytic
Stomatal density	88.1 / mm^2	168.5 / mm^2
Pollen		
Size	31.1-36.7 μm	37.5-56.8 μm
Width between two adjacent pores	8.2-12.3 μm	11.2-17.4 μm
Lumen number across the diameter	9-12	6-9
Lumen size	3.8-4.7 μm	6.4-10.1 μm
Granules in a lumen	Obscure, undeveloped	25-40, well developed
Number of coarse granules of murus in cross-section under LM	18-22	14-17
Width between two adjacent granules under LM	3.9-5.3 μm	8.2-11.3 μm

Additional specimens examined: CHINA. Guangdong Province: Zhaoqinq City, at the boundary of the Lankeshan Provincial Nature Reserve, 23°08′15″N, 112°34′58″E, alt. 15 m, 8 Oct 2010, *Bo Li and Zhu-Qiu Song* 201002 (IBSC).

Conservation status: *Persicaria lankeshanensis* is only recorded from the Lankeshan Provincial Nature Reserve, Guangdong, China. Given the species was observed only in two nearby localities in a about 5 km^2 area at the nature reserve, we propose to categorize it as Vulnerable (VU) on the IUCN Red List, according to the criteria D2 (IUCN, 2001).

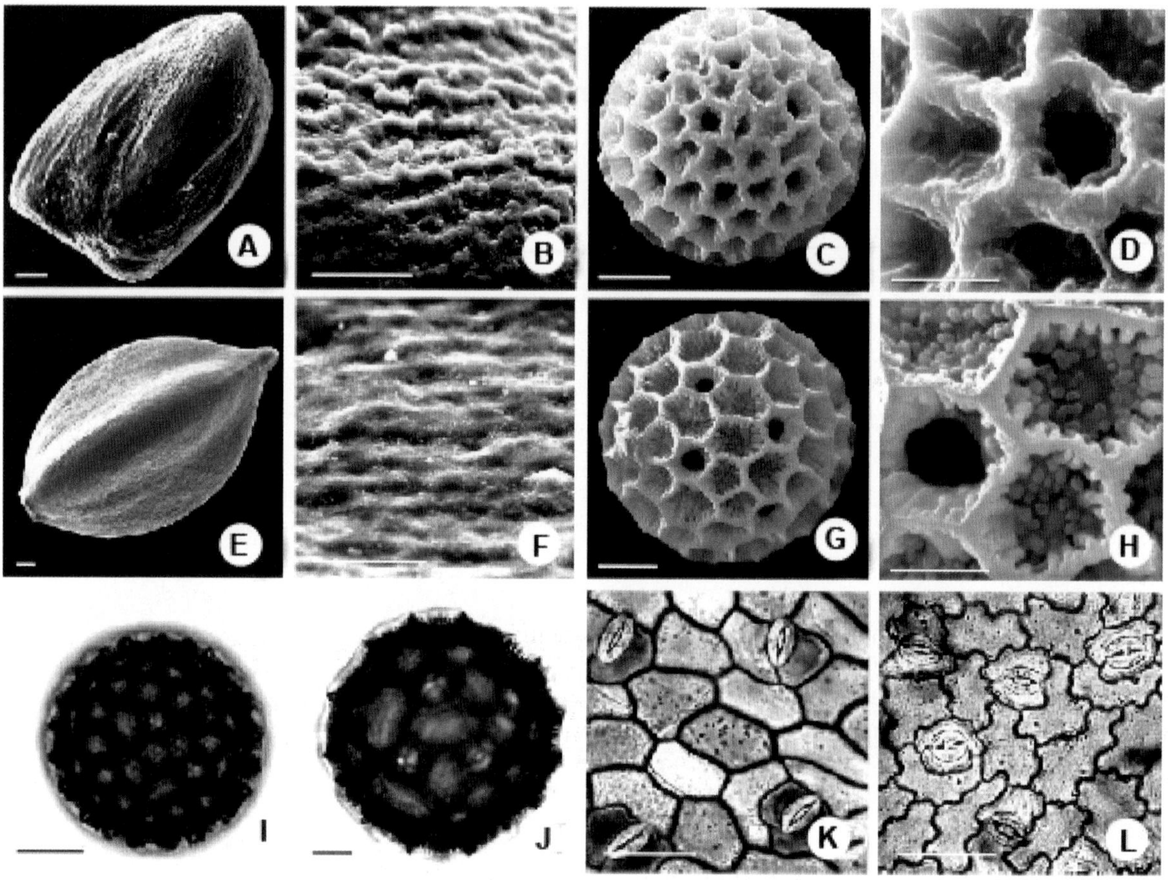

Fig. 3. Micromorphology of *Persicaria lankeshanensis* (A–D, I, K) and *P. taquetii* (E–H, J, L). A,B,E,F. Achene morphology under scanning electron microscopy (SEM); A,E. Lateral view; B,F. Epicarp sculpturing, C,D,G,H. Pollen morphology under SEM; C,G. Whole view; D,H. Lumen and granules inside; I,J. Pollen morphology under light microscopy (LM); K,L. Leaf abaxial epidermis under LM. Scale bars: A,E = 100 μm; B,F,K,L = 50 μm; C,G,I,J = 10 μm; D,H = 5 μm.

Acknowledgements

The authors are grateful to Prof. An-Jen Li of the Institute of Botany, Chinese Academy of Science, for identifying specimens, to Ms. Yun-Xiao Liu and Ms. Xiao-Ying Hu, South China Botanical Garden, Chinese Academy of Sciences, for the illustrations and SEM observations, respectively, and to Mr. Zhu-Qiu Song for field assistance.

References

Brandbyge, J. 1993. Polygonaceae. *In*: Kubitzki, K. and Bittich, V. (Eds), The Families and Genera of Vascular Plants. Vol. **2**. Springer Verlag, Berlin, pp. 531-544.

Freeman, C.C. 2005. Polygonaceae. *In*: Flora of North America Editorial Committee, Flora of North America. Vol. **5**. Oxford University Press, New York, pp. 574-594.

Haraldson, K. 1978. Anatomy and taxonomy in Polygonaceae subfam. Polygonoideae Meisn. emend Jaretzky. Symb. Bot. Upsal. **22**: 1-95.

Hedberg, O. 1946. Pollen morphology in the genus *Polygonum* L. *s.l.* and its taxonomical significance. Sven. Bot. Tidskr. **40**: 371-414.

Hong, S.P., Ronse Decraene, L.P. and Smets, E. 1998. Systematic significance of tepal surface morphology in tribes Persicarieae and Polygoneae (Polygonaceae). Bot. J. Linn. Soc. **127**: 91-116.

Hou, Y.T. 2006. Systematic studies on the tribe Polygoneae of China. Shandong Normal University, Jinan, 306 pp.

IUCN 2001. IUCN red list categories and criteria, v. 3.1. IUCN Species Survival Commission, Switzerland.

Meng, R.X., Zhou, Z.Z. and Wang, S.L. 1997. A study on foliar epidermis of genus *Polygonum* in Anhui. Anhui Univ. Nat. Sci. Edit. **21**: 81-93.

Qaiser, M. 2001. Polygonaceae. *In*: Nasir, E. and Ali, S.I. (Eds), Flora of Pakistan. Vol. **205**. Department of Botany, Karachi University & Missouri Botanical Garden Press, St. Louis, pp. 1-190.

Qu, C.Y., Hou, Y.T., Li, A.L., Lu, F.J. and Li, F.Z. 2006. Fruit shape and pericarp micromorphological characteristics of *Polygonum* section *Persicaria* from China. Bull. Bot. Res. **26**: 275-285.

Ronse Decraene, L.P. and Akeroyd, J.R. 1988. Generic limits in *Polygonum* and related genera (Polygonaceae) on the basis of floral characters. Bot. J. Linn. Soc. **98**: 321-371.

Ronse Decraene, L.P., Hong, S.P., and Smets, E. 2000. Systematic significance of fruit morphology and anatomy in tribes Persicarieae and Polygoneae (Polygonaceae). Bot. J. Linn. Soc. **134**: 301-337.

Wang, J.X. and Feng, Z.J. 1994. A study of the pollen morphology of the genus *Polygonum* L. in China. Acta Phyt.. Sin. **32**: 219-231.

Yang, J., Li, Y.H. and Wang, J.W. 1991. A primary study on the microstructural features of achene surfaces and its taxonomic importance to *Polygonum*. Acta Bot. Boreali-Occidentalia Sin. **11**: 1-7.

Yasmin, G., Khan, M.A., Shaheen, N. and Hayat, M.Q. 2010a. Taxonomic significance of leaf epidermal anatomy of selected *Persicaria* Mill. species of family Polygonaceae from Pakistan. Afr. J. Biotechnol. **9**: 3759-3768.

Yasmin, G., Khan, M.A. and Shaheen, N. 2010b. Pollen morphology of selected *Polygonum* L. species (Polygonaceae) from Pakistan and its taxonomic significance. Pak. J. Bot. **42**: 3693-3703.

Zhang, X.P. and Zhou, Z.Z. 1998. A study on pollen morphology and its phylogeny of Polygonaceae in China. University of Science and Technology of China Press, Hefei, 235 pp.

Zhu, L.T., Lu, F.J. and Hou, Y.T. 2007. Micro-morphology of leaf epidermis and its taxonomical significance of *Polygonum* section *Cephalophilon* in China. J. Wuhan Bot. Res. **25**: 136-142.

11

A NEW *RANUNCULUS* SPECIES (RANUNCULACEAE) FROM SHAANXI, CHINA

MIN-RONG LUO[1] AND LIANG ZHAO[2]

College of Life Sciences, Northwest A&F University, Yangling, Shaanxi, 712100, China

Keywords: Ranunculus shanyangensis; New species; China.

Abstract

Ranunculus shanyangensis M.R. Luo & L. Zhao, a new species of Ranunculaceae from Shaanxi, China, is described and illustrated. *Ranunculus shanyangensis* is morphologically similar to *R. repens*, however, it differs from the latter due to the following characters: longer adventitious roots, wholly creeping stems with no branches, longer rachis, fewer floral organs and pollen wall sculpturing. SEM micrography of the pollens for *R. shanyangensis* and *R. repens* are presented along with distribution map.

Introduction

Ranunculus is the largest genus within Ranunculaceae including about 600 species, primarily distributed in temperate to arctic or subantarctic zones, but a few species are also found in high montane regions of the tropics (Tamura, 1995). Wang and Gilbert (2001) revised the genus *Ranunculus* in China and recognized 125 species in China, of which 66 are endemic.

When we were carrying out a fieldwork survey in May 2009 in Shaanxi Province, China, we came across a distinct population of *Ranunculus.* This population grew in slightly moist region in the Qinling Mountains. Notable morphological characteristics that distinguish this population from other species include longer adventitious roots, longer rachis, wholly creeping stems that lack branches and fewer floral organs. Based on literature review (Davis, 1965; Timokhina, 1993; Tutin, 1993; Whittemore, 1997; Wang and Gilbert, 2001; Kadoto, 2006; Tzvelev, 2007) and examination of specimens from several herbaria, we have determined that individuals of this population are similar to *R. repens* L. However, combination of characteristics distinguishes this population as new species, *R. shanyangensis*. Here we describe this new species.

Materials and Methods

The specimens of *Ranunculus repens* from several herbaria (PE, WUK, NAS, IFP, LBG, IBSC, IBK) have been examined representing collections made in China as well as Europe and North America. Flowers and fruits were rehydrated and measurements were taken for each character.

Palynological investigation was conducted to determine whether unique characteristics were present in *R. shanyangensis*. Pollen samples were obtained from dried herbarium specimens and suspended in distilled water, after acetolysis, fixed on SEM specimen holders, which had been cleaned with acetone. The specimens were sputter-coated with gold for 1.5 min and photographed using a Hitachi S-4800 scanning electron microscope. Palynological terminology follows Santisuk (1979).

[1]College of Horticulture, Northwest A&F University, Yangling, Shaanxi, 712100, China
[2]Corresponding author. Email: biology_zhaoliang@126.com

Ranunculus shanyangensis M.R. Luo & L. Zhao **sp. nov.** **(Figs 1, 3)**.

Diagnosis: *Species nova R. repenti L. similis est, a quo caule toto tranversaliter repente, simplice, flaccido, ad nodos radicante, radicibus tenuibus, 5-10 cm longis, pedunculo longiore 13-15 cm longo, staminibus 35-40 per florem, granis pollenis pantocolpatis, carpellis 15-20 differt.*

Type: Shanyang County, Shaanxi Province, North-west China; altitude 1100 m; 33°25′20"N, 110°08′00"E, 15 May 2009, Liang Zhao 20090501 (*Holotype*: WUK; *Isotype*: WUK, PE).

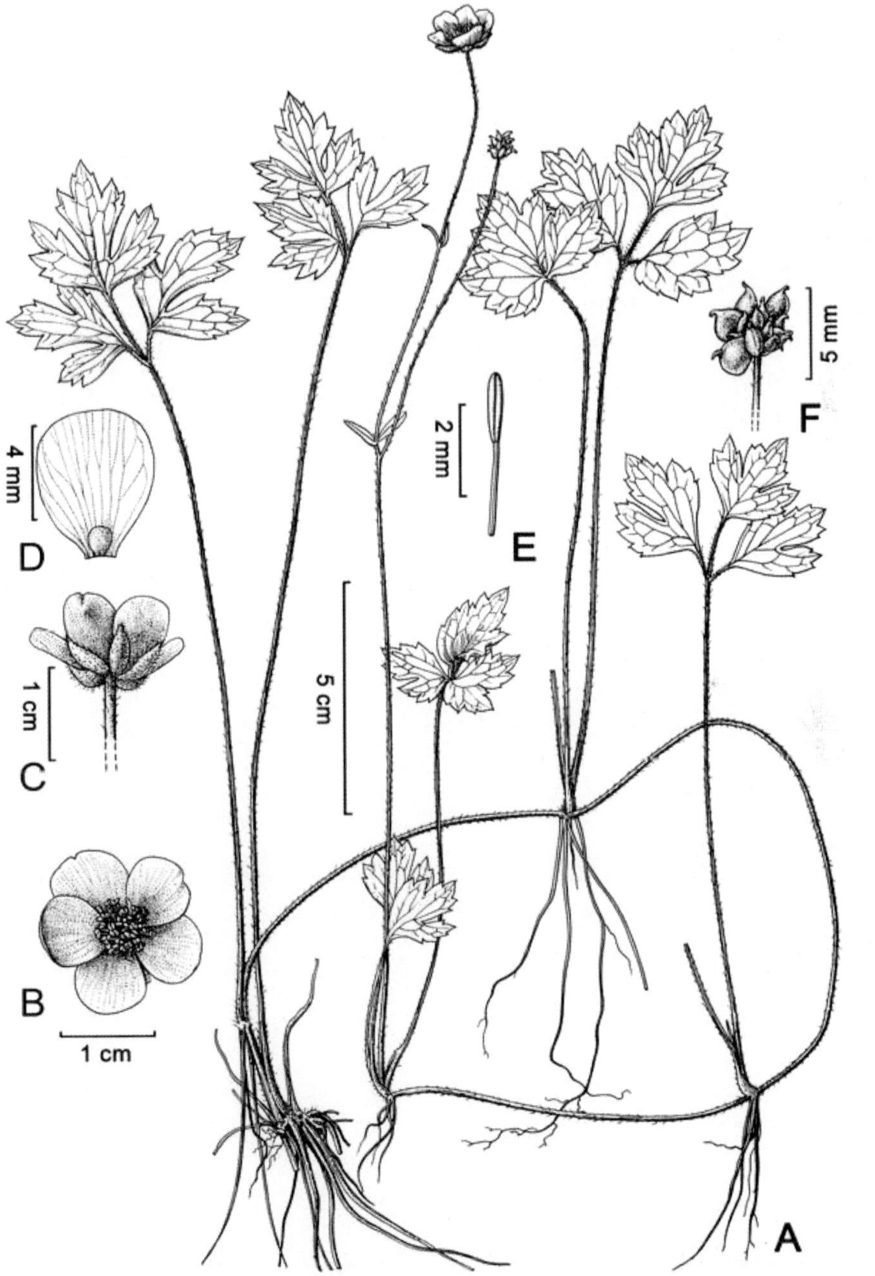

Fig. 1. *Ranunculus shanyangensis* M.R. Luo & L. Zhao **sp. nov.** (A) Habit, (B) Flower - apical view, (C) Flower - lateral view, (D) Petal, (E) Stamen, (F) Fruit.

Perennial herbs. Adventitious roots 5-10 cm long. Stems wholly creeping, 20-50 cm long, slender, c. 1.0-1.2 mm in diameter, hirsute, rooting at the nodes, old petiole persistent on nodes. Leaves mostly basal, or in nodes of stolons (2-4); petiole 5-18 cm long; blade ternate, reniform to pentagonal; both surfaces hirsute; central leaflet broadly rhombic, 2.0-2.6 × 2.0-2.5 cm; base broadly cuneate, secondary lobes lobed or partite; lateral leaflets oblique, unequally 2- or 3-lobed or partite; petiolule 0.5-2.0 cm long. Cauline leaves smaller. Flowers (1-2) in loose cyme, axillary; rachis 13-15 cm long; bracts 3-lobed or undivided, lanceolate to linear. Flowers 1.4-1.7 cm in diameter; pedicels 5-7 cm long, hirsute. Receptacle hirsute. Sepals 5.5-6.0 × 2-4 cm, navicular at anthesis, hirsute on abaxial surface. Petals 5-8 × 4-5 mm, obovate or widely obovate; claw 1 mm long; nectary covered by a scale, c. 1 mm long. Stamens 35-40; anthers c. 1.5 mm long; filaments c. 2.0 mm long. Carpels 15-20. Achene bilaterally compressed, obliquely obovate, c. 2.2×2.0 mm, glabrous, narrowly marginate; beak c. 0.5 mm long.

Phenology: It sprouts in late February, and the aboveground parts die in early November. Flowering occurs in May - August and fruiting in June - September.

Specimens examined: Shaanxi, Shanyang County, in slightly moist areas, 1100 m, 15 May 2009, Liang Zhao 20090502 (SANU); *ibid.,* 1105 m, Liang Zhao 20090503 (SANU), Liang Zhao 20090504 (SANU), Liang Zhao 20090505 (SANU), Liang Zhao 20090506 (SANU), Liang Zhao 20090507 (SANU), Liang Zhao 20090508 (SANU); *ibid.,* Liang Zhao 20090509(PE), Liang Zhao 20090510 (PE).

Distribution: Ranunculus shanyangensis is found in the moist area in Shanyang County, Shaanxi Province, North-west China (Fig. 2). Recent observations suggest that the population of this new species has a narrow geographic distribution, and possibly endemic to this site. After intensive field investigation in Qinling Mountains from 2009 to 2010, we only found one population with about 200 individuals.

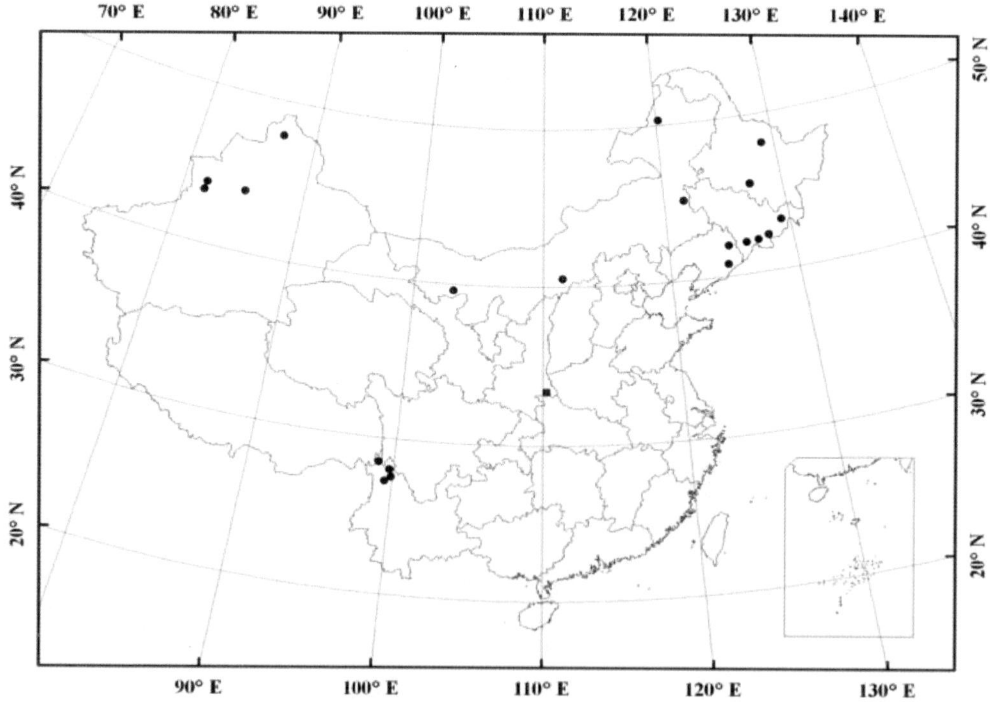

Fig. 2. Distribution of *Ranunculus repens* (●) and *R. shanyangensis* (■) in China.

Conservation status: Using the IUCN categories and criteria (IUCN, 2001), we suggest ranking of *Ranunculus shanyangensis* as 'Critically Endangered' (CR), which is based on the size of this population and its close proximity to urban and agricultural areas.

Etymology: The specific epithet refers to Shanyang County where the type collections were made.

Notes: Ranunculus shanyangensis is morphologically similar and possibly closely related to *R. repens* L. Both have creeping stems, rooting at the nodes; smooth, bilaterally compressed and narrowly marginate achenes; and the petal nectary pit is covered by a scale. However, *R. shanyangensis* differs from *R. repens* in adventitious root (5-10 cm vs. 1-3 cm), stem (wholly creeping, no branched vs. ascending or suberect, branched above), rachis (13-15 cm vs. 5-8 cm) , stamens (35-40 vs. c. 50), carpels (15-20 vs. 30-35) and pollen (stephanocolpate vs. tricolpate) characters.

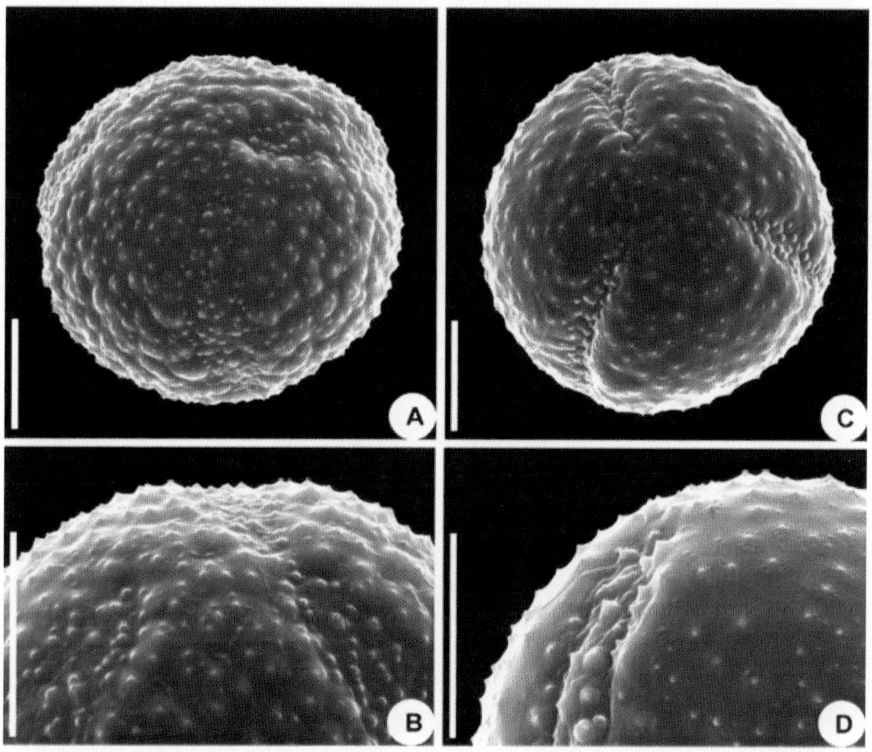

Fig. 3. Pollen morphology under scanning electron microscopy (SEM). (A, B) *Ranunculus shanyangensis*, (C, D) *R. repens.* Scale bar: A, B, C, D = 10 μm

Pollen morphology:

In our examination of the pollen in *R. repens* from China, Europe and North America, we note that all of them are tricolpate (Fig. 3C) with small and sparsely spaced spinules (Fig. 3D). This character is in accord with the results of Erdtman *et al.* (1961), whose material was collected from Scandinavia. However, the pollen of *Ranunculus shanyangensis* is pantocolpate (Fig. 3A) and bears densely spaced spinules (Fig. 3B).

Acknowledgements

We are grateful to Professor Wen-Tsai Wang, Institute of Botany, Chinese Academy of Sciences, China for his help. We thank Dr. Mare Nazaire, Washington State University, United

States, and Dr. Jeremy Lundholm, Saint Mary's University, Canada for their helpful comments and correcting the English. We also thank Mr. Yin-Hou Xiao for his assistance in taking SEM photographs. This project was supported by the Fundamental Research Funds for the Central Universities (No. QN2012020), the Specialized Research Fund for the Doctoral Program of Higher Education of China (No. 20120204120032) and the National Natural Science Foundation of China (31300158 and 31100141).

References

Davis, P.H. 1965. *Ranunculus* L. *In:* Davis, P.H. (Ed.), Flora of Turkey. Vol. **1.** Edinburgh University Press, pp. 146-197.

Erdtman, G., Berglund, B. and Praglowski, J. 1961. An introduction to a Scandinavian pollen flora. Grana **2**: 1-92.

IUCN 2001. IUCN red list categories and criteria, version 3.1. IUCN Species Survival Commission, Switzerland.

Kadoto, Y. 2006. *Ranunculus* L. *In:* Iwatsuki, K., Boufford, D.E. and Ohba, H. (Eds), Flora of Japan. Vol. **2**(a). Kodansha, Tokyo, pp. 310-322.

Santisuk, T. 1979. A palynological study of the tribe Ranunculeae. Opera Bot. **48**: 1-74.

Tamura, M. 1995. Ranunculaceae. *In:* Hiepko, P. (Ed.), Die Natürlichen Pflanzenfamilien, 2nd ed. Vol. **17**a (4). Duncker und Humblot, Berlin, pp. 1-556.

Timokhina, S.A. 1993. *Ranunculus* L. *In:* Malyschev, L.I. and Peschkova, G.A. (Eds), Flora of Siberia. Vol. **6**. Science Publishers, Inc., Enfield, NH, USA, pp. 162-195.

Tutin, T.G. 1993. *Ranunculus* L. *In:* Tutin, T.G., Burges, N.A., Chater, A.O., Edmondson, J.R., Heywood, V.H., Moore, D.M., Valentine, D.H., Walters, S.M. and Webb, D.A. (Eds), Flora Europaea. Vol. **1**. 2nd edition. Cambridge University Press, pp. 269-186.

Tzvelev, N.N. 2007. *Ranunculus* L. *In:* Tzvelev, N.N. (Ed.), Flora of Russia. Vol. **10**(a). Taylor and Francis / Balkema. AK Leiden, The Netherlands. pp. 116-188.

Wang, W.T. and Gilbert, M.G. 2001. *Ranunculus* L. *In:* Wu, Z.Y., Raven, P.H. and Hong, D.Y. (Eds), Flora of China. Vol. **6**. Science Press & Missouri Botantical Garden Press. pp. 391-431.

Whittemore, A.T. 1997. *Ranunculus* L. *In:* Flora of North America Editorial Committee, Flora of North America. Vol. **3**. Oxford University Press, pp. 88-134.

12

NEW RECORDS OF SEVEN FUNGAL SPECIES FOR BANGLADESH

AMENA KIBRIA, K.S. HOSSAIN[1], N. AKHTAR[2], M.A.A. JAHAN[2], MD. A.M. SARKER[2] AND MST. N. BEGUM[2]

Department of Botany, Jagannath University, Dhaka 1100, Bangladesh

Keywords: Fungi; New records; Bangladesh.

Abstract

From three farming stages of *Arthrospira platensis* (Nordstedt) Gomont [*Spirulina platensis* (Gomont) Geitler], seven fungal species, namely, *Cladosporium varians* Braun, Melnik & K. Schub., *Fusarium trichothecioides* Wollenw., *Geotrichum candidum* Link, *Mucor circinelloides* Tiegh., *M. hiemalis* Wehmer, *Penicillium frequentans* Westling and *Verticillium albo-atrum* Reinke & Berthold were recorded for the first time from Bangladesh.

Introduction

Fungi are the second largest kingdom, with about 70,000−100,000 known species (Kendrick, 2000). But knowledge on the mycodiversity is still deficient and incomplete, making inventory ponderous due to the inadequate number of researchers in this area (Mueller *et al*., 2004). A total of 275 fungal species under 125 genera have been reported from Bangladesh (Siddiqui *et al*., 2007). *Spirulina*, the most widely exploited food microalga, usually expose under open air during commercial production. Air of the earth contains various fungal propagules. These airborne propagules can be settled on culture, slurry and powder of *Spirulina*. Eleven fungi species were isolated and identified from culture, slurry and powder of *Spirulina* at Bangladesh Council of Scientific and Industrial Research (BCSIR), Dhaka. Out of 11 species seven were recorded for the first time from Bangladesh. This study presents taxonomic description of these seven fungal species under five different genera.

Materials and Methods

Fungi were isolated from culture, slurry and powder of *Arthrospira platensis* (Nordstedt) Gomont [*Spirulina platensis* (Gomont) Geitler; common name: Spirulina] at Bangladesh Council of Scientific and Industrial Research (BCSIR), Dhaka following Agar and/or Dilution Plate techniques during the last week of December 2012.

Colony character of pure culture of each isolate was studied. They were examined microscopically after preparation of glass slides with cover slip or sticky tap by using lactophenol-cotton blue solution (0.05 gm cotton blue in 100 ml lacto-phenol). Microscopic images of each isolates were captured with the aid of Olympus-DP20 digital camera attached with Olympus CX41 compound microscope at 400x. Isolates were identified up to species level with the help of standard mycological literature (Bensch *et al.,* 2012; Booth, 1971; Crous and Groenewald, 2013; Ellis, 1971; Ellis *et al.,* 2007; Harvey, 1965; Jabnoun *et al.,* 2010; Kurtzman *et al.,* 2011; Raper and Thom, 1949; Schipper, 1973, 1976, 1978; Zare, 2003).

---

[1]Corresponding author. Email: ksh1968@gmail.com
[2]Biological Research Division, Bangladesh Council of Scientific and Industrial Research (BCSIR), Dhaka-1000, Bangladesh.

Pure cultures of the described species are preserved at the Department of Botany, Jagannath University, Dhaka, Bangladesh.

Results and Discussion

A total of 11 fungal species belonging to eight different genera were studied and identified. Among these, seven species *viz.*, *Cladosporium varians* Braun, Melnik & K. Schub., *Fusarium trichothecioides* Wollcnw., *Geotrichum candidum* Link, *Mucor circinelloides* Tiegh., *M. hiemalis* Wehmer, *Penicillium frequentans* Westling and *Verticillium albo-atrum* Reinke & Berthold are found new records for Bangladesh. Four others were *Arthrinium phaeospermum* (Corda) Ellis, *Cladosporium sphaerospermum* Penz., *Cunninghamella bertholletiae* Stadel and *Nigrospora oryzae* (Berk. & Br.) Petch.

Taxonomic descriptions of the newly recorded species are as follows:

1. **Cladosporium varians** U. Braun, V.A. Melnik, and K. Schubert, Mikologiya I Fitopatologiya 42(3): 215 (2008). **(Figs 1 & 8)**.

Colonies on PDA reaching 29.5 mm in diameter after 14 days, dark grey-olivaceous, olivaceous or iron-grey, reverse grey olivaceous, velvety to powdery, growth regular, flat to low curved, margin white, extensive. Mycelium mainly immersed, pale olivaceous grey; consisting of branched or unbranched, 4.1–5.0 µm wide hyphae. Conidiophores macronematous or micronematous, solitary, straight or rising, unbranched or branched or dichotomous branched, terminally or laterally formed from hyphae, septate, 58.1–87.2 × 2.8–5.5 µm. Conidiogenous cells are cylindrical oblong, integrated, terminal, 15.6–17.5 × 6.9–7.5 µm. Conidia broadly ellipsoidal to ellipsoidal or subglobose, aseptate conidia 5.4–12.9 × 3.9–5 µm, single septate conidia 10.0–17.5 × 3.8–5.6 µm, ramoconidia 11.4 × 5.0 µm.

Specimen examined: Isolate no. K.S. Hossain 267, 30 December 2012, from spirulina powder, BCSIR, Dhaka, collected by Amena Kibria.

Notes: *Cladosporium varians* belongs to the *C. cladosporioides* complex, but differs from *C. cladosporioides* by its long, frequently branched conidiophores. Furthermore, the tips of the conidiogenous cells are often somewhat swollen or unilaterally swollen, and the ramoconidia have up to four septa, and subglobose conidia are not abundant. *C. tenuissimum*is another comparable species, which is, however, easily distinguishable by its setiform, usually unbranched conidiophores (Bensch *et al.*, 2012).

2. **Fusarium trichothecioides** Wollenw., J. Wash. Acad. Sci. 2:147 (1912). **(Figs 2 & 9)**.

Colonies attaining 5 cm in diameter after 4 days on PDA, 6.5 cm in diameter after 6 days on SNA, pale pinkish-white, and reverse light brownish pink. It grows sparingly at 37° C, negative on urease reaction. It's able to grow without vitamin supplement. Mycelium cottony, extensive, 2.5–4.4 µm wide. Conidiophores comprise of a basal cell bearing 2 or 3 apical phialides, 18.0 × 3.8 µm. Macroconidia 14–24 × 2.5–5 µm, 3 to 5 septate or 8–16 × 2.5–5.0 µm, 1 to 2 septate, formed frequently from lateral conidiophores. Microconidia are not present. Chlamydospores are smooth walled, terminal, intercalary, spherical, 8–15 µm in diameter. Sporodochia, ascus and ascospore absent.

Specimen examined: Isolate no. K.S. Hossain 251, 24 December 2012, from spirulina slurry, BCSIR, Dhaka, collected by Amena Kibria.

3. **Geotrichum candidum** Link, Magazin der Gesellschaft Naturforschenden Freunde Berlin 3(1): 3–42 (1809). **(Figs 3 & 10)**.

Colonies on PDA reaching 4.5 cm in diameter after 6 days, grows rapidly, white to cream colored, powdery, flat with aerial mycelium. Mycelium consisting of septate, hyaline, dichotomously branched and 3–5 µm wide hyphae. Arthroconidia hyaline, one-celled, smooth, slimy, subglobose to cylindrical, 4.7–10.1 × 2.7–5.4 µm, released by the separation of a double septum. Disjunctor cells, blastoconidia, budding and ascospore not found. It grows sparingly at 37° C, negative on urease reaction. It's able to grow without vitamin supplement.

Specimen examined: Isolate no. K.S. Hossain 262, 24 December 2012, from spirulina powder, BCSIR, Dhaka, collected by Amena Kibria.

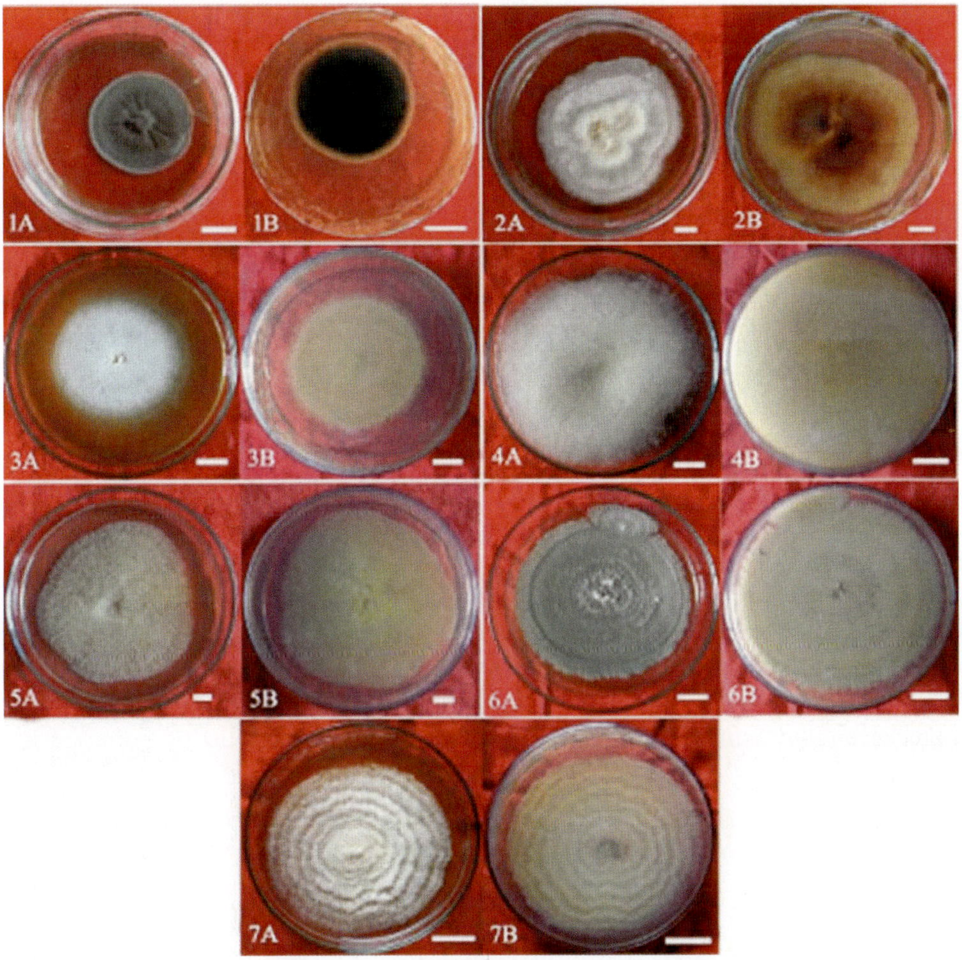

Figs 1-7. Colonies of the seven fungal species on PDA medium at different incubation periods, A. Front views; B. Reverse views. 1. *Cladosporium varians*, 2. *Fusarium trichothecioides*, 3. *Geotrichum candidum*, 4. *Mucor circinelloides*, 5. *Mucor hiemalis*, 6. *Penicillium frequentans*, 7. *Verticillium albo-atrum*. [Scale bars = 10 mm]

4. **Mucor circinelloides** Tiegh., Annales des Sciences Naturelles Botanique 1: 94 (1875). **(Figs 4 & 11)**.

Colonies on PDA attaining 6.1 cm in diameter, up to 13 mm high at room temperature; 6.9 cm in diameter at 30° C and 7.8 cm in diameter and 13 mm high at room temperature on Beer Wort Agar after 4 days. Colonies pale smoke gray without any zonation, reverse off white. Sporangiophores branched, 6.2–16.2 µm in diameter. Sporangia 18.75–53.75 × 19.38–50.63 µm, globose or spherical. Columellae globose or spherical, 16.25–41.25 × 18.13–31.88 µm, collars present. Sporangiospores 5.6–13.1 × 2.25–3.13 µm, broadly ellipsoidal to ellipsoidal, few fusiform. Chlamydospores present, 7.50–16.88 µm thick.

Specimen examined: Isolate no. K.S. Hossain 255, 24 December 2012, from spirulina culture, BCSIR, Dhaka, collected by Amena Kibria.

5. **Mucor hiemalis** Wehmer, Annales Mycologici 1(1): 37 (1903). **(Figs 5 & 12)**.

Colonies on PDA attaining 8.3 cm in diameter at room temperature; on Beer Wort Agar 4.4 cm in diameter at 30° C and 5.2 cm in diameter and 5 mm high at room temperature after 4 days. Colonies cottony to fluffy, pale yellow-brown colour, reverse light yellowish to ocherish. Sporangiophores branched, 2.5–13.7 µm in diameter. Sporangia globose, 40.63–65 × 37.5–62.5 µm. Columellae globose to subglobose, 12.5–36.88 × 12.5–32.5 µm, collars present. Sporangiospores 4.4–12.5 × 2.8–8.1 µm, subglobose to ellipsoidal, few ovals. Chlamydospores 5.63–16.88 µm thick.

Specimen examined: Isolate no. K. S. Hossain 259, 24 December 2012, from spirulina culture, BCSIR, Dhaka, collected by Amena Kibria.

6. **Penicillium frequentans** Westling, Arkiv før Botanik 11(1): 133 (1911). **(Figs 6 & 13)**.

Colonies on PDA 5.2 cm in diameter after 10 days at room temperature, spreading rapidly, radiately wrinkled, broadly zonate, margin thin. Mycelium consisting of hyaline, aseptate, branched hyphae. Conidiophores arise from hyphae velvety, crowded, short, septate, heavily sporing, up to 200.6 µm long and 2.6–3.6 µm wide, with smooth or finely roughed walls, and apices enlarged up to 4 µm or more in width. Sterigmata mostly 8.0–12.0 × 3.0–3.5 µm, formed in crowded clusters numbering 10 to 12 or more, produce conidial chain. Conidia mostly 2.6–3.1 µm in diameter, smooth, globose to subglobose, thin-walled.

Specimen examined: Isolate no. K.S. Hossain 263, 27 December 2012, from spirulina slurry, BCSIR, Dhaka, collected by Amena Kibria.

7. **Verticillium albo-atrum** Reinke & Berthold, Die Zersetzung der Kartoffel druch Pilze 1: 75 (1879). **(Figs 7 & 14)**.

Colonies on PDA reaching 4.1 cm in diameter after 14 days at room temperature, white, velvety, thinning at margin, several centric weavy zone present, margin weavy. Reverse light brownish yellow. Mycelium septate, hyaline. Conidiophores branched, hyaline, more or less erect, and verticilliate, c. 4.25 µm thick. Phialides 2–4, 2 µm thick. Conidia broadly ellipsoidal to ellipsoidal, few oval, end tapered, ellipsoidal to sub cylindrical, hyaline, aseptate, 3.9–5.5 × 1.7–2.4 µm.

Specimen examined: Isolate no. K.S. Hossain 266, 29 December 2012, from spirulina culture, BCSIR, Dhaka, collected by Amena Kibria.

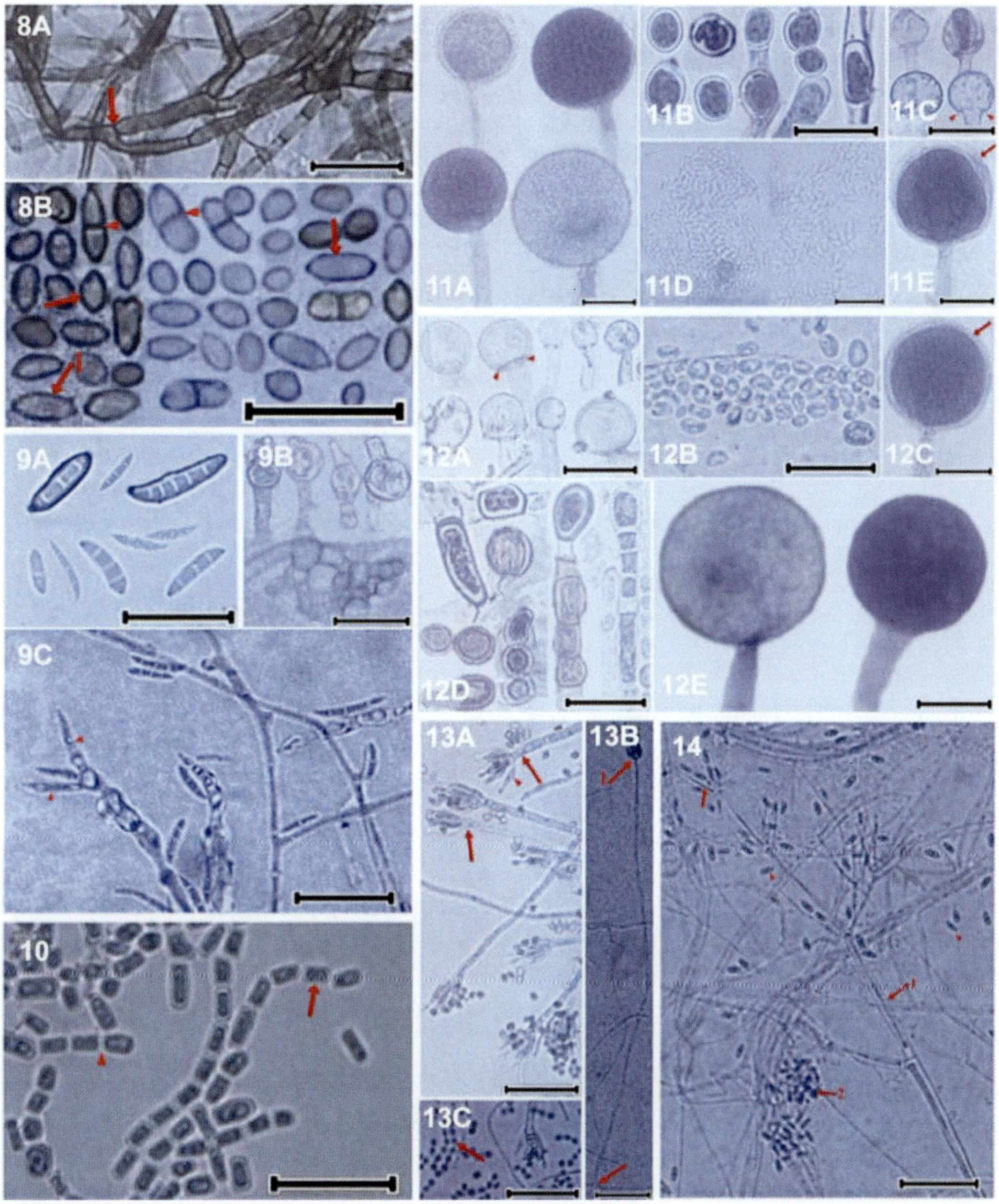

Figs 8–14. Microscopic images of the seven fungi. 8. *Cladosporium varians*, A. Dichotomous branching of conidiophores, B. Different aseptate (arrow), single septate (arrowhead) conidia and ramoconidia (arrow-1); 9. *Fusarium trichothecioides*, A. Macroconidia, B. Chlamydospores, C. Conidiophores bearing conidia on apical phialides (arrowhead); 10. Double septation (arrowhead) and arthroconidial chain of *Geotrichum candidum* (arrow); 11. *Mucor circinelloides*, A. Sporangia, B. Chlamydospores, C. Columellae with collarette (arrowheads), D. Sporangiospores, E. Sporangia with incrusted wall (arrow); 12. *Mucor hiemalis*, A. Columellae with collarette (arrowheads), B. Sporangiospores, C. Sporangia with incrusted wall (arrow), D. Chlamydospores, E. Sporangia; 13. *Penicillium frequentans*, A. Sterigmata (arrowhead) on width apex of conidiphore (arrow), 13B. Long monoverticillate conidiophore arise from mycelium (arrow to arrow-1), C. Chain of conidia; 14. Conidiophores (arrow-1), whorl of phialides (arrow), conidia (arrowhead) and a cluster of conidia (arrow-2) of *Verticillium albo-atrum*. [Scale bars = 25 μm]

Acknowledgements

The authors express their sincere thanks and gratitude to Professor Dr. Robert L Wick, University of Massachusetts, USA for his great support to fulfill the lacking of literature regarding fungal identification. Microscopic photographic equipments facility was provided by the Director, National Mushroom Development & Extension Center, Savar, Dhaka, Bangladesh, whose cooperation is greatly appreciated. The authors are grateful to Mohammad Anwar Hossain of the same institute for his cooperation in microscopic photography.

References

Bensch, K., Braun, U., Groenewald, J.Z. and Crous, P.W. 2012. The genus *Cladosporium*. Studies in mycology No. 72, CBS-KNAW Fungal Biodiversity Centre, Utrecht, The Netherlands, pp. 1–401.

Booth, C. 1971. The Genus *Fusarium*. Commonwealth Mycological Institute, Kew, Surrey, England, pp. 1–237.

Crous, P.W. and Groenewald, J.Z. 2013. A phylogenetic re-evaluation of *Arthrinium*. IMA Fungus **4**(1): 133–154.

Ellis, M.B. 1971. Dematiaceous Hyphomycetes. Common wealth Mycological Institute, Kew, Surrey, England, pp. 1–608.

Ellis, D., Davis, S., Alexiou, H., Handke, R. and Bartley, R. 2007. Descriptions of medical fungi. 2nd ed. Mycology Unit, Women's and Children's Hospital, North Adelaide, Australia, pp. 1–198.

Harvey, C.S. 1965. The morphology of *Verticillium albo-atrum*, *V. dahliae*, and *V. tricorpus*, New Zeal. J. Agr. Res. **8**(3): 450–478.

Jabnoun-Khiareddine, H., Daami-Remadi, M., Barbara, D.J., and El Mahjoub, M. 2010. Morphological variability within and among *Verticillium* species collected in Tunisia. Tunisian J. Plant Prot. **5**: 19–38.

Kendrick, B. 2000. The Fifth Kingdom. 3rd ed. Focus Publishing, R Pullins Company, Newburyport, USA, pp. 1–373.

Kurtzman, C., Fell, J.W., and Boekhout, T. (Eds). 2011. The Yeasts: A Taxonomic Study, Vol. **2**, 5th ed. Elsevier Science, USA, pp.1–2354.

Mueller, G.M., Bills, G.F. and Foster, M.S. (Eds). 2004. Biodiversity of Fungi: Inventory and monitoring methods. Elsevier Academic Press, London, pp. 1–777.

Raper, K.B. and Thom, C. 1949. A Manual of the Penicillia. The Willams & Wilkins company, Baltimore, USA, pp. 1–875.

Schipper, M.A.A. 1973. A study on variability in *Mucor hiemalis* and related species. Studies in Mycology No. 4, CBS-KNAW Fungal Biodiversity Centre, Utrecht, The Netherlands, pp. 1–40.

Schipper, M.A.A. 1976. On *Mucor circinelloides, Mucor racemosus* and related species. Studies in Mycology No.12, CBS-KNAW Fungal Biodiversity Centre, Utrecht, The Netherlands, pp.1–40.

Schipper, M.A.A. 1978. On certain species of *Mucor* with a key to all accepted species. Studies in Mycology No. 17, CBS-KNAW Fungal Biodiversity Centre, Utrecht, The Netherlands, pp. 1–70.

Siddiqui, K.U., Islam, M.A., Ahmed, Z.U., Begum, Z.N.T., Hassan, M.A., Khondker, M., Rahman M.M., Kabir, S.M.H., Ahmed, M., Ahmed, A.T.A., Rahman, A.K.A. and Haque, E.U. (Eds). 2007. Encyclopedia of Flora and Fauna of Bangladesh. Vol. **2**. Cyanobacteria, Bacteria and Fungi. Asiatic Society of Bangladesh, Dhaka, pp. 1–415.

Zare, R. 2003. A revision of plant-associated *Verticillium* species. Rostaniha **4**: 29–54.

13

A NEW SPECIES OF *ARUNDINELLA* RADDI (POACEAE) FROM THE WESTERN GHATS, INDIA

C. N. SUNIL, M. K. RATHEESH NARAYANAN[1], PRAJEESH PARAMESWARAN[2], M. SIVADASAN[3,4] AND A. H. ALFARHAN[3]

Department of Botany, S.N.M. College, Maliankara P.O., Ernakulam–683 516, Kerala, India

Keywords: Arundinella thirunelliensis; New species; India; Kerala; Wayanad; Western Ghats.

Abstract

Arundinella thirunelliensis (Poaceae), a new species from the Western Ghats in Kerala, India is described and illustrated. It closely resembles *A. nepalensis* but differs by being a rheophyte with large culms, glabrous culm nodes, leaf sheath and peduncles, effuse panicles with highly branched drooping branches, large, 3-nerved, deeply bifid smooth lemma with short twisted column of awn, and the lower glume distinctly shorter than the lower lemma.

Introduction

The genus *Arundinella* Raddi (1823) (Poales: Poaceae) is with about 55 species, having distribution in the tropical regions of the world, mainly in Asia (Clayton and Renvoize, 1986). Watson and Dallwitz (1992 onwards: http://delta-intkey.com/grass/www/arundine.htm) also accounted 55 species for the genus. Sun and Phillips (2006) estimated the present approximate number of species to be 60. As per Clayton *et al.* (2006 onwards: http://www.kew.org/data/grasses-db.html) there are 68 species of *Arundinella*. The genus is represented by 21 species in India, of which 11 are found in Kerala (Hooker, 1896; Fischer, 1934; Bor, 1960; Sreekumar and Nair, 1991; Shaju and Mohanan, 2004; Kabeer and Nair, 2009).

During a recent floristic exploration in the foot hills of Brahmagiri, Wayanad district, Kerala, India some interesting specimens belonging to the genus *Arundinella* Raddi were collected. Critical study and analysis revealed that it is distinct from the hitherto known taxa of the genus and is described here as a new species *Arundinella thirunelliensis*.

Arundinella thirunelliensis Sunil, Ratheesh & Sivadasan, **sp. nov**. **(Figs 1 & 2)**.

Diagnosis: *Arundinella thirunelliensis* is similar to *A. nepalensis* Trin., but differs in being a rheophyte having large culms, glabrous culm nodes, leaf sheath and peduncles, effuse panicles with highly branched drooping branches, large, 3-nerved, deeply bifid smooth lemma with short twisted awn-column, and lower glume distinctly shorter than the lower lemma.

Type: India, Kerala: Wayanad District, Thirunelli, Kalindi river bed, 900 m, 8 Dec 2012, *Sunil & Ratheesh Narayanan* 4854 (*Holotype*: CAL!; *Isotypes*: MH!).

Paratypes: India, Kerala: Wayanad District, Thirunelli, Kalindi, 12 Jan 2014, *Sunil, Ratheesh Narayanan & Prajeesh* 1872 (CABC-MSSRF Herbarium!, Wayanad).

[1]Department of Botany, Payyanur College, Edat P.O., Payyanur, Kannur–670 327, Kerala, India
[2]M. S. Swaminathan Research Foundation, Puthoorvayal P.O., Kalpetta, Wayanad–673 121, Kerala, India
[3]Department of Botany & Microbiology, College of Science, King Saud University, P.O. Box 2455, Riyadh–11451, Kingdom of Saudi Arabia
[4]Corresponding author: E-mail: drmsivadasan@gmail.com

Tufted, perennial rheophytes; clumps up to 60 cm wide; roots wiry. Culms 80-200 cm high, 3-5 mm across; nodes glabrous. Leaves: sheath up to 27 cm long, glabrous, ribbed; ligule a narrow rim with c. 4 mm long fimbriate hairs at apex and back; blade linear or linear-lanceolate, flat, stiff, 18-42 × 0.4-1.0 cm, base rounded, margin scaberulous or smooth, acuminate at apex, glabrous or upper surface with few scattered hairs. Inflorescence terminal panicles, 16-42 cm long, 7-20 cm wide, pyramidal; branches lax to effuse, alternate, spreading, lowermost up to 25 cm long; rachis

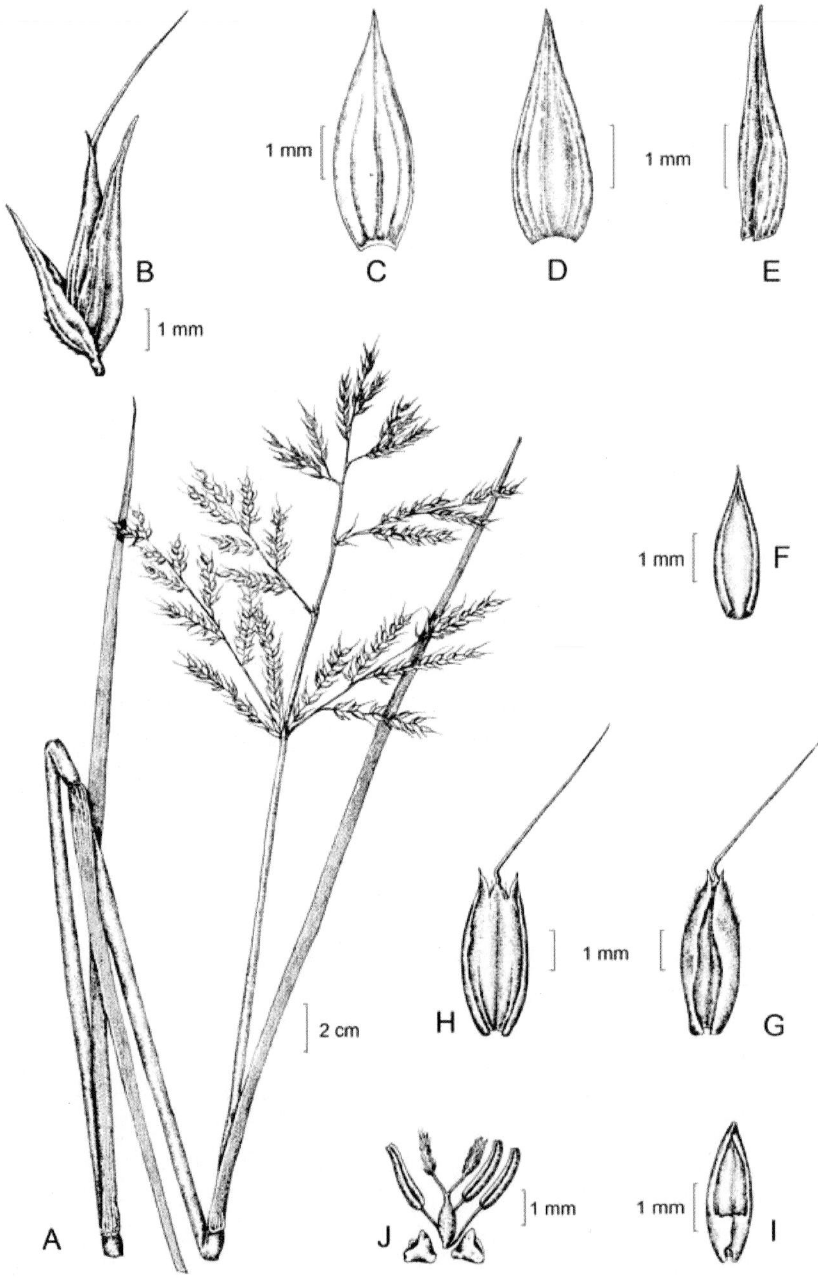

Fig. 1. *Arundinella thirunelliensis* Sunil, Ratheesh & Sivadasan, **sp. nov.** A. Habit; B. Spikelet; C. Lower glume; D. Upper glume; E. Lower lemma; F. Lower palea; G. Upper lemma; H. Upper lemma - spread open; I. Upper palea; J. Lodicules, stamens and pistil.

Fig. 2. *Arundinella thirunelliensis* Sunil, Ratheesh & Sivadasan, **sp. nov.** A. Habitat; B. A portion of inflorescence; C. Lower glume; D. Upper glume; E. Lower lemma – ventral view; F. Lower lemma – dorsal view; G. Lower palea; H. Upper lemma; I. Upper lemma showing bifid apex and twisted column; J. Upper palea; K. Caryopsis.

angular, scaberulous. Spikelets elliptic-lanceolate, 4.0-4.5 mm long, pale to dark purple; pedicels 0.5-4.0 mm long, angular, scabrous. Lower glume ovate-lanceolate, 3.0-3.5 × 1.0-1.5 mm, acute to acuminate at apex, 3 or 5-nerved, scabrid on mid-nerve, pale purple with green nerves, chartaceous. Upper glume ovate-lanceolate, 4.0-4.5 × 1.0-1.5 mm, acuminate at apex, margin hyaline, 5 or 7-nerved, pale purple with green nerves, chartaceous. Lower floret male or neuter. Upper floret bisexual. Callus glabrous or sparsely hairy. Lower lemma ovate-lanceolate, 3.5-4.0 × 1.2-1.5 mm, margin hyaline and ciliate towards apex, apex acute, very rarely with an up to 2 mm long awn, chartaceous, 3 or 5-nerved, green. Lower palea lanceolate to oblong-elliptic, 3.0-3.5 × 0.5-1.0 mm, acute at apex, 2-keeled, 2-nerved, membranous, hyaline. Upper lemma ovate or ovate-oblong, 2.5-3.0 × 1.0-1.3 mm, margin involute, apex 2-lobed, awned from the sinus, sub-chartaceous, hyaline, 3-nerved, puberulous, whitish in fruits, smooth; lobes c. 0.5 mm long, ciliate along margin, acute or obtuse at apex; awn 4.0-4.5 mm long, geniculate with a chestnut-coloured, flat, twisted, 0.5-1.0 mm long column. Upper palea ovate-lanceolate, 2.5-3.0 × 0.7-1.0 mm, ciliate along margin, acute at apex, hyaline, membranous, 2-keeled, 2-nerved. Lodicules 2, triangular, c. 0.3 mm long, hyaline, apex irregularly lobed. Stamens 3, anthers 1.0-1.5 mm long, oblong, yellowish-brown. Ovary ovoid, 0.3-0.5 mm long; styles 2, c. 1 mm long; stigma c. 1 mm long, feathery, purple. Caryopsis ellipsoid, 1.0-1.2 × 0.5-0.6 mm, light brown.

Phenology: Flowering and fruiting occurs during October to February.

Etymology: The specific epithet denotes the type locality – Thirunelli in Wayanad district, which is one of the floristically rich regions on southern Western Ghats.

Ecology and distribution: Arundinella thirunelliensis was collected from a perennial stream bed. The new species has a distribution restricted to perennial stream beds that are occasionally subjected to flash floods. The plants are often associated with *Homonoia riparia* Lour., *Rotula aquatica* Lour., *Cyperus* sp., *Equisetum* sp., etc.

Comparison of characters of *Arundinella thirunelliensis* with *A. nepalensis* is provided in Table 1.

Table 1. Comparison of characters of *Arundinella thirunelliensis* with *A. nepalensis*.

Characters	*Arundinella thirunelliensis* **sp. nov.**	*Arundinella nepalensis*
Habit	Rheophytic grass	Terrestrial grass
Culm node	Glabrous	Pubescent
Leaf sheath	Glabrous	Glabrous or tuberculate hairy
Peduncle	Glabrous below panicle	Pubescent below panicle
Panicle	Effuse, highly branched, branches alternate and drooping	Densely spiculate, branches simple and sub-verticillate, secund
Lower glume	Distinctly shorter than the lower lemma	About as long as the lower lemma
Upper lemma	2.5-3.0 mm long, smooth, apex deeply 2-lobed; lobes c. 0.5 mm long, acute or obtuse	1.0-1.5 mm long, granulose, apex emarginate or shallowly 2-lobed; lobes c. 0.05 mm long, obtuse
Awn	Column of awn 0.5-1.0 mm long when dry	Column of awn 2.2-3.2 mm long when dry

Acknowledgements

The authors are grateful to Dr. J. F. Veldkamp, Leiden, for a critical review of the manuscript. The encouragement provided by the Manager and Head of the Department of Botany, Sree Narayana Mangalam College, Maliankara, the Manager and Head of the Department of Botany, Payyanur College, Payyanur, and the Director, Community Agrobiodiversity Centre, M. S.

Swaminathan Research Foundation, Wayanad are gratefully acknowledged. The last two authors thankfully acknowledge the encouragements and support extended by the Deanship of Scientific Research, King Saud University, through the research group project No. RGP-VPP-135. Sincere thanks are expressed towards Mr. Jayesh P. Joseph for the illustration and towards Mr. T. S. Rajeev, Dr. Sanal Kumar and Dr. Baiju, Department of Botany, S. N. M. College, Maliankara, Ernakulam for various help.

References

Bor, N.L. 1960. *Arundinella. In*: The Grasses of Burma, Ceylon, India and Pakistan. Pergamon Press, Oxford, pp. 417-428.

Clayton, W.D. and Renvoize, S.A. 1986. Genera Graminum: Grasses of the World. Kew Bull. Additional Ser. **13**: 1-389.

Clayton, W.D., Vorontsova, M.S., Harman, K.T. and Williamson, H. 2006 (onwards). GrassBase - The Online World Grass Flora. http://www.kew.org/data/grasses-db.html. Accessed on 17 April 2014.

Fischer, C.E.C. 1934. *Arundinella. In*: Gamble, J.S., Flora of the Presidency of Madras. Adlard & Sons Ltd., London, pp. 1798-1802.

Hooker, J.D. 1896. *Arundinella. In*: Hooker, J.D., The Flora of British India, Vol. **7**. L. Reeve & Co. Ltd., London, pp. 68-77.

Kabeer, K.A.A. and Nair, V.J. 2009. *Arundinella* Raddi. *In*: Flora of Tamil Nadu - Grasses. Botanical Survey of India, Kolkatta, pp. 334-344.

Raddi, G. 1823. *Agrostographia brasiliensis*. Bertini, Lucca, 58 pp.

Shaju, T. and Mohanan, N. 2004. A new species of *Arundinella* Raddi (Poaceae) from Kerala, India. *Rheedea* **14**: 47-50.

Sreekumar, P.V. and Nair, V.J. 1991. *Arundinella* Raddi. *In*: Flora of Kerala Grasses. Botanical Survey of India, Culcutta, pp. 332-346.

Sun, B. and Phillips, S.M. 2006. *Arundinella. In*: Wu, Z.-Y., Raven, P.H. and Hong, D.-Y. (Eds), Flora of China, Vol. **22**. Science Press, Beijing. Peoples Republic of China and Missouri Botanical Garden Press, St. Louis, Missouri, USA, pp. 563-570.

Watson, L. and Dallwitz, M.J. 1992 onwards. The grass genera of the world: descriptions, illustrations, identification, and information retrieval; including synonyms, morphology, anatomy, physiology, phytochemistry, cytology, classification, pathogens, world and local distribution, and references. Version: 5th February 2014. http://delta-intkey.com/grass/www/arundine.htm. Accessed on 17 April 2014.

A NEW VARIETY AND TWO NEW RECORDS OF *ELATOSTEMA* J. R. FORSTER & G. FORSTER (URTICACEAE) FROM VIETNAM

YUN LIN, LIN DONG DUAN[1] AND HAI YAN BI[2]

Hunan Food and Drug Vocational College, Changsha 410014, P. R. China

Keywords: Elatostema; New record; New variety; Urticaceae; Vietnam.

Abstract

Based on examination of specimens of *Elatostema* J. R. Forster & G. Forster (Urticaceae) from Vietnam, *Elatostema albopilosum* W.T. Wang var. *vicinum* L. D. Duan & Y. Lin, a new variety collected from Bac Can Province and Tuyen Quang Province, Vietnam, is described and photographed. This new variety has glabrous stems, glabrous leaves and glabrous peduncles which differs from var. *albopilosum* that having sparsely puberulent leaf blade on abaxial surface along veins and sparsely strigillose leaf blade on adaxial surface, and short pilose peduncles. *E. integrifolium* (D. Don) Wedd. and *E. pseudodissectum* W. T. Wang, two species previously known from Vietnamese adjacent countries, are newly recorded from Ninh Binh Province and Tam Dao National Park, Vietnam, respectively. The vouchers are kept in the Herbarium (PE) of Institute of Botany, Chinese Academy of Sciences, Beijing, China, and their descriptions and photographs are also provided.

Introduction

The genus *Elatostema* J. R. Forster & G. Forster is the largest genus in the family Urticaceae, consists of c. 500 species and is distributed in tropical and subtropical regions of Africa, Asia and Oceania, 30 species of which are found in Vietnam (Fu *et al.*, 2013; Gagnepain, 1929; Ho, 2003; Lin, 2008; Lin *et al.*, 2003, 2011; Wang, 1995, 2012). Based on examination of *Elatostema* specimens housed at the Herbarium (PE) of Institute of Botany, Chinese Academy of Sciences, Beijing, China, and after thoroughly consulting the literature (Gagnepain, 1929; Lin, 2008; Bi *et al.*, 2011; Lin *et al.*, 2003, 2011; Wang, 1980, 1995, 2012; Duan and Lin, 2013; Wei *et al.*, 2013), a new variety was found from Vietnam, and two species previously known from Vietnamese adjacent countries, were discovered to be new to Vietnam, and were reported here.

1. Elatostema albopilosum W. T. Wang var. **vicinum** L. D. Duan & Y. Lin, **var. nov. (Figs 1, 2).**

Diagnosis: The new variety *Elatostema albopilosum* W. T. Wang var. *vicinum* L. D. Duan & Y. Lin is morphologically similar to *E. albopilosum* W. T. Wang var. *albopilosum*, but differs in glabrous stems, glabrous leaves and glabrous peduncles; the latter pilose stems, sparsely puberulent leaf blade on abaxial surface along veins, sparsely strigillose leaf blade on adaxial surface, and short pilose peduncles.

Type: Vietnam. Tuyen Quang Province, Na hang, 400 m a.s.l., 28 October 1998*, H. N. Qin, H. Zhu, M. L. Zhang & T. G. Gao 319* (*Holotype*: PE). Bac Can Province, Cho Don District, Xuan Lac Commune, 105°32'95"E, 22°21'91"N, 200 m a.s.l., 27 October 2004, *L. Q. Li, N. T. Hiep, Z. Y. Zhang, X. C. Zhang, T. G. Gao, Z. T. Wang, N. S. Khang & N. X. Tam 0432* (*Paratype:* PE).

[1]Shaoyong University, Shaoyang 422004, P. R. China
[1]Beijing Museum of Natural History, Beijing 100050, P. R. China. Corresponding author. Email: rubybi@126.com

Fig. 1. **Elatostema albopilosum** W. T. Wang var. **vicinum** L. D. Duan & Y. Lin, **var. nov.** A) male habit (*H. N. Qin et al. 319*, PE), B) staminate inflorescence (*L. Q. Li et al.* 0432, PE), C) pistillate inflorescence, dorsal view. (*L. Q. Li et al.* 0432, PE).

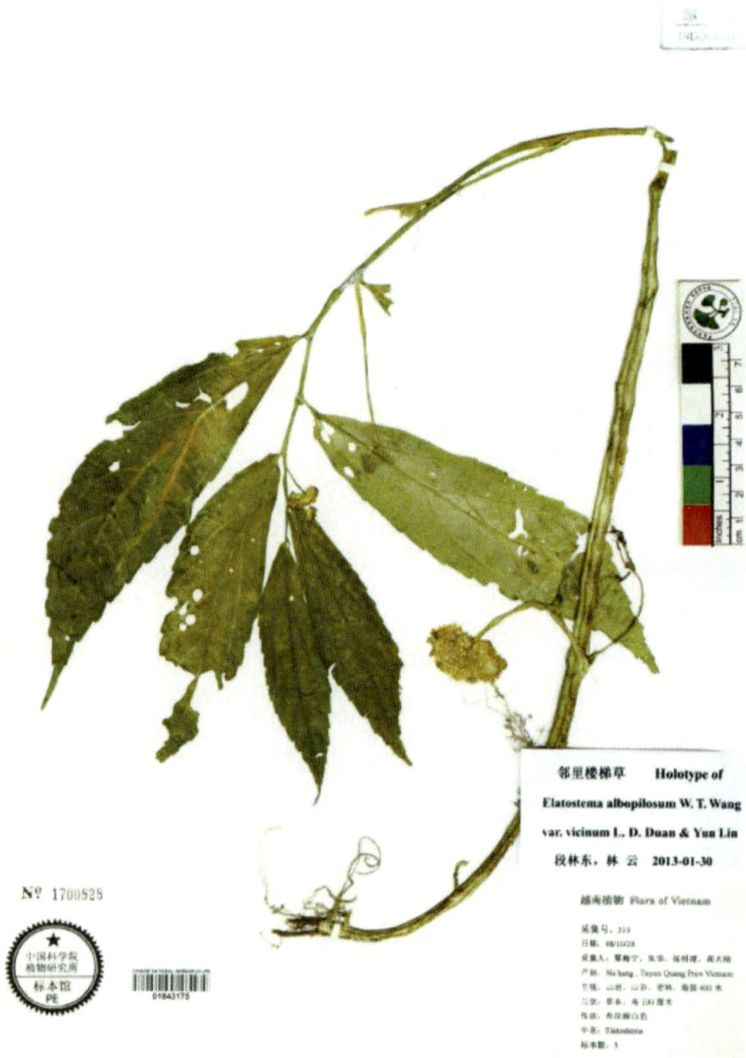

Fig. 2. Holotype of **Elatostema albopilosum** W. T. Wang var. **vicinum** L. D. Duan & Y. Lin (from *H. N. Qin et al. 319*, PE)

2. Elatostema integrifolium (D. Don) Wedd. in DC., Prodr. **16**(1): 179 (1869). *Procris integrifolia* D. Don, Prodr. Fl. Nepal. : 61 (1825). **(Fig. 3).**

Type: Nepal. Precise locality not known, *Wallich s. n.* (? K).

Perennial herbs or subshrubs, monoecious or dioecious, 60-200 cm tall. Stems erect, branched, glabrous. Leaves alternate, glabrous; nanophylls absent; stipules narrowly lanceolate, 6-10 × 1.5-2.0 mm, with cystoliths; petiole 1.5-6.0 (-10) mm long; leaf blade obliquely elliptic or obliquely oblanceolate, 5-19 × 2-6 cm, papery or herbaceous, major basal lateral veins asymmetric, 1 basal, the other arising above base, cystoliths conspicuous, dense, 0.2-0.4 (-0.6) mm long; base obliquely cuneate, margin entire or c. 2-crenate, apex acuminate or caudate. Male inflorescence solitary or in pairs, simple, 10-18 × 6-12 mm, sessile; receptacle 4-6 mm in diam.; bracts 4, triangular, equal, connate, c. 1 mm long; bracteoles cymbiform, c. 1.2 mm long. Female inflorescence solitary or in pairs, 5-8 mm in diam.; peduncle 0-1 mm long; receptacle very small; bracts 20-30, triangular, nearly equal, 0.8 mm long; bracteoles narrowly linear, 0.8-1.2 mm long.

Male flowers 4-merous. Achenes ellipsoid, 0.7 mm long, 8-ribbed.

Phenology: Flowering from February to May.

Distribution: Bhutan, China, India, Indonesia, Myanmar, Nepal and Thailand. New record to Vietnam.

Habitat: The species grows in valley forests, streamsides at altitudes of 460-760 m in Vietnam and 900-1600 m in other countries.

Specimens examined: Vietnam. Ninh Binh Province: Bong, Biology Station, 460-760 m a.s.l., 10.2.1965, *Sino-Vietnam Exped. 2332* (PE).

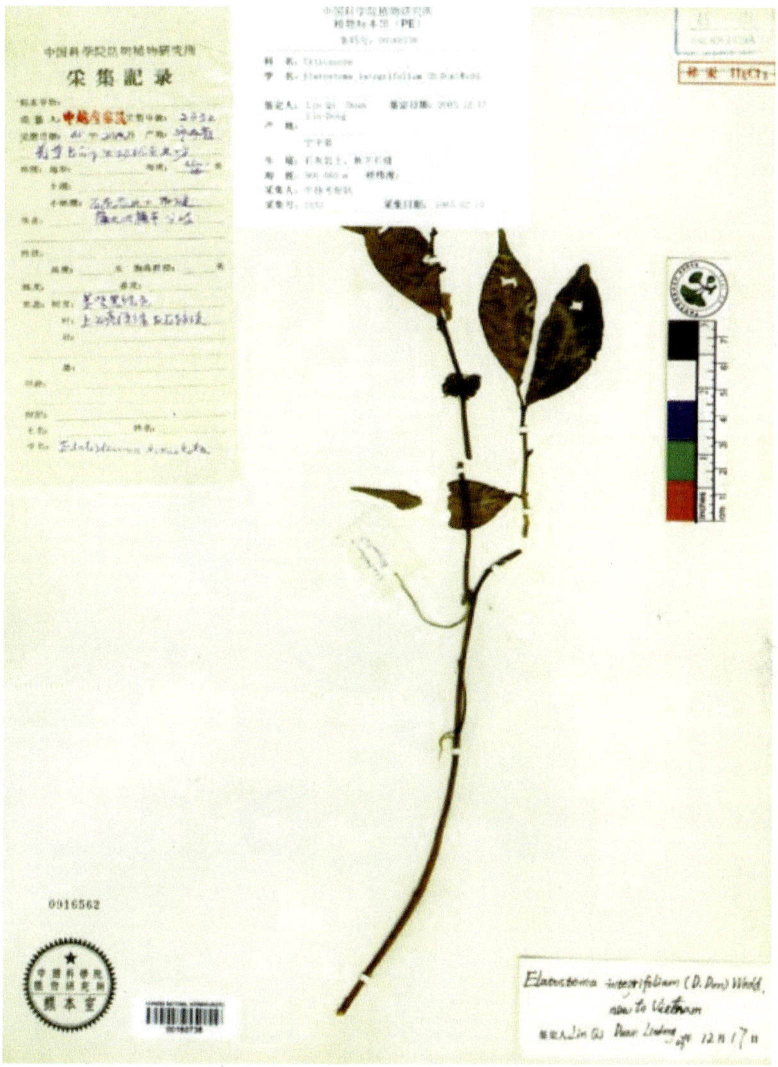

Fig. 3. **Elatostema integrifolium** (D. Don) Wedd. (from *Sino-Vietnam Exped. 2332*, PE).

Note: This newly recorded species is similar to *E. laevissimum* W.T. Wang (Wang, 1980) in Vietnam, but differing in stipules narrowly lanceolate, 6-10 × 1.5-2.0 mm, with cystoliths, leaves glabrous, male inflorescence 10-18 × 6-12 mm, female inflorescence 5-8 mm in diam., the latter with stipules narrowly triangular, 2-5 × 0.8-1.0 mm, without cystoliths, leaf blade sparsely strigillose on adaxial surface, male inflorescence 3-4 mm in diam. and female inflorescence 1.5-2.5 mm in diam.

3. Elatostema pseudodissectum W. T. Wang, Bull. Bot.. Lab. N.-E. For. Inst. **7**(7): 55 (1980). **(Fig. 4).**

Type: China. Guangxi: Baise, 15 August 1928, *R. C. Ching 7427* (*Holotype*: PE!).

Perennial herbs, monoecious, 20-40 cm tall, glabrous. Stems ascending or erect, branched. Leaves alternate; nanophylls absent; stipules narrowly triangular, 0.7-1.5 × 0.2-0.4 mm, without cystoliths; petiole 1-3 mm long; leaf blade obliquely ovate or elliptic-ovate, 4-11 × 1.5-4.0 cm, membranous or herbaceous, major basal lateral veins both arising at base of leaf blade, cystoliths conspicuous, dense, 0.2-0.6 mm long; base obliquely cuneate, apex acuminate or cuspidate, margin denticulate. Male inflorescence solitary, simple, 4-7 mm in diam.; peduncle 12-40 mm long; receptacle small; bracts 3-4, connate, equal, broadly ovate. Female inflorescence solitary, 30-flowered or more, 3-6 mm in diam., sessile; receptacle 3-5 mm in diam.; bracts 11-20, narrowly ovate, nearly equal, c. 1.2 mm long; bracteoles spathulate-linear, 0.3-0.5 mm long. Male flowers 4-merous. Achenes ellipsoid, c. 0.6 mm long, smooth.

Fig. 4. **Elatostema pseudodissectum** W. T. Wang (from *L. Q. Li et al. 559*, PE).

Phenology: Flowering from September to October.

Distribution: South-west China. New record to Vietnam.

Habitat: This species grows in valley wet forests, streamsides at altitudes of 1000 m in Vietnam and 1100-2200 m in China.

Specimens examined: Vietnam. Tam Dao National Park, 105°38'68"E, 21°27'47"N, 1000 m a.s.l., 31.10.2004, *L. Q. Li, N. T. Hiep, Z. Y. Zhang, X. C. Zhang, T. G. Gao, Z. T. Wang, N. S. Khang & N. X. Tam 559* (PE). Precise locality not known, *Ke Can,* 150 m a.s.l., 5.1.1965, *Sino-Vietnam Exped. 1750* (PE).

Note: This newly recorded species is similar to *E. dissectum* Wedd. (Weddell, 1856) in Vietnam, differing in stipules narrowly triangular, 0.7-1.5 × 0.2-0.4 mm, without cystoliths, leaf blade obliquely ovate or elliptic-ovate, male inflorescence 4-7 mm in diam., the latter having stipules narrowly linear or subulate, 3-5 × 0.1-0.3 mm, with cystoliths, leaf blade obliquely oblong or obliquely oblong-lanceolate and male inflorescence 10-12 × 8-10 mm.

Acknowledgments

Thanks are due to the Curator of herbarium (PE) for permission to examine specimens. This work was supported by the Project of the Education Department in Hunan Province (11A109).

References

Bi, H.Y., Yang, Z.R. and Lin, Q. 2011. New taxa of *Elatostema* (Urticaceae) from Thailand and India. Bangladesh J. Plant Taxon. **18**(2) : 149-152.

Duan, L.D. and Lin, Y. 2013. *Elatostema biopposítum* (Urticaceae), a new species from Guangxi, China. Bangladesh J. Plant Taxon. **20**(2) : 179-183.

Fu, L.F., Do, V.T., Wen, F., Liu, S.Y. and Wei, Y.G. 2013. New records of *Elatostema* and *Pellionia* (Urticaceae) from Vietnam. Guihaia **33**(6): 801-803.

Gagnepain, F. 1929. *Pellionia* Gaudich. Flore Generale de L'Indo-Chine **5**(9): 892-921.

Ho, P.H. 2003. An Illustrated Flora of Vietnam. Vol. **2.** NXB Tre Press, Ho Chi Minh, pp. 589-593.

Lin, Q. 2008. A revision of *Elatostema* section *Weddelia* series *Salvinioida* (Urticaceae). Bot. J. Linn. Soc. **158**(1): 62-66.

Lin, Q., Friis, I. and Wilmot-Dear, C.M. 2003. *Elatostema* (Urticaceae). *In:* Wu, Z.Y. and Raven, P.H. (Eds), Flora of China. Vol. **5**. Science Press, Beijing, and Missouri Botanical Garden Press, St. Louis, USA, pp. 127-163.

Lin, Q., Yang, Z.R., Duan, L.D. and Gao, T.G. 2011. Miscellaneous taxonomic notes on *Elatostema* (Urticaceae) from China and its adjacent area. Nordic J. Bot. **29**(5): 590-597.

Wang, W.T. 1980. Classificatio specierum sinicarum *Elatostematis* (Urticaceae). Bull. Bot. Lab. N.-E. For. Inst. **7**(7): 1-96.

Wang, W.T. 1995. *Elatostema* J.R. Forster & G. Forster. *In:* Wang, W.T. and Chen, C.J. (Eds), Flora Reipublicae Popularis Sinicae, Tomus **23**, No. 2. Science Press, Beijing, pp. 187-317.

Wang, W.T. 2012. Nova Classificatio Specierum Sinicarum *Elatostematis* (Urticaceae) *In:* Fu, D.Z. (Ed.), Paper Collection of W.T. Wang. Vol. **2**. Higher Education Press, Beijing, pp. 1061-1178.

Weddell, H.A. 1856. Monograhpie de la famille des Urticées. Archives du Muséum D'Histoire Naturelle **9**: 1-332.

Wei, Y.G., Wen, F., Fu, L.F. and Wang, W.T. 2013. Three new species of *Elatostema* J.R. Forst. & G. Forst. (Urticaceae) in limestone caves from Guangxi and Guizhou, China. Bangladesh J. Plant Taxon. **20**(1): 1-8.

15

PRIMULINA LECHANGENSIS (GESNERIACEAE), A NEW SPECIES FROM A LIMESTONE CAVE OF NORTHERN GUANGDONG, CHINA

SHOU-BIAO ZHOU[1], XIN HONG[2], WEN MA AND FANG WEN[2,3]

College of Life Sciences, Anhui Normal University, CN-241000, Wuhu, China

Keywords: Primulina lechangensis; New species; Gesneriaceae; Karst cave flora; China.

Abstract

Primulina lechangensis (Gesneriaceae), a new species from northern Guangdong province in southern China is described and illustrated. Its morphology suggests affinities to *P. longicalyx,* but can be distinguished from similar taxon by several distinct characters including: smaller leaf blade (2.0-3.5 × 1.0-1.2 cm), base cuneate-attenuate; smaller bracts (1.0-1.5 × c. 0.15 cm); shorter calyx lobes (0.8-0.9 cm long) tuberculate inside; smaller corolla tube (c. 2 cm long), inflated in the middle and contracted near the mouth; shorter glabrous filaments (0.3-0.4 cm long), inserted 1.3-1.5 cm from base of corolla; stamens and staminodes glabrous; pistil 1.4-1.5 cm long, and capsule c. 1.5 cm long. This species is further exemplified by photographs, and a distribution map is presented.

Introduction

The expanded *Primulina* Hance (1883) (Gesneriaceae) comprises more than 150 species, of which the distribution and differentiation center is located from Southwest China to North Vietnam (Wei *et al.*, 2010). In late autumn of 2011, we collected some *Primulina* specimens in fruits from crevices of a carbonate sedimentary rocky hill along a roadside near Lechang, Shaoguang, Guangdong of Southern China. The living plants were introduced and cultivated in the nursery of the Gesneriad Conservation Center of China (GCCC) for future floral examination. After a year of careful cultivation, plants flowered in spring of 2013. The shape of the corolla tube (inflated in the middle, contracted near the mouth) is unusual in this genus, which can be readily distinguished from other species in Northern and Northwestern Guangdong. After consulting the local Flora and relevant literature (Wang *et al.*, 1990,1998; Li and Wang, 2004; Wei *et al.*, 2010; Hong *et al.*, 2012; Chung *et al.*, 2013; Liang *et al.,* 2013; Lu *et al.,* 2013; Ning *et al.,* 2013; Wen *et al.*, 2013; Xu *et al.*, 2013; Zhao *et al.*, 2013; Li *et al.,* 2014; Wen and Wei, 2014; Zheng and Deng, 2014; Zhou *et al.,* 2014) and herbarium specimens deposited at ANU, BJFU, CDBI, CTC, GH, HN, IBK, IBSC, KUN, NU, PH, PE, US and VMN, we identified the collected specimens as a new species, *Primulina lechangensis*. The new species is described and illustrated along with its distribution map.

Primulina lechangensis X. Hong, F. Wen & S.B. Zhou, **sp. nov**. **(Figs 1 & 2)**.

Diagnosis: *Primulina lechangensis* resembles *P. longicalyx* but differs from the latter by having leaf blades with cuneate to attenuate bases, smaller inflorescence bracts, the corolla tube inflated in the middle and contracted near the mouth, and glabrous filaments.

---

[1]The Key Laboratory of Conservation and Employment of Biological Resources of Anhui, CN-241000, Wuhu, China

[2]Gesneriad Conservation Center of China, Guangxi Institute of Botany, Guangxi Zhuang Autonomous Region and the Chinese Academy of Sciences, CN-541006, Guilin, China

[3]Corresponding author. Email: wenfang760608@139.com

Type: China. Guangxi Province, cultivated in the nursery of Gesneriad Conservation Center of China (GCCC), introduced from Lechang, Shaoguang City, 430 m a.s.l., growing in damp crevices of limestone rocks in a limestone cave, 6 May 2013, WF13050601 (*Holotype:* IBK!, *Isotype:* ANU!).

Acaulescent perennial herb; rhizome internodes inconspicuous. Leaves 6-8, basal, opposite, petiole cylindrical, 1-2 × c. 0.1 cm, pubescent; leaf blade slightly oblique, ovate to elliptic, 2.0-3.5 × 1.0-1.2 cm, chartaceous when dry, apex acute to obtuse, margin crenate from base, base cuneate to attenuate, adaxially appressed puberulent, abaxially densely velutinous. Lateral veins 3-4 on each side of midrib, adaxially impressed, abaxially conspicuous. Cymes axillary, 1-2 or more, unbranched, commonly 4-flowered; peduncle 2.5-3.0 cm long, c. 0.1 cm wide, densely glandular puberulent. Bracts 2, free, 1.0-1.5 × c. 0.15 cm, lanceolate-ovate, entire, densely puberulent. Pedicels c. 0.2 cm long, densely puberulent, intermixed with glandular puberulent. Calyx 5-partite from base, segments equal, lanceolate-linear, 0.8-0.9 × c. 0.1 cm, densely puberulent outside, tuberculate inside, margin entire, apex attenuate. Corolla whitish lilac to bluish purple, with mauve lines, c. 2 cm long, puberulent or glandular puberulent outside, glandular puberulent only on veins inside where filaments and staminodes fused to a tube, tube swollen, 1.4-1.5 cm long, contracted near the mouth, orifice c. 1.4 cm in diameter, inflated in the middle, c. 1.6 cm in diameter, base constricted; limb distinctly 2-lipped, adaxial lobs 2-partite to the base, divaricate, c. 0.5 × 0.3-0.4 cm, lobes oblong; abaxial lobes 3-partite to the base, divaricate, c. 0.6 cm long, c. 0.3 cm in diameter, oblong. Stamens 2, adnate to corolla tube c. 0.3 cm above the base; filaments white, geniculate about 0.2 cm above point of attachment, 0.3-0.4 cm long, glabrous; anthers c. 0.2 cm long, dorsified, glabrous. Staminodes 3, translucent, the central one capitate, c. 0.5 mm long, inserted c. 0.1 cm from base of corolla, lateral ones short linear, glabrous, c. 0.2 cm long, inserted c. 0.3 cm from base of corolla. Disc annular, margin entire, 0.1-0.2 mm high. Pistil nearly as long as corolla tube; the transition between ovary and style conspicuous. Ovary linear, c. 2 mm long, c. 1 mm in diameter, densely glandular-puberulent and puberulent; style 7-9 mm long, c. 0.5 mm in diameter, densely puberulent; stigma pale green, obtrapeziform, apex 2-lobed to about half of the stigma, c. 1 mm long, lobes ligulate. Capsules linear, c. 1.5 cm long.

Flowering period: April - May.

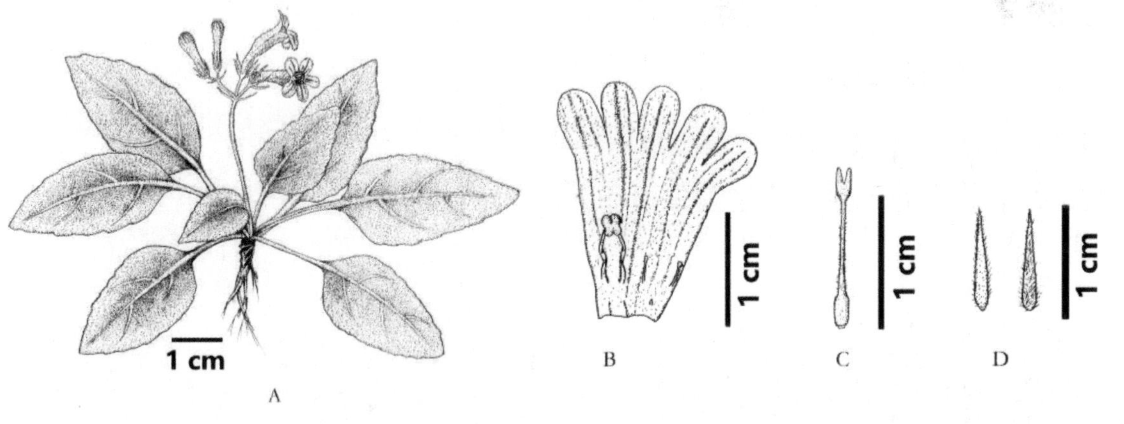

Fig. 1 A-D. *Primulina lechangensis* X. Hong, F. Wen & S.B. Zhou, **sp. nov.** A. Habit; B. Opened corolla showing stamens and staminodes; C. Pistil; D. Adaxial and abaxial calyx lobes (All from the type, WF13050601, IBK).

Distribution: Only known from the type locality Lechang, Shaoguang city, northern Guangdong Province, China (Fig. 3).

Habitat: Primulina lechangensis grows in damp, shady cervices at the foot of a carbonate sedimentary rock hill in the north of Guangdong, China, at 430 m a.s.l. It is locally abundant and grows in subtropical evergreen seasonal rain forest.

Etymology: The species is named after the type locality Lechang in Guangdong Province.

Vernacular name: China: Lechang baochunjutai.

Fig. 2 A-K. *Primulina lechangensis* X. Hong, F. Wen & S.B. Zhou, **sp. nov.** A. Habitat; B. Plant with flowers; C. Cyme; D. Frontal view of cyme; E. Frontal view of a flower; F. Lateral view of flower; G. Opened corolla; H. Anthers; I. Pistil; J. Stigma; K. Capsule.

Additional collections: CHINA. Guangdong Province: Shaoguang, Lechang, c. 430 m, 27 November 2012, *Wen & Hong 20121116* (IBK!).

Note: Primulina lechangensis is morphologically close to *P. longicalyx* (J.M. Li & Y.Z. Wang) Mich. Möller & A. Weber based on the shape of leaf and flower, but it can be easily distinguished by some additional characteristics. A detailed comparison of the diagnostic characters between *P. lechangensis* and *P. longicalyx* is shown in Table 1.

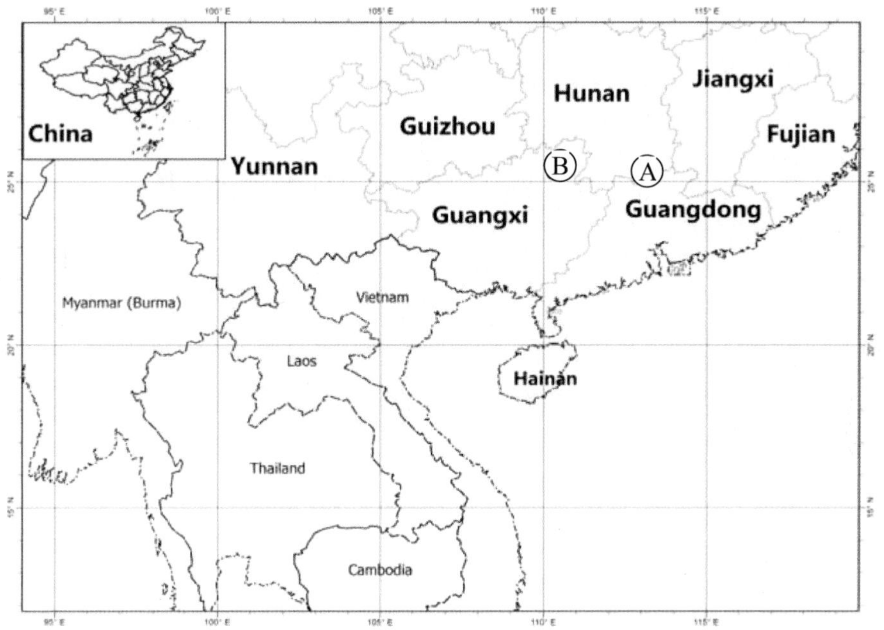

Fig. 3. Distribution of *Primulina lechangensis* X. Hong, F. Wen & S.B. Zhou, **sp. nov.** (A) and its related species, *P. longicalyx* (J.M. Li & Y.Z. Wang) Mich. Möller & A. Weber (B) in China.

Table 1. Diagnostic morphological characters of *Primulina lechangensis* and *P. longicalyx*.

Characters	*Primulina lechangensis*	*P. longicalyx*
Leaf blade	2.0-3.5 × 1.0-1.2 cm, base cuneate-attenuate	8-25 × 3.5-8.0 cm, base broadly cuneate
Bracts	1.0-1.5 cm long	2.5-3.0 cm long
Calyx	0.8-0.9 cm long, tuberculate inside, glabrous	1.5-2.5 cm long, inside with dense eglandular hairs
Corolla size	c. 2 cm long	5.0-5.6 cm long
Tube shape	Swollen tubular, inflated in the middle, contracted near the mouth	Tubular
Tube size	1.4-1.5 cm long, orifice c. 1.4 cm in diameter	3.8-4.0 cm long, orifice c. 1.8 cm in diameter
Filaments	0.3-0.4 cm long, inserted c. 0.3 cm from base of corolla, glabrous	c. 1.4 cm long, inserted 1.3-1.5 cm from base of corolla, slightly puberulent
Indumentum of anthers	Glabrous	Densely bearded
Indumentum of staminodes	Glabrous	Laterals puberulent
Pistil	1.4-1.5 cm long	3.9-4.3 cm long
Capsule	c. 1.5 cm long	2.2-2.8 cm long

Acknowledgements

The authors are grateful to Prof. Yi-Gang Wei for checking the specimens. The authors also thank Ms. Xiao-Ming Xu for the illustration. This study was supported by Key Foundation of Education Department of Anhui Province (KJ2011A129), Provincial Key Laboratory of Biotic Environment and Ecological Safety in Anhui (2004sys003), Project of National College Students Innovation and Entrepreneurship of Anhui Normal University, the Guangxi Natural Science Foundation (2013GXNSFAA019071), National Natural Science Foundations of China (31270236) and International S & T Cooperation Projects of Guangxi (Guikehe 1347004-4) and Guilin (20130412).

References

Chung, K.F., Huang, H.Y., Peng, C.I. and Xu, W.B. 2013. *Primulina mabaensis* (Gesneriaceae), a new species from a limestone cave of northern Guangdong, China. Phytotaxa **92**: 40-48.

Hance, H.F. 1883. *Primulina tabacum* Hance. J. Bot. **21**: 165-170.

Hong, X., Zhou, S.B. and Wen, F. 2012. *Primulina chizhouensis* sp. nov. (Gesneriaceae), a new species from a limestonecave in Anhui, China. Phytotaxa **50**: 13-18.

Li, J., Hong, X., Wen, F. and Yan, L.J. 2014. *Primulina pseudoroseoalba* (Gesneriaceae), a new species from a karst cave in Guangxi, China. Ann. Bot. Fenn. **51**: 86-89.

Li, Z.Y. and Wang, Y.Z. 2004. *Primulina*, *Chirita* and *Chiritopsis*. *In*: Li, Z.Y. and Wang, Y.Z. (Eds), Plants of Gesneriaceae in China. Henan Sci. & Technol. Publ. House, Zhengzhou, pp. 170-282.

Liang, H.L., Kong, D.X., Shi, Y.C., Zhao, B. and Wen, F. 2013. *Primulina tsoongii* sp. nov. (Gesneriaceae) from a limestone area in north Guangxi, China. Nord. J. Bot. **32**: 75-79.

Lu, S.N., Fu, L.F., Liang, G.Y. and Wen, F. 2013. *Primulina bullata*, a new species of *Primulina* (Gesneriaceae) from Guangxi. Guihaia **33**: 42-45.

Ning, Z.L., Li, G.F., Wang, J., Smith, J.F., Rasolonjatovo, H. and Kang, M. 2013. *Primulina huaijiensis* (Gesneriaceae), a new species from Guangdong, China. Ann. Bot. Fenn. **50**: 119-122.

Wang, W.T., Pan, K.Y. and Li, Z.Y. 1990. Gesneriaceae. *In*: Wang, W.T. (Ed.), Flora Reipublicae Popularis Sinicae **69**. Science Press, Beijing, pp. 125-581.

Wang W.T., Pan, K.Y. and Li, Z.Y. 1998. Gesneriaceae. *In*: Wu, Z.Y. and Raven, P.H. (Eds), Flora of China, Vol. **18**. Science Press, Beijing and Missouri Bot. Garden Press, St. Louis, 322 pp.

Wei, Y.G., Wen, F., Möller, M., Monro, A., Zhang, Q., Gao, Q., Mou, H.F., Zhong, S.H. and Cui, C. 2010. Gesneriaceae of South China. Guangxi Science and Technology Publishing House, Nanning, 777 pp. (in Chinese and English).

Wen, F. and Wei, Y.G. 2014. *Primulina dongguanica* F. Wen, Y.G. Wei & R.Q. Luo (Gesneriaceae), a new species from South China. Candollea **69**(1): 9-19

Wen, F., Zhao, B., Liang, G.Y. and Wei, Y.G. 2013. *Primulina lutvittata* (Gesneriaceae), a new species from a limestone cave in Guangdong, China. Ann. Bot. Fenn. **50**: 87-90.

Xu, W.B., Liu, Y., Kono, Y., Chang, H., Peng, C.I. and Chung, K.F. 2013. *Primulina cardaminifolia* (Gesneriaceae), a rare new species from limestone areas in Guangxi, China. Bot. Stud. **54**: 19-27.

Zhao, B., Hong, X., Huang, S.L. and Wen, F. 2013. *Primulina crassirhizoma* sp. nov. (Gesneriaceae) from a limestone area along the boundary of Sino-Vietnam. Bangladesh J. Plant Taxon. **20**: 129-134.

Zheng, Y.L. and Deng, Y.F. 2014. A new species of *Primulina* (Gesneriaceae) from Guangdong, China. Phytotaxa **163**(1): 48-53.

Zhou S.B., Hong, X., Li, Z.L. and Wen, F. 2014. *Primulina diffusa* (Gesneriaceae), a new species endemic to the karst limestone area in southwestern Guangxi, China. Ann. Bot. Fenn. **51**: 212-216.

A NEW INFRA-SPECIFIC TAXON OF *ACALYPHA FIMBRIATA* SCHUM. & THONN. FROM NIGERIA

AKEEM BABALOLA KADIRI

Department of Botany, University of Lagos, Akoka, Nigeria

Keywords: Acalypha fimbriata; New variety; Nigeria.

Abstract

This study shows that the populations of *Acalypha fimbriata* Schum. & Thonn. occurring in Nigeria are divisible into two recognizable varieties, viz., *A. fimbriata* Schum. & Thonn. var. *fimbriata* and *A. fimbriata* var. *robusta* Kadiri var. nov. Generally, the populations have a suite of similar characters in their vegetative and reproductive organs and the ultra-structures of the leaf cuticle using scanning electron microscope showed uniformity of characters like sunken stomata, wide-narrow stomatal opening and conspicuous periclinal wall. However, the two varieties are discriminated based on 1 vs. 2 fruits per bract, slim vs. robust stature, scanty vs. clumsy inflorescence on the axillary spike, leaf apex character as well as smooth and rough unicellular trichomes. The observation is supported by a T-test at 95% similarity confidence while the 14 populations consisting of 280 individuals used for the analysis are separated into two by the scatter diagram. Neighbour joining analysis using Euclidean measure and UPGMA dendrogram reveal the level of relationships. Therefore, one of the varieties is based on the description of the earlier described type specimen and the other variety being different is considered as a new taxon *Acalypha fimbriata* var. *robusta* Kadiri.

Introduction

Acalypha L. (Euphorbiaceae) is pantropical in distribution, comprising about 450 species worldwide, 80 species in Africa and 15 species in West Africa with 9 species represented in Nigeria (Thonner, 1915; Hutchinson and Dalziel, 1958; Kadiri *et al.*, 2009). They grow in a wide range of disturbed terrestrial habitats such as gutter-walls, farm lands, road sides, forest and savannah. Hutchinson and Dalziel (1958) and Burkill (1994) reported that they have high medicinal and aesthetic properties. *Acalypha fimbriata* Schum. & Thonn. is one of the well-spread species of the genus belonging to the series Polygynae-Pleurogynae Mull.-Arg. and section Ciliatae Mull-Arg. It is characterized by simple lanceolate leaves that is 4-10 x 2.5-4.0 cm in size. The inflorescence is arranged on axillary spike and the female flower is located within a boat-shaped bract having 3-chambered fruit that contains one seed per chamber. *Acalypha fimbriata* is medicinally important in respiratory and digestive problems, rheumatism and for sore and wound dressing (Burkill, 1994 and *personal communication* with natives). The bioactive principles are obtainable through the common practices of pulverisation, cooking, boiling and grinding.

De Candolle (1866) and Hepper (1976) considered the species as a synonym of *A. ciliata* Forsk. However, Radcliffe-Smith (1989, 1996) regarded them as distinct taxa because of differences in some morphological characters and added that *A. ciliata* is common in the north, while *A. fimbriata* is restricted to the south of Nigeria. Nevertheless, the two species can be contiguous in distribution but there is no doubt about the constancy of their distinctness (Radcliffe-Smith, 1989). *A. fimbriata* is well distributed in wet areas; it is usually found throughout rainy season and disappears as soon as dry season begins. During this period of growth, it was observed that individuals of the same population and locality have robust and slim stature, clumsy and sparse inflorescence distribution, different number of ovary in a bract and some other morphological numerical differences. In view of this, the present study was carried out in order to scrutinize the heterogeneity of the populations across their entire range in the country, define the species properties as accurate and recognize the possible infra-specific taxa among the populations. The outcome of the study is expected to be valuable to individual plant recognition especially in medicine.

Materials and Methods

A total of 14 out of 18 populations were focused because of loss of assessable data in four of them, each containing 20 individuals were studied in the herbaria of Universities of Lagos (LUH) and Ibadan (UIH), Forestry Research Institute of Nigeria, Ibadan (FHI). The fresh samples collected from the fields were examined critically. Herbarium abbreviations are as reported by Holmgren *et al.* (1990) and provenances of the specimens used are shown in Table 1. Samples were collected at every 3rd, 4th or 5th step depending on size and degree of variability within the population, in line with the approach of Anderson (1941, 1943), Clausen (1960) and Olowokudejo (1995). They were examined both unaided and with the aid of x10 magnifying hand lens for exomorphological characteristics.

For scanning electron microscopy, approximately 8 mm^2 of the preserved leaves was hydrated in a solution of water Aerosol OT and chloral hydrate. They were later dehydrated in graded ethanol series: 50%, 70%, 90% and 100%; then bisected and mounted on stubs with adaxial and abaxial surfaces facing upward before sputter coating with Palladium Gold alloy. Samples were observed at 20KeV using Phillips 505T SEM and photographs were taken with Polaroid 55P/IN film. The approach followed Kadiri *et al.* (2009). Voucher specimens have been deposited in the herbarium and representative samples maintained in cultivation in the botanical garden of University of Lagos, Nigeria.

Principal component and neighbor joining analyses based on Euclidean similarity measure were performed with three characters (leaf apex length, number of inflorescence per plant and number of ovary in the bract), and a UPGMA dendogram based on the nine characters studied was generated using the statistical program of Hammer *et al.* (2001) to reflect the degree of relationships among the populations.

Results and Discussion

Acalypha fimbriata Schum. & Thonn. var. **fimbriata**

The taxon is representative of the type already described (Akobundu and Agyakwa, 1998). It was compared with the described *Acalypha ciliata* (Hutchinson and Dalziel, 1958), a very close ally, as there is no record of it in the flora (Hutchinson and Dalziel, 1958). Therefore, the attention given to it is for the purpose of comparison with the new taxon.

Acalypha fimbriata Schum. & Thonn. var. **robusta** Kadiri **var. nov.**

Diagnosis: *Folia apex acutatus-acuminatus, basis attenuatus-truncatus, margine serratus, forma ovatus-lanceolatus. Venatio craspedodroma, Folia magnitudo 4.8 – 6.3 × 3.0 – 3.6 cm, folia apex longitudo 9 -11 mm, petiolo longitudo 3.1 – 4.6 cm, inflorescentia numero 27 – 35, planta altitude 39.3 – 48.2 cm, pistillum numero per bractea 2, cuticularis, stomata depressus cum tenuis, stomata margine et latus - longust stomata apetura et conicus trichoma verruca.*

Type: Collected from c. 250 m from community primary school, Molade area, Iwo road Ibadan, 14 July 200. (*Holotype*: LUH; *Isotype*: FHI).

Table 1. Provenances of the specimens of *Acalypha fimbriata* used for the study.

No.	Localities	Voucher specimens
1	Ibadan	P. Wit, 17.8.1971, 27563 (FHI)
2	Ibadan	A. P. D. Jones, 23.10.1945, 13737 (FHI)
3	Igboora	O. B. B., 15.10.1981, 96310 (FHI)
4	Cross-River	Ibhanesbor and Oguntayo, 8.6.1972, 65249 (FHI)
5	Kwara	Olorunfemi and Ibhanesbor, 14.5.1973, 70013 (FHI)
6	Ibadan	M. C. Ejiofor, 7.9.1949, 24560 (FHI)
7	Abeokuta	Oduwo and Binuyo, 30.8.1984, 102011 (FHI)
8	Lagos	Oguntayo and Olorunfemi, 7.11.1977, 91921 (FHI)
9	Owena	Oyayomi and Osanyinlusi, 21.6.1977, 84421 (FHI)
10	Ago-Are	Olorunfemi *et al.*, 19.10.1981, 96370 (FHI)
11	Ipake	Magbagboeola & Co., 1.2.1981, 94854 (FHI)
12	Otukpa	Emwiogbon and Oguntayo, 30.6.1978, 103402 (FHI)
13	Ohumbe	Oyayomi and Osanyinlusi, 13.6.1977, 82984 (FHI)
14	Cross-River	Ekwuno & Co, 14.3.1979, 89054 (FHI)
15	Cross-River	W. Punt, 22.5.1959, 51931 (FHI)
16	Abeokuta	C. Owechi, 22.5.1959, 52381 (FHI)
17	Lagos	J. G. Adam, 18.4.1959, 52381 (FHI)
18	Oshogbo	C. Owehi, 1.8.1964, 4485 (FHI)
19	Ibadan (Kadiri 2)	Kadiri, 10.9.1999, 2025 (LUH)
20	Ibadan	Kadiri, 8.7.2000, 2026 (LUH)
21	Ibadan	Kadiri, 26.10.2000, 2027 (LUH)
22	Ibadan	Kadiri, 6.7.2001, 2028 (LUH)
23	Iwo (Kadiri 55)	Kadiri, 9.8.2000, 2029 (LUH)
24	Ilorin (Kadiri 20)	Kadiri, 12.9.2000, 2030 (LUH)
25	Ijebu-Ode (Kadiri 4)	Kadiri, 26.10.2000, 2031 (LUH)
26	Ife (Kadiri 8)	Kadiri, 7.8.2000, 2032 (LUH)
27	Ibadan (Kadiri 15)	Kadiri, 9.8.2000, 2032 (LUH)
28	Ijebu- Ode (Kadiri 30)	Kadiri, 26.10.2000, 2033 (LUH)
29	Ijebu- Ode (Kadiri 35)	Kadiri, 3.12.2000, 2034 (LUH)
30	Iwo (Kadiri 29)	Kadiri, 6.7.2001, 2035 (LUH)
31	Ibadan (Kadiri 56)	Kadiri, 6.7.2001, 2036 (LUH)
32	Ibadan (Kadiri 57)	Kadiri, 14.7.2001, 2037 (LUH)
33	Igboora	Kadiri, 6.7.2001, 2036 (LUH)
34	Abeokuta	Kadiri, 14.7.2001, 2037 (LUH)
35	Ijebu-Ode	Kadiri, 22.10.2008, 2335 (LUH)
36	Ijebu-Igbo (Kadiri 28)	Kadiri, 18.11.2008, 2345 (LUH)
37	Shagamu	Kadiri, 12.12.2008, 2435 (LUH)
38	Ibadan (Kadiri 59)	Kadiri, 14.12.2008, 2467 (LUH)
39	Ibadan (Kadiri 12)	Kadiri, 12.2.2009, 2535 (LUH)

An annual herb, 30-70 mm long. Stem woody at the base and few to profusely branching. Leaves simple, alternate, 41-71 × 19-43 mm, usually lanceolate, margin serrate, base oblique, apex acuminate, petiole slender, 21-52 mm long. Inflorescence on axillary spikes, shorter than the petioles, inflorescence number 17-36 per individual. Flowers greenish, male flower is above the female, clumsy, arranged alternately in boat-shaped bracts. Fruit a 3-chambered capsule, usually 2 (very rarely 1) per bract, 1-2 mm long. Seeds brown, c. 1-3 mm long. Leaf cuticular characters include thin stomata rim, wide-long stomatal aperture and warty and long unicellular trichomes.

Phenology: Flowering: April to August; Fruiting: May to August.

Habitat: Farmland and roadside.

Distribution: It is very common in the south western to the south southern Nigeria.

Etymology: The name is based on the overall appearance of the plant.

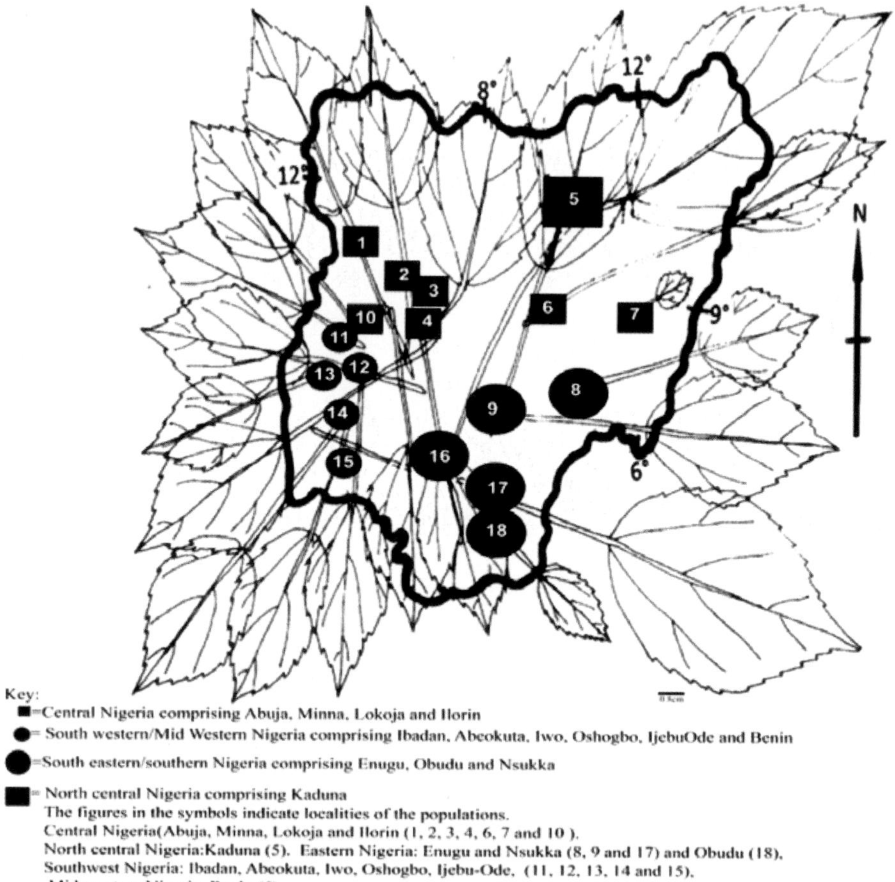

Fig. 1. Map of Nigeria showing distribution and leaf variations among the populations of *Acalypha fimbriata* across Nigeria.

The two varieties can be distinguished by number of ovaries per bract which is usually 1 in var. *fimbriata* and 2 in var. *robusta* (Fig. 2B, C, F, G) and mean inflorescence number on the axillary spike that is usually varied from 2-16 in the former and 27-35 in the latter (Table 2). Another important feature for distinction is the length of leaf apex; it varies from 5-8 mm in var. *fimbriata* to 9-11 mm in var. *robusta* (Fig. 2D, H; Table 2).

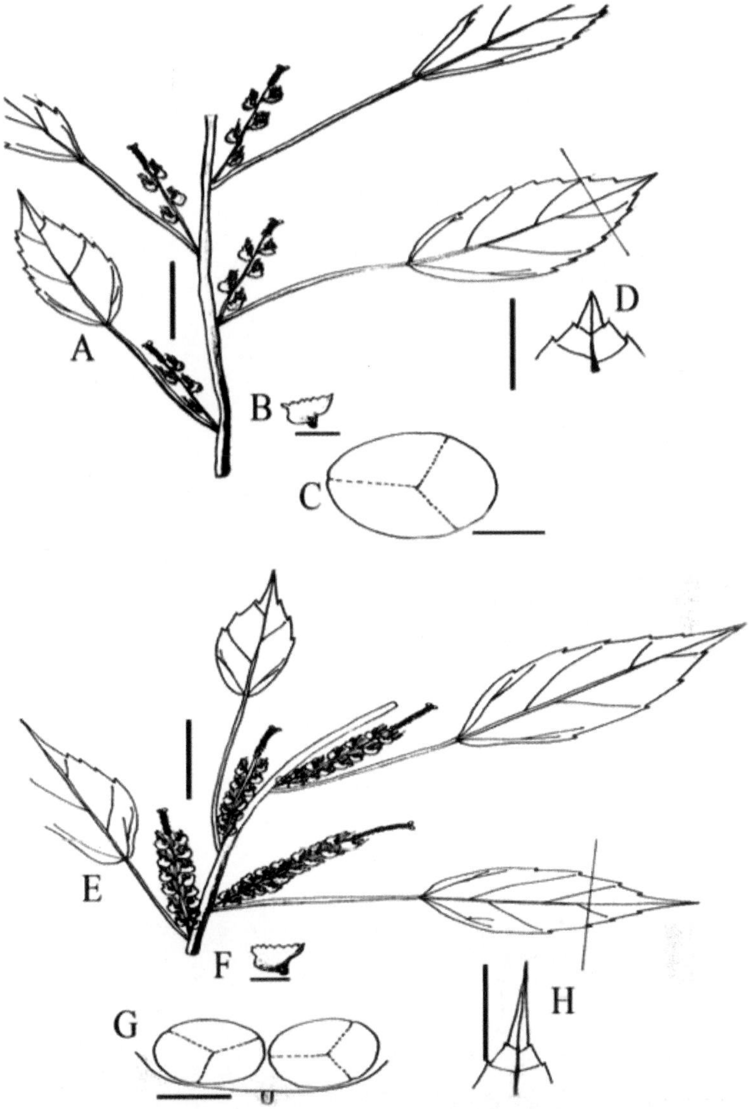

Fig. 2. Line drawings showing morphological features of the two infra-specific taxa of *Acalypha fimbriata*. A-D: *A. fimbriata* var. *fimbriata,* E-H: *A. fimbriata* var. *robusta* Kadiri **var. nov**. Kadiri. D and H: leaf apex; B and F, bracts partially sectioned to show fruits, C and G, number of fruits per bract. Scale bars: A, C, D, E, G and H = 1 cm; B and F = 0.6 cm.

Other features are stomatal complex configuration (thin-wide rim and narrow-wide plus long-short stomatal aperture) and the trichome ornamentation which is smooth in var. *fimbriata* and warty in var. *robusta* (Figs 3G; 4E, G). It was observed that individuals of the same populations often have different appearances. Some may have slim stature and scanty inflorescence distribution, while others have robust stature and clumsy inflorescence distribution (Fig 2). However, the leaf blade and petiole length reasonably overlap among the populations (Table 2). The present study also reveals that the populations of the species are clearly divided into the two recognized varieties using principal component analysis (Fig. 5A) and the relationship patterns are

Table 2. The quantitative characters of the populations of *Acalypha fimbriata* used for group recognition

Pop. Acc. No	Leaf width (mm)	Leaf length (mm)	Petiole length (mm)	Fruit length (mm)	Seed diameter (mm)	Plant height (cm)	Leaf apex (mm)	Pistil No.	Inflorescence no. per plant
Kadiri 2	23(37±2)39	52(65±2)70	40(49±2)52	1.0(1.5±0.1)2.0	1.0(2.0±0.2)2.0	35.0(43.4±5.0)47.6	5.0(6.0±2)7.0	1	6(8±2)9
Kadiri 8	26(39±3)43	50(58±3)68	32(37±3)41	1.0(0.9±0.2)1.5	1.0(2.0±0.2)2.0	28.3(34.5±3.5)39.8	7.0(8.0±2.0)8.5	1	14(16±5)18
Kadiri 30	24(30±1)32	53(58±3)62	28(33±2)40	1.6(2.3±0.2)2.5	1.0(2.0±0.2)2.0	19.4(22.3±2.8)26.4	5.5(6.0±2)7.0	1	6(8±2)10
Kadiri 4	25(38±2)42	54(63±2)67	38(42±2)47	1.0(2±0.3)2.2	1.0(3.0±0.2)3.0	24.4(30.0±3.1)42.0	7.0(8.0±2)9.0	1	2(2±0)2
Kadiri 12	22(32±2)35	48(55±3)58	10(16±3)23	0.8(1.0±0.1)1.0	1.0(2.0±0.2)2.0	18.9(27.7±2.9)31.4	4.7(5.0±2)6.0	1	6(8±2)9
Kadiri 55	18(31±2)34	48(58±2)63	32(44±3)53	1.0(2.0±0.1)2.0	1.0(3.0±0.2)3.0	29.3(40.3±5.1)49.4	6.5(7.0±2)8.0	1	12(13±3)15
Kadiri 56	20(36±2)39	60(63±4)69	22(30±2)34	1.0(2.0±0.1)2.0	1.0(2.0±0.2)2.0	19.2(27.9±3.2)38.3	7.0(8.0±2)9.0	1	11(12±2)15
Kadiri 28	21(25±3)30	46(53±2)61	26(32±3)41	1.0(2.0±0.1)2.0	1.0(3.0±0.2)3.0	24.4(39.3±3.3)46.2	9.5(10.0±3)11.0	2	34(35±5)37
Kadiri 29	25(31±4)36	43(55±3)64	21(31±2)43	1.0(2.0±0.1)2.0	1.0(3.0±0.2)3.0	35.1(48.2±5.3)70.0	8.0(9.0±3)10.0	2	26(29±4)32
Kadiri 35	30(38±3)43	48(54±2)65	34(46±2)54	1.0(2.0±0.1)2.0	1.0(2.0±0.2)2.0	33.1(44.4±5.7)52.0	8.5(9.0±3)10.5	2	17(19±3)21
Kadiri 20	28(32±2)36	52(61±5)71	23(32±3)42	1.0(2.0±0.1)2.0	1.0(3.0±0.2)3.0	30.2(39.3±3.3)43.6	9.0(10.0±3)11.0	2	28(30±6)34
Kadiri 57	21(31±2)34	43(55±2)65	30(43±2)52	1.0(2.0±0.1)2.0	1.0(2.0±0.2)2.0	31.1(40.4±5.1)52.3	8.5(9.0±2)10.0	2	27(32±4)36
Kadiri 15	23(30±3)34	41(50±3)60	22(31±3)40	1.0(2.0±0.1)2.0	1.0(3.0±0.2)3.0	34.2(48.0±5.2)55.8	8.0(9.0±2)10.0	2	24(27±3)31
Kadiri 59	19(31±3)35	58(48±2)62	25(33±2)42	1.0(2.0±0.1)2.0	1.0(3.0±0.2)3.0	31.2(46.0±5.2)54.4	10.0(11.0±3)12.0	2	26(30±4)33

reflected by the dendrograms of neighbour joining analysis and unweighted pair group method with arithmetic mean (UPGMA) (Figs 5B & 5C). The statistical interpretations (i.e. principal component analysis, neighbour joining and UPGMA dendrograms) of the findings reasonably support the division of the populations of the species into two and recognition of them as different varieties. One taking after the type specimen already described for *A. fimbriata* (Akobundu and Agyakwa, 1998) and deposited in the herbarium and the other being newly described for the first time in the present work as *A. fimbriata* var. *robusta*.

The distinctive morphological characterstics of *Acalypha fimbrita* var. *fimbriata* and *A. fimbriata* var. *robusta* are presented in Table 3.

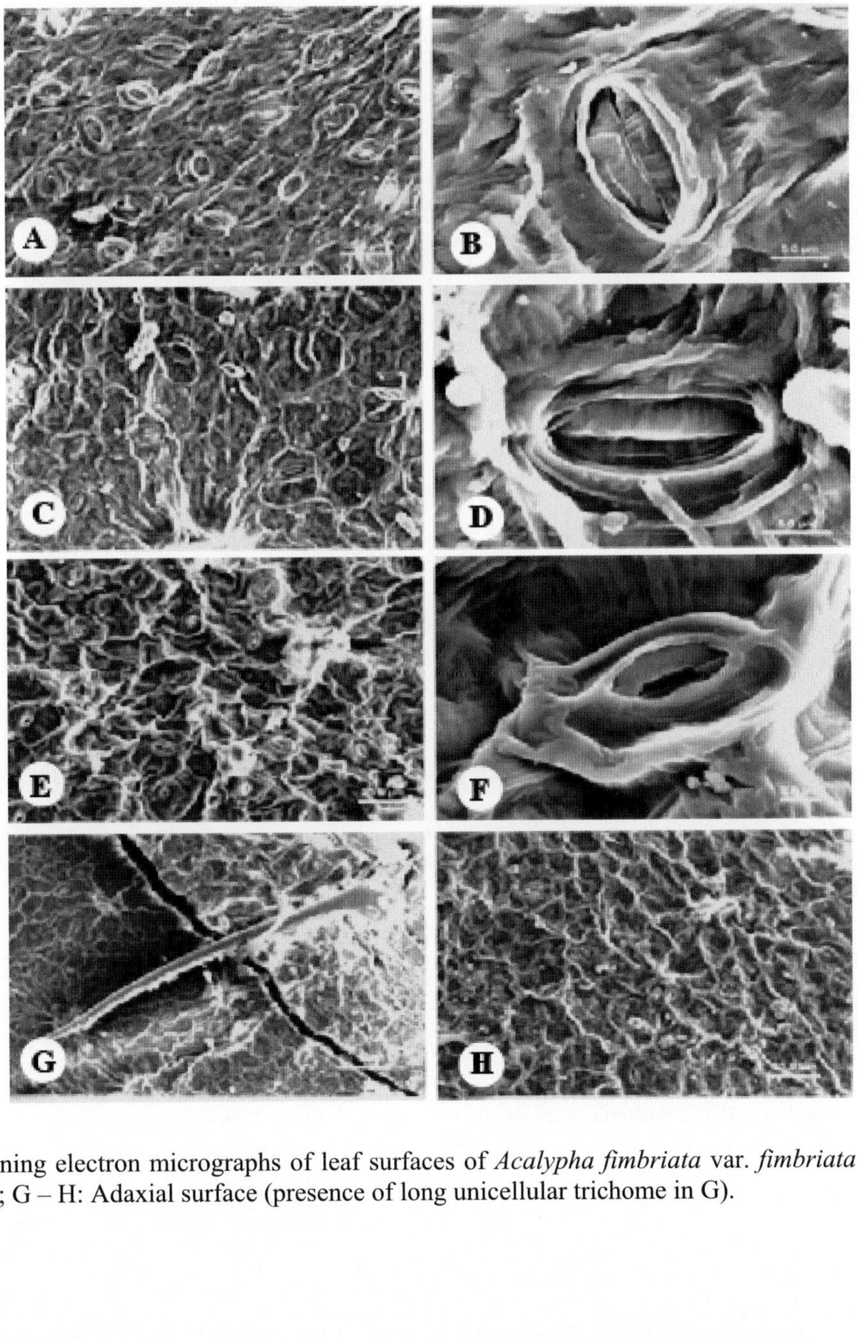

Fig. 3. Scanning electron micrographs of leaf surfaces of *Acalypha fimbriata* var. *fimbriata* A – F: Abaxial surface; G – H: Adaxial surface (presence of long unicellular trichome in G).

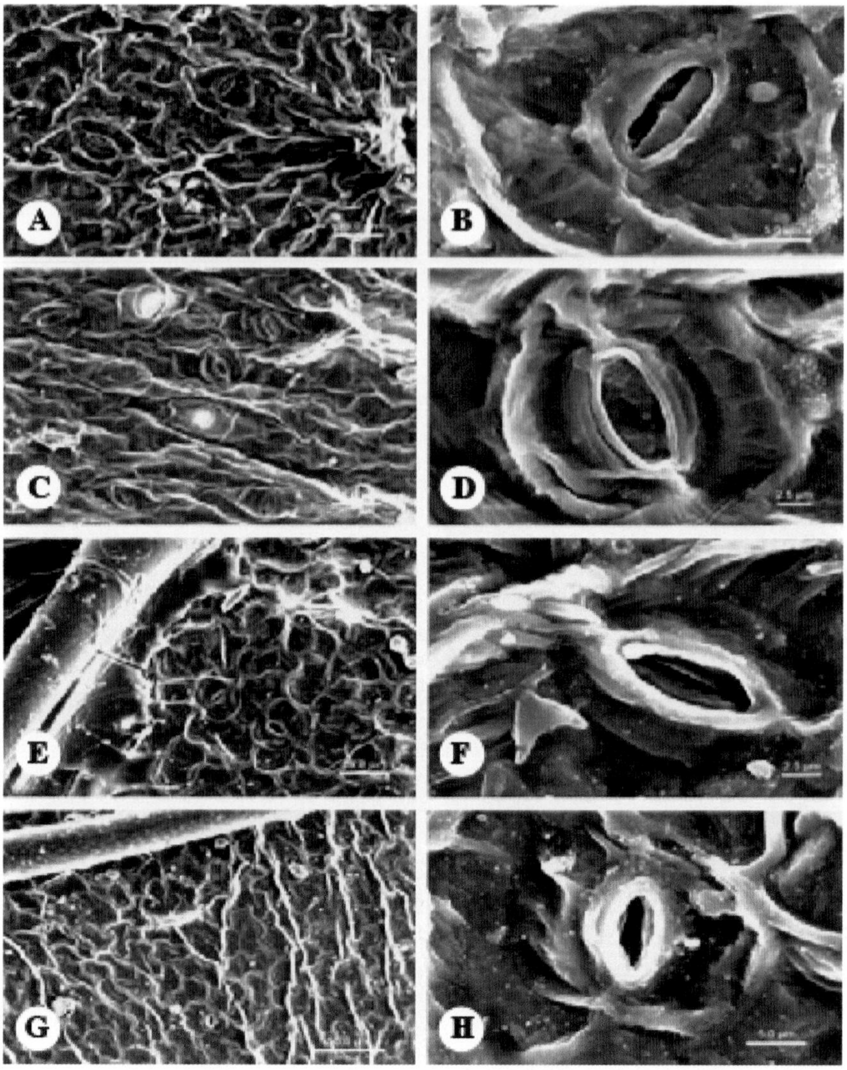

Fig. 4. Scanning electron micrographs of leaf surfaces of *Acalypha fimbriata* var. *robusta* Kadiri **var. nov.** A – F: Abaxial surface; G – H: Adaxial surface (warty surface of trichomes are in E and G).

Table 3. Distinctive morphological characteristics of the recognized infra-specific taxa of *Acalypha fimbriata* in Nigeria.

Characters	*Acalypha fimbriata* var. *fimbriata*	*Acalypha fimbriata* var. *robusta* **var. nov.**
Inflorescence number	2 – 16	27 - 35
Pistil number per bract	1	2
Cuticular characters	Wide stomatal rim	Thin stomatal rim
	Narrow-short stomatal aperture	Wide-long stomatal aperture
	Smooth unicellular conical trichome	Warty unicellular conical trichome

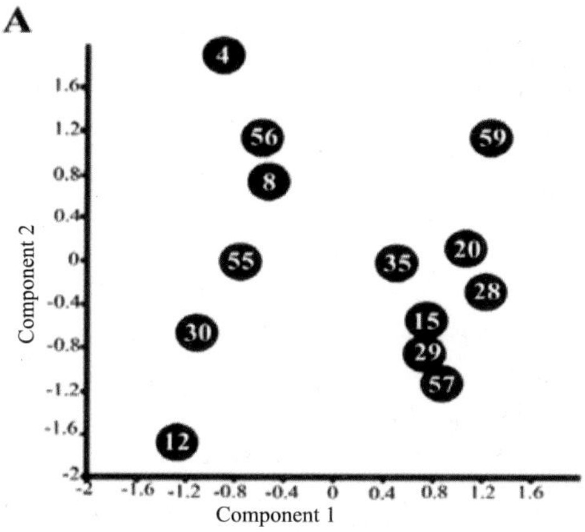

Principal component analysis showing the two varieties identified.
The figures are the populations' accession numbers.

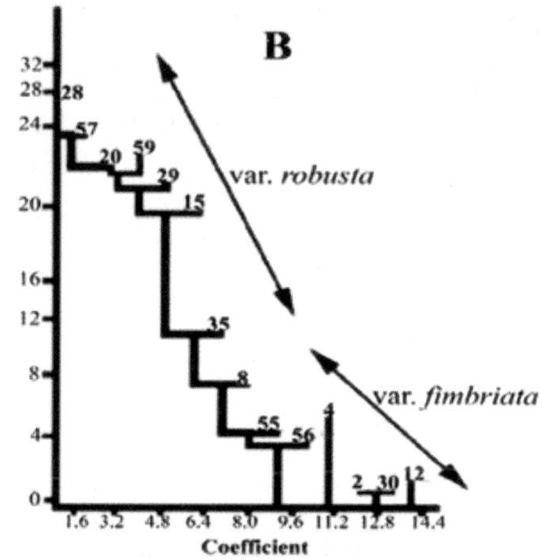

Neighbour joining analysis using
Euclidean similarity measures.

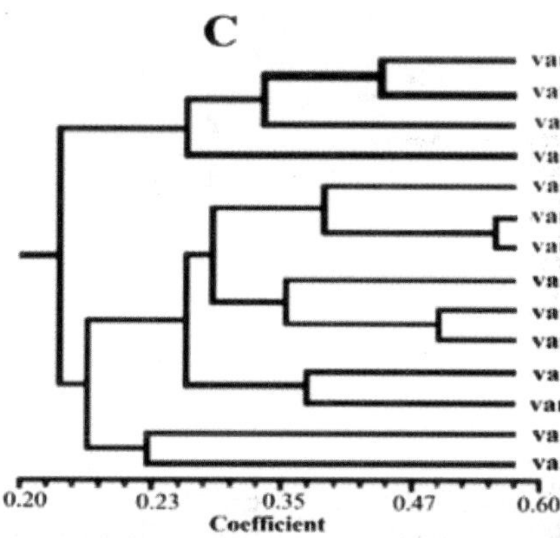

A UPGMA dendrogram showing relationships
among the populations.

Fig. 5. Relationships among the studied populations of *Acalypha fimbriata*. 5A. Division of the populations supported by principal component analysis; 5B. Neighbour joining dendrogram showing the separation of the populations; 5C. UPGMA tree showing the pattern of relationships among the populations of *Acalypha fimbriata*.

Most of the populations used in the study are from southwest Nigeria where Hutchinson and Dalziel (1958) implicated as the most probable centre of genetic diversity for most species of *Acalypha* in West Africa. The characters used for analysis have been reported by Hutchinson and Dalziel (1958) and Agnew (1974) to be taxonomically useful in the genus. In corroboration, Davis and Heywood (1963) stated that morphology is a pedestal upon which taxonomic decisions are based. Although, cuticular characters of the leaf surface are not discriminating enough to support the groups recognized, however, presence of warty and long unicellular trichome and thin stomatal rim recorded in the var. *robusta* Kadiri var. nov. distinguish it from the var. *fimbriata*. Stomatal and trichome characters are useful for delimitation of taxa (Stace, 1965; Inamdar and Gangdhara, 1977; Metcalfe and Chalk, 1979; Devi *et al.,* 2013), which has been supported by the present study. The statistical analyses of principal component analysis, neighbour joining analysis and UPGMA dendrograms also reflect the level of relationships among the populations and support recognition of the two distinct groups i.e. varieties in *Acalypha fimbriata.*

Acknowledgement

The support of the University of Lagos, Nigeria through 2001-2005 postgraduate fellowship stipends is acknowledged, without which the field trips and visits to herbaria would have been impossible. I am also grateful to Professors J.D. Olowokudejo and O.T. Ogundipe for their encouragement. The work is a part of the author's Ph. D. Thesis.

References

Agnew, A.D.Q. 1974. Upland Kenya Wild Flowers. Oxford University Press, London, pp. 213-217.

Akobundu, I. A. and Agyakwa, C.W. 1998. A Handbook of West African Weeds. International Institute of Tropical Agriculture, Nigeria, 564 pp.

Anderson, E. 1941. The technique and use of mass collections in plant taxonomy. Anna. Missour. Bot. Gard. **23**: 287-292.

Anderson, E. 1943. Mass collections. Chron. Bot. **7**: 378-380.

Burkill, H.M. 1994. The Useful Plants of West Tropical Africa. Royal Botanic Gardens, Kew, 636 pp.

Clausen, J. 1960. A simple method for the sampling of natural populations. Reports of Scottish Plant Breeding Station, pp. 69-75.

Davis, P.H. and Heywood, V.H. 1963. Principles of Angiosperm Taxonomy. Oliver & Boyd. Ltd., Edinburgh, 559 pp.

De Candolle, A.P. 1866. Prodromus Systematis **15**(2): 491.

Devi, N.J., Padma, Y., Narasimhudu, C.L. and Raju, R.R.V. 2013. Diversity of stomata and trichomes in *Euphorbia* L. – I. Bangladesh J. Plant Taxon. **20**(1): 27-38.

Hammer, O., Harper, D.A.T. and Ryan, P.D. 2001. PAST: Paleontological statistics software package for education and data analysis. Palaeontologia Electronica **4**(1): 1-9.

Hepper, F.N. 1976. The West African Herbaria of Isert and Thonning. Robert MacLehose & Company Ltd., Great Britain, 227 pp.

Holmgren, P.K., Holmgren, N.H. and Banner, T.L.C. 1990. Index Herbariorum. Part I: The Herbaria of the World [Regnum Veg. Vol. **120**]. New York Botanical Garden, New York.

Hutchinson, J. and Dalziel, J.M. 1958. Flora of West Tropical Africa. Crown Agents for Overseas Governments and Administrations, London, 828 pp.

Inamdar, J.A. and Gangadhara, M. 1977. Studies on the trichomes of some Euphorbiaceae. Feddes Repertorium **88**: 103-111.

Kadiri, A.B., Olowokudejo, J.D. and Ogundipe, O.T. 2009. Leaf epidermis morphology of West African species of the genus *Acalypha* (Euphorbiaceae). Botanica Lithuanica **15**(2): 65-78.

Metcalfe, C.R. and Chalk, L. 1979. Anatomy of the Dicotyledons. Oxford University Press, Oxford, 276 pp.

Olowokudejo, J.D. 1995. Taxonomic study of the *Biscutella variegata* complex (Cruciferae). Willdenowia **25**: 25-38.

Radcliffe-Smith, A. 1989. Revised conspectus of the Euphorbiaceae. Euphorbiaceae Newsletter **2**: 4-9.

Radcliffe-Smith, A. 1996. Euphorbiaceae, subfamilies Phyllanthoideae, Oldfieldioideae, Acalyphoideae, Crotonoideae, and Euphorbioideae, tribe Hippomaneae. In Pope G. V. (Editor). Flora Zambesica **9**(4): 1-357.

Stace, C.A. 1965. Cuticular studies as an aid to plant taxonomy. Bull. Brit. Mus. Nat. Hist. Bot. **4:** 1-78.

Thonner, F.R. 1915. Flowering Plants of Africa. Dulan & Co. Ltd., London, 647 pp.

17

A NEW SPECIES OF *POLYGONATUM* MILL. (ASPARAGACEAE) FROM GUIZHOU, CHINA

MING-TAI AN, YUN LIN[1, 2], JIA-GUO WANG, JIANG-HUA WU[3] AND MIN MENG[4]

Forestry College, Guizhou University, Guiyang 550025, Guizhou, P. R. China

Keywords: Polygonatum sinopubescens; Asparagaceae; Yinjiang County; SW China.

Abstract

A new species *Polygonatum sinopubescens* from Yinjiang County, Guizhou Province, South-west China is described and illustrated. This species was found growing in evergreen broad-leaved forests or mixed needle-leaved and evergreen broad-leaved forests on slopes at altitudes of 870-930 m. It is related to *Polygonatum filipes* Merr. *ex* C. Jeffrey & Mc Ewan, but differs from the latter by pubescent stems, petiole, peduncle and pedicel, leaf blade shortly dense pubescent on abaxial surface, 2-3-flowered inflorescences and filaments 7-11 mm long.

Introduction

Polygonatum Mill. (Asparagaceae) comprises of ca. 60 species and distributed in temperate regions of the North hemisphere, mainly from the Himalayas to Japan having 39 species in China (Chen and Tamura, 2000).

During our expeditions in Yinjiang County, northeast of Guizhou Province, southwest China in May 2014 and May 2015, we collected previously unknown specimens of *Polygonatum* from evergreen broad-leaved forests or mixed needle-leaved and evergreen broad-leaved forests on slopes at altitudes of 870-930 m above sea level, in latitude 27°36′50.6″-27°57′28″ N, and longitude 108°26′15″-108°36′31.9″ E. After critical examination of the specimens deposited at GZAC, HGAS, IBK, IBSC, KUN and PE, and carefully consulting relevant literature (Baker, 1875; Li, 1966; Tang, 1978; Chen and Liu, 1984; Tu, 1986; Liang, 1987; Wan and Gao, 1990; Zhu, 1992; Chen and Tamura, 2000; Fu *et al.*, 2002), it was identified as a new species of *Polygonatum*. This paper describes and illustrates the new species as *Polygonatum sinopubescens* M. T. An, Yun Lin & J. G. Wang.

Polygonatum sinopubescens M. T. An, Yun Lin & J. G. Wang, **sp. nov.** (**Figs 1 & 2**).

Diagnosis: *Polygonatum sinopubescens* is morphologically similar to *Polygonatum filipes* Merr. *ex* C. Jeffrey & Mc Ewan based on moniliform or terete-moniliform rhizome, alternate leaves, oblong-lanceolate to elliptic leaf blade, very slender peduncle, 1.5-2.0 cm long perianth. However, it differs from the latter in stems pubescent, petiole pubescent, peduncle pubescent and pedicel pubescent (vs glabrous in *P. filipes*), leaf blade shortly dense pubescent on abaxial surface (vs shortly pubescent on abaxial veins in *P. filipes*), filaments 7-11 mm long (vs 4 mm long in *P. filipes*).

---

[1]Hunan Medication Vestibule School, Changsha 410208, Hunan, P. R. China.
[2]Corresponding author. Email: leoliny@foxmail.com
[3]Guizhou Normal College, Guiyang 550018, Guizhou, P. R. China.
[4]Yinjiang Forestry Bureau, Yinjiang 555200, Guizhou, P. R. China.

Type: China. Guizhou Province: Yinjiang County, Yangxi Nature Reserve, alt. 870-930 m in evergreen broad-leaved forests or mixed needle-leaved and evergreen broad-leaved forests on slopes, 16 May 2011, *J.G. Wang & X.F. Li YJ* 2014-0110 (*Holotype*: GZAC, GZAC Herb. Bar Code No. 0026320; *Isotypes*: HUFD, PE); same locality, 21 May 2015, *M.T. An & J.G. Wang* 2015-0627 *(Paratypes*: GZAC, HUFD).

Rhizome moniliform or terete-moniliform, 1.0-1.8 cm thick, up to 30 cm long. Stem erect or ascending, 30-60 cm long, pubescent. Leaves 5-9, alternate; petiole short or indistinct, pubescent; leaf blade ovate-elliptic, elliptic or oblong-lanceolate, 8.0-11.5 cm long, 2.5-4.0 cm wide, base broadly cuneate to rounded, margin entire, apex obtuse to shortly acuminate, shortly dense pubescent on abaxial surface. Inflorescence axillary, 2-3-flowered; peduncle slender, 3.0-5.5 cm long, dense pubescent; bracts lanceolate, 2-4 mm long, greenish white, caducous. Flowers pendulous; pedicel 1-2 cm long, pubescent. Perianth yellowish green or greenish white, cylindric, 1.5-2.0 cm long; lobes 4-5 mm long. Stamens 6, 1.5-1.8 cm long; filaments 7-11 mm long, shortly cottony; anthers 3 mm long. Ovary obovoid, 4 mm long; style 1.3-1.5 cm long. Young berries obovoid, 4-5 mm long, 3-4 mm in diameter.

Phenology: Flowering from May to June, and fruiting from July to September.

Etymology: *Polygonatum sinopubescens* is named after shortly dense pubescent on abaxial surface of leaf blade, and this species is from China.

Vernacular name: *Roumao Huangjing*

Habitat: This species grows in evergreen broad-leaved forests or mixed needle-leaved and evergreen broad-leaved forests on slopes at altitudes of 870-930 m above sea level., latitude 27°36′50.6″-27°57′28″ N, longitude 108°26′15″-108°36′31.9″ E, comprises about 200 individuals growing in five populations within the nature reserve.

Distribution: Polygonatum sinopubescens is only known from its type locality, Yangxi Nature Reserve, Yinjiang County, northeast Guizhou Province, southwest China.

Pharmaceutical value: In traditional Chinese medicine the rhizome is used to moisten the lung, nourish the kidney and invigorate the spleen.

A comparison between the new species *Polygonatum sinopubescens* and its closely related is appended in Table 1.

Table 1. Comparison of morphological characteristics between *Polygonatum sinopubescens* sp. nov. and *P. filipes*.

Characters	*P. sinopubescens* **sp. nov.**	*P. filipes*
Stem	pubescent	glabrous
Leaf	petiole pubescent; leaf blade shortly dense pubescent on abaxial surface	petiole glabrous; leaf blade shortly pubescent on abaxial veins
Inflorescence	2-3-flowered; peduncle 3.0-5.5 cm long, pubescent	2-7-flowered; peduncle 3.0-8.0 cm long, glabrous
Pedicel	pubescent	glabrous
Filament	7-11 mm long	c. 4 mm long
Berry	obovoid	spherical

Conservation status: Polygonatum sinopubescens is known only from the type locality, comprises about 200 individuals growing in five populations within the nature reserve, and is therefore given the assessment of Data Deficient (DD) according to IUCN (2001) criteria.

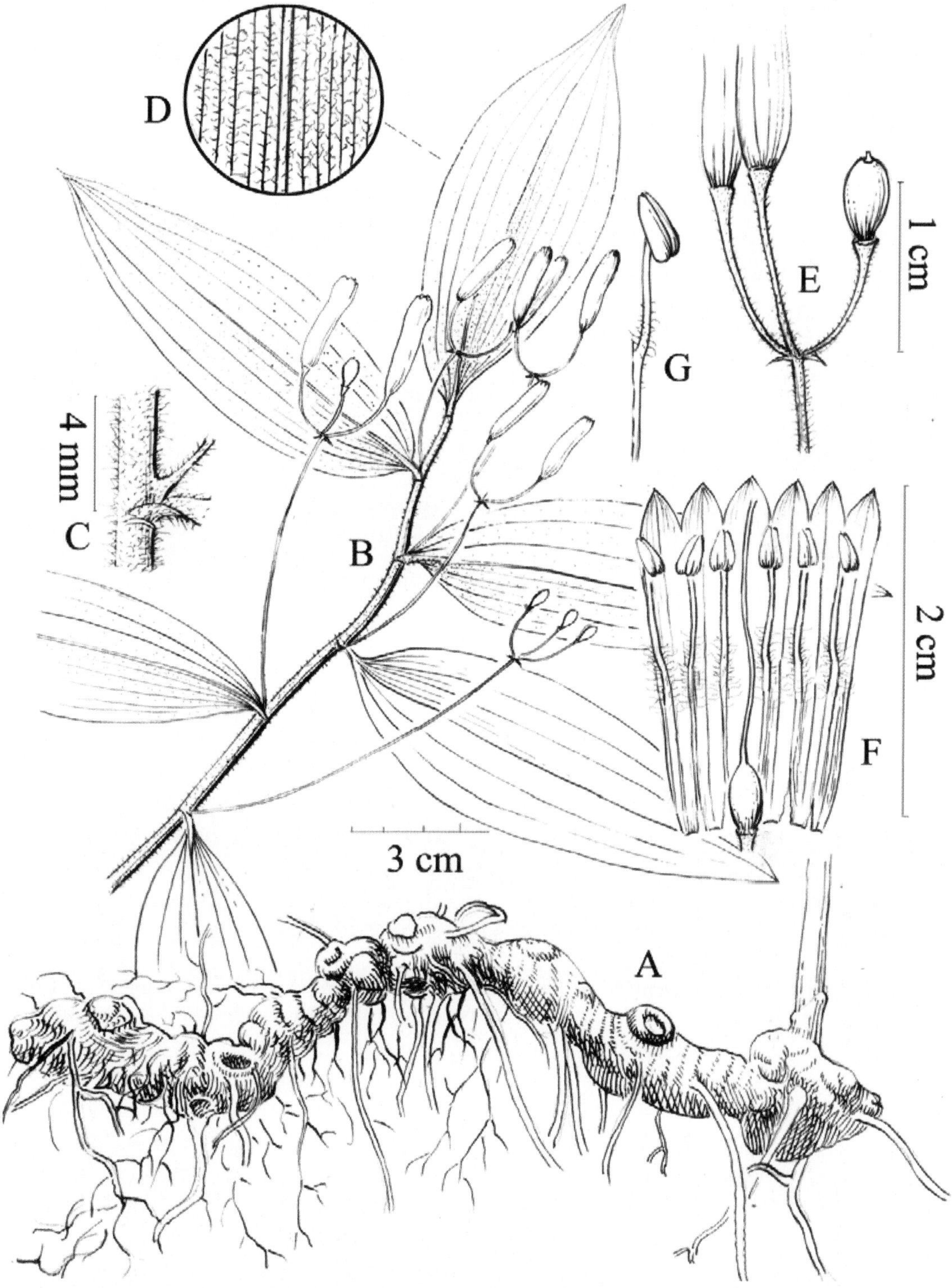

Fig. 1. **Polygonatum sinopubescens** M.T. An, Yun Lin & J.G. Wang, **sp. nov.** A. Rhizome; B. Habitat of flowering plant; C. Branch with node; D. Leaf blade pubescence on abaxial surface; E. Inflorescence; F. Dissected flower; G. Stamen. (*J. G. Wang & X. F. Li* YJ-2014-0110).

Fig. 2. **Polygonatum sinopubescens** M.T. An, Yun Lin & J.G. Wang, **sp. nov.** A. Habitat; B. Rhizome; C. Stem with pubescent; D. Leaf blade, showing shortly dense pubescent on abaxial surface; E. Inflorescence; F. Berries with bracts.

Acknowledgements

Thanks are due to the curators of herbaria, namely GZAC, HGAS, IBK, IBSC, KUN and PE for permission to examine their specimens. This work was supported by Application Fundamentals Major Special Projects: Ecological restoration and its optimal regulation of eco-economic system in Karst Rocky Desert Area grant number Qian-Ke-He JZ-NO. [2014] 2002,,Plant Specimen Digitization and Chinese Virtual Herbarium Establishment (grant number 2005DKA21401), and the Second Investigation on National Key Protected Wild Plant Resources in Guizhou Province (grant number Lin-Hu-Fa-No. [2012] 87 and Qian-Lin-Hu-Tong-No. [2013] 251). We also thank Mr. Hua Xie for the drawing.

References

Baker, J.G. 1875. Revision of the genera and species of Asparagaceae. J. Linn. Soc. Bot. **14**: 508–630.

Chen, S.C. and Liu, D.Q. 1984. Two new species of Liliaceae from China. Acta Phytotaxon. Sin. **22**(5): 417–419.

Chen, X.Q. and Tamura, M.N. 2000. *Polygonatum* (Liliaceae). *In*: Wu, Z.Y. and Raven, P.H. (Eds), Flora of China. Vol. **24**. Science Press, Beijing & Missouri Botanical Garden Press, St. Louis, pp. 223–232.

Fu, L.G., Chen, T.Q., Lang, K.Y., Hong, T., Lin, Q. and Li, R. 2002. Higher Plants of China. Vol. **13**. Qingdao Publishing House, Qingdao, pp. 204–217.

IUCN. 2001. IUCN Red List Categories and Criteria, Version 3.1. Prepared by the IUCN Species Survival Commission. IUCN, Gland, Switzerland, and Cambridge, United Kingdom.

Li, P.Y. 1966. Some new plants of Liliaceae from Tsinling. Acta Phytotaxon. Sin. **11**(3): 251–253.

Liang, S.Y. 1987. Two new species of *Polygonatum* Mill. (Liliaceae) from China. Acta Phytotaxon. Sin. **25**(1): 64–66.

Tang, Y.C. 1978. *Polygonatum* (Liliaceae). *In*: Wang, F.T. and Tang, T. (Eds), Flora Reipublicae Popularis Sinicae. Vol. **15**. Science Press, Beijing, pp. 52–80.

Tu, Y.L. 1986. *Polygonatum* (Liliaceae). *In*: Chang, S.S. (Ed), Flora Guizhouensis. Vol. **3**. Guizhou People's Publishing House, Guiyang, pp. 384–390.

Wan, Y. and Gao, C.Z. 1990. A new species and two varieties from Guangxi. Guihaia **10**(3): 177–180.

Zhu, Z.Y. 1992. A new species of *Polygonatum* from Emeishan. Bull. Bot. Res., Harbin **12**(3): 267–269.

18

TWO NEW RECORDS AND ONE REDISCOVERY OF ANGIOSPERMS FOR BANGLADESH

KHANDAKAR KAMRUL ISLAM[1] AND NAIMUR RAHMAN

Bangladesh National Herbarium, Chiriakhana Road, Mirpur-1, Dhaka-1216, Bangladesh

Keywords: New records; Rediscovery; Angiosperm; Bangladesh

Abstract

Two species of angiosperms, *Spiradiclis bifida* Kurz (Rubiaceae) and *Strobilanthes urticifolia* Wall. *ex* Kuntze (Acanthaceae) are being reported as new records for Bangladesh. One species, *Corymborkis veratrifolia* (Reinw.) Blume belonging to the Orchidaceae has been rediscovered from Hazarikhil Wildlife Sanctuary after about 90 years of last record. These species are described with updated nomenclature, synonyms, taxonomic description, ecology and geographical distribution. Illustrations with field photographs are also provided.

Introduction

During recent botanical explorations in Hazarikhil Wildlife Sanctuary, under Chittagong Forest Division, Chittagong District the authors collected three plant specimens belonging to the families Rubiaceae, Acanthaceae and Orchidaceae. On examining the specimens under Rubiaceae, Acanthaceae and Orchidaceae housed at Bangladesh National Herbarium (DACB), Bangladesh Council of Scientific and Industrial Research, Chittagong (BCSIRH), Herbarium of Bangladesh Forest Research Institute (BFRIH), Dhaka University Salar Khan Herbarium (DUSH) and Herbarium of Chittagong University (HCU)) the specimens did not matched with any specimens stored at those herbaria. Later, these were identified as *Spiradiclis bifida* Kurz (Rubiaceae), *Strobilanthes urticifolia* Wall. *ex* Kuntze (Acanthaceae) and *Corymborkis veratrifolia* (Reinw.) Blume (Orchidaceae).

Encyclopedia of Flora and Fauna of Bangladesh described a total of 107 taxa of the family Acanthaceae (Begum *et al.*, 2008) and a total of 170 taxa of the family Rubiaceae (Rahman and Das, 2009) for the flora of Bangladesh. Das *et al.* (2009, 2012 and 2013) and Das and Rahman (2010, 2011) added another 16 species to the account of Rubiaceae. Rahman and Das (2009) recorded only one species under the genus *Spiradiclis* and Begum *et al.* (2008a) recorded fourteen species under the genus *Strobilanthes.*

These two species, *Spiradiclis bifida* Kurz (Rubiaceae) and *Strobilanthes urticifolia* Wall. *ex* Kuntze (Acanthaceae) were not previously recorded from the territory of Bangladesh in the relevant works of Hooker (1880, 1884), Kurz (1877), Prain (1903), Heinig (1925), Cowan (1926), Cowan and Cowan (1929), Kanjilal *et al.* (1939), Raizada (1941), Datta and Mitra (1953), Sinclair (1956), Rahman and Das (2009), Begum *et al.* (2008), Islam (2009), Uddin and Rahman (2015), and Rahman *et al.* (2015). These were recorded by Hooker (1880, 1884) from Khasia mountains, upper Assam and Cachar in India and from Pakistan, Nepal and India respectively.

On the other hand, Huda (2008) described a total of 179 taxa of the family Orchidaceae for the flora of Bangladesh. *Corymborkis veratrifolia* (Reinw.) Blume belonging to the Orchidaceae was recorded from Chittagong and the Chittagong Hill Tracts by Hooker (1890), Prain (1903) and Heinig (1925). Since then it has neither been collected nor recorded to occur else where in

[1] Corresponding author. E-mail: orchidcu.islam@gmail.com

Bangladesh. Recently it has been rediscovered from Hazarikhil Wildlife Sanctuary by the authors after about 90 years of last record.

Hence, the species *Spiradiclis bifida* Kurz and *Strobilanthes urticifolia* Wall. *ex* Kuntze are new records and *Corymborkis veratrifolia* (Reinw.) Blume is a rediscovery for Bangladesh.

Materials and Methods

Specimens of each species encountered in flowering and fruiting condition were collected and preserved at Bangladesh National Herbarium (DACB). The photographs of fertile specimens in natural habitat were taken during the field trips. Both fresh materials and herbarium specimens were studied and examined by using long arm steriomicroscope. All available taxonomic resources *viz.* literatures, herbarium specimens and botanical illustrations were taken under consideration to identify the species. Digital images of the species are also used to supplement plant identification and document their habitats. Unnamed specimens were identified and described by consulting relevant floristic literatures of Hooker (1880, 1884), Prain (1903), Rahman and Das (2009), Begum *et al.* (2008), Huda (2008) and Wu *et al.* (2011).

Taxonomic enumeration of these two new records and one rediscovery are prepared. In the enumeration, each species is cited with updated nomenclature, commonly known synonyms, taxonomic description, ecology, geographical distribution and citation of voucher specimens deposited at DACB.

Results

1. Spiradiclis bifida Kurz, J. Asiat. Soc. Beng., Pt. 2, Nat. Hist.. 41(2): 310 (1872). Hook. f., Fl Brit. India 3: 76 (1880). **(Figs. 1a & 1b)**

Herbs, up to 48 cm tall, apparently perennial. Stems slender. Leaves 10–19× 3–7 cm., blade drying thinly papery, elliptic or elliptic-lanceolate; apex acute or shortly acuminate or subcaudate, cuneate at base, adaxially sparsely pilose to subglabrous, abaxially pubescent along veins, secondary veins 15–19 pairs. Petiole 1.5–3.0 cm long, pubescent. Stipules subulate. Inflorescence paniculate, up to 18 cm, simple or bifid axes numerous, 1.0 –1.5 cm, scorpioid. Flowers 5-merous, pale creamy, 0.3–0.5 cm long. Calyx glabrous or subglabrous. Ovary 2 locules. Fruits a capsule, c 0.2 cm in diam., glabrous or subglabrous, valves 4. Seeds black.

Flowering and fruiting: January to December.

Habitat: Grows in wet places in deep forest at the slope of the hill about 76–152 m altitude.

Distribution: India (Khasia mountains, upper Assam, Naga hills and Cachar), southern China, Java and Indonesia.

Specimen examined: **Chittagong:** Hazarikhil Wildlife Sanctuary; 5.11.2015, K.K. Islam and N. Rahman, KKI573 (DACB).

Note: This newly recorded species is nearer to *S. cylindrica* Wall. *ex* Hook. f., having leaves 3.0–6.5 cm long with 7–9 pairs of secondary veins, and white flowers with glabrous sepals.

2. Strobilanthes urticifolia Wall. *ex* Kuntze, Revis. Gen. Pl. 2: 499 (1891). *Strobilanthes alatus* Nees, in DC. Prodr. 11:194(1847) *(non* Blume 1826); Hook. f., Fl Brit. India 4: 464 (1884). **(Figs. 2a & 2b)**

A small erect shrub, up to 1.2 m tall, with sticky hairy branches, round to 4-angled. Leaves on 2.3-7.9 cm long branchlets, upper leaves sessile and lower ones long petioled, petiole 2.5–6.4 cm linear or winged upwards, lamina ovate, 7.5–15 × 4.0–7.5 cm, base cordate, rounded or attenuate, acuminate, serrate. Flowers 5-merous, 2.7–3.8 cm long, dark-blue, glandular-pubescent

Figs 1-3: 1a. *Spiradiclis bifida*, 1b. *Spiradiclis bifida* (Illustration), 2a. *Strobilanthes urticifolia*, 2b. *Strobilanthes urticifolia* (Illustration), 3a. *Corymborkis veratrifolia*, 3b. (A-G). *Corymborkis veratrifolia:* A. Habit; B. Leaf; C. Inflorescence; D. Flower; E. Lip with column; F. Front view of column; G. Fruit.

paniculate spikes, c. 2 cm across, in lax, upper flowers soon in distant pairs, not capitate, tube cylindrical below as long as calyx, limb petals open. Bracts ovate, glandular, deciduous. Calyx 0.6–1.0 cm, deeply divided, linear, obtuse, green, sepals 5, unequal, ligulate, blunt, viscous-hairy, enlarged and accrescent in fruit, one much larger than the rest. Stamens 4, glabrous,staminal filaments dilated. Petals 5, about 1.0–1.3 cm long, ventricose, mouth curved. Ovary glandular,hairy;style very remotely minutely hairy; stigma linear c. 0.8-0.9 cm. Capsule c. 1.9 cm long, narrowly elliptic-oblong, glandular-hairy, 4-seeded. Seeds about 0.9 cm long, hairy ,ovate.

Flowering and fruiting: November to January.

Habitat: Grows as undergrowth at the top of the hill about 275–305 m altitude.

Distribution: Pakistan, Afganistan, Nepal and India. It is also found in the Himalayas at altitudes of 2,000–3,500 m.

Specimen examined: **Chittagong**: Hazarikhil Wildlife Sanctuary; 2.12.2015, K.K. Islam and N. Rahman, NR29 (DACB).

Note: This newly recorded species is nearer to *S. dalhousieanus* (Nees) C.B. Clarke, having elliptic or ovate-elliptic leaves, flowers in capitate heads or condensed spikes and bracts equal to calyx.

3. **Corymborkis veratrifolia**(Reinw.) Blume, Coll. Orch. Arch. Ind.: 125(1859). Huda, Ency. Fl. Fa. Bangladesh 12: 32 (2008). *Corymbis veratrifolia* (Reinw.) Rchb. f., Flora 48: 184 (1865), Hook. f., Fl Brit. Ind. 6: 91 (1890), Prain, Beng. Pl. 2: 772 (1903), *Hysteria veratrifolia* Rein., Syll. Pl. Nov. 2: 5 (1826). **(Figs. 3a & 3b)**

A rigid terrestrial herb about 1 m tall. Stem erect, thick, subterete, 0.7–1.3 cm in diameter. Leaves large, 26–40 × 5–10 cm, narrowly elliptic or elliptic-lanceolate, caudate-acuminate, abaxially often brown scurfy, apex long acuminate, scattered with sheathing bases, basal sheaths 4–10 cm long. Inflorescence a panicle, 10–29 or more flowered, 8–12 cm long, 2–6 branched; bracts at branch base lanceolate, 1.0–1.5 cm, floral bracts ovate, 0.2–0.4 cm. Flowers fragrant, axillary, 2–3 cm across, tubular or campanulate, greenish-white, c. 0.4–0.6 cm. Sepals and petals subsimilar, oblanceolate-linear or narrowly, 3.0–3.7 cm, with the upper one-third deflexed. Lip similar in size with the sepals and other petals, narrowly clawed which encloses the column and flares into an ovate blade with crenulate margin, blade of lip lanceolate, acuminate, apical lobe orbicular or broadly ovate-elliptic with 2 longitudinal ridges. Column c. 3 cm long, subcylindric, shorter than lip. Pollinia 2, soft, elongated. Rostellum bifid, c. 0.3 cm.

Flowering and fruiting: May to November.

Habitat: Near the streams bank in the dense shade of the deep forest about 15–76 m altitude.

Distribution: Bhutan, Cambodia, India, Indonesia, Japan, Laos, Malaysia, N. Australia, Sri Lanka, SW. Pacific Islands, Taiwan, Thailand and Vietnam. In Bangladesh, the plant was recorded from Chittagong and Chittagong Hill Tracts.

Specimen examined: **Chittagong**: Hazarikhil Wildlife Sanctuary, 4.11.2015, K.K. Islam and N. Rahman, KKI463 (DACB).

Acknowledgements

The authors express their gratitude and sincere thanks to Dr. Sarder Nasir Uddin, Principal Scientific Officer, Bangladesh National Herbarium for his kind cooperation to identify these plant specimens. The authors are also grateful to Dr. Mostafa Kamal Pasha, former Professor, Department of Botany, University of Chittagong for his kind cooperation for article preparation.

References

Begum, M., Afroz, S., Sultana, N. and Hassan, M.A. 2008. Acanthaceae. *In*: Ahmed, Z.U., Begum, Z.N.T., Hassan, M.A., Khondker, M., Kabir, S.M.H., Ahmad, M., Ahmed, A.T.A., Rahman, A.K.A. and Haque, E.U. (Eds.). Encyclopedia of Flora and Fauna of Bangladesh, Vol. **6**. Angiosperms: Dicotyledons (Acanthaceae-Asteraceae). Asiatic Society of Bangladesh, Dhaka, pp. 1–75.

Cowan, J.M. 1926.The flora of the Chakaria Sundarbans. Rec. Bot. Survey India. **11**(2): 197–225.

Cowan, A.M. and Cowan, J.M. 1929. The trees of Northern Bengal- including shrubs, woody climbers, bamboos, palms and tree ferns. Bengal Secretariat Book Dept., Calcutta. pp. 1–178.

Das, S.C., Rashid, M.H. and Rahman, M.A. 2009. Taxonomic revision of the genus *Pavetta* L. (Rubiaceae) of Bangladesh. Plant Archives **9**(2): 813–820.

Das, S.C. and Rahman, M.A. 2010. Notes on the Rubiaceae 3: Five new records for Bangladesh. Bangladesh J. Bot. **39**(2): 215–222.

Das, S.C. and Rahman, M.A. 2011. Taxonomic revision of the genus *Morinda*L. (Rubiaceae) in Bangladesh. Bangladesh J. Bot. **40**(2): 113–120.

Das, S.C., Dev, P.K. and Rahman, M.A. 2012. Notes on the Rubiaceae 4: Five new records forBangladesh. Bangladesh J. Bot. **41**(1): 21–28.

Das, S.C., Dev, P.K. and Rahman, M.A. 2013. Notes on the Rubiaceae 5: Five new records for Bangladesh. Bangladesh J. Bot. **42**(2): 257–264.

Datta, R.M. and Mitra, J.N. 1953. Common plants in and around Dacca. Bull.Bot. Soc. Beng. **7**(1&2): 1-110.

Heinig, R.L.1925. List of Plants of the Chittagong Collectorate and Hill Tracts. The Bengal Government Branch Press, Darjeeling, India. pp. 1–89.

Hooker, J.D. 1890. Orchidaceae. *In*: Hooker, J.D. (Ed.),The Flora of British India, Vol. **6**. The Oast House, Brooke, Ashford, Kent, England. p. 91.

Hooker, J.D. 1880. Rubiaceae. *In*: Hooker, J.D. (Ed.), The Flora of British India, Vol. **3**. The Oast House, Brooke, Ashford, Kent, England. p. 76.

Hooker, J.D. 1884. Acanthaceae, *In*: Hooker, J.D. (Ed.), The Flora of British India. Vol. **4**. The Oast House, Brooke, Asford, Kent, England. p. 464.

Huda, M.K. 2008. Orchidaceae. *In*: Ahmed, Z.U., Begum, Z.N.T., Hassan, M.A., Khondker, M., Kabir, S.M.H., Ahmad, M., Ahmed, A.T.A., Rahman, A.K.A. and Haque, E.U. (Eds.). Encyclopedia of Flora and Fauna of Bangladesh, Vol. **12**. Angiosperms: Monocotyledons (Orchidaceae-Zingiberaceae). Asiatic Society of Bangladesh, Dhaka. p. 32.

Islam, K.K. 2009. Diversity, Ecology, Ethnobotany of Orchids of Madhupur Sal Forest and their conservation. Bangladesh. The University of Chittagong, Chittagong. M.S. Thesis (unpublished). pp. 1-218.

Kanjilal, U.N., Das, A., Kanjilal, P.C., and De, R.N. 1939 (Repr. 1982). Flora of Assam, Vol. **3**. A Von Book Company, Ajmeri Gate, Delhi, India. pp. 11–95 and 408–457.

Kurz, S. 1877 (Repr. 1974). Forest Flora of British Burma, Vol. **2**. Bishen Singh Mahendra Pal Singh, Dehra Dun, India. pp. 4–76 and 239–248.

Prain, D. 1903 (Repr. ed. 1963). Bengal Plants, Vol. **1&2**. Bishen Singh Mahendra Pal Singh, Dehra Dun, India.

Rahman, M.A. and Das, S.C. 2009. Rubiaceae. *In*: Ahmed, Z.U., Hassan M.A., Begum Z.N.T., Khondaker, M., Kabir, S.M.H., Ahmed, M. and Ahmed, A.T.A. (Eds.). Encyclopedia of Flora and Fauna of Bangladesh, Vol. 10. Angiosperms: Dicotyledons (Ranunculaceae-Zygophyllaceae). Asiatic Society of Bangladesh, Dhaka. pp. 38–159.

Rahman, M.S., Hossain, G.M., Khan, S.A. and Uddin, S.N. 2015. An annotated checklist of the vascular plants of Sundarban mangrove forest of Bangladesh. Bangladesh J. Plant Taxon. **22**(1): 17–41.

Raizada, M.B. 1941. On the Flora of Chittagong. Indian Forester **67**(5): 245–254.

Sinclair, J. 1956. The Flora of Cox's Bazar, East Pakistan. Bull. Bot. Soc. Beng. **9**(2): 84–116.

Uddin, S.N., and Rahman, N. 2015. Notes on occurrence of the genus *Mycetia* Reinwardt (Rubiaceae) in Bangladesh. Bull. Bangladesh National Herb. **4:**103–110.

Wu, Z. Y., Raven, P. H. and Hong, D. Y. (Eds.) 2011. Flora of China. Vol. **19**. Science Press, Beijing and Missouri Botanical Garden Press, St. Louis.

19

A NEW SPECIES OF *LAPORTEA* GAUDICH. (URTICACEAE) FROM HIMALAYA, INDIA

BACHAN LAL BHELLUM[1] AND BIKARMA SINGH[2]

Department of Botany, Government College for Women, Parade, Jammu, J & K State, India

Keywords: Conservation status; Endangered; *Laportea stolonifera;* New species; Taxonomy.

Abstract

A new herb species of Urticaceae, *Laportea stolonifera* B. L. Bhellum & B. Singh, is described and illustrated from a restricted habitat of subtropical forest of Northwest Himalaya, India. The new taxon is vegetatively similar to *Laportea ovalifolia* (Schumach. & Thonn.) Chew, an African endemic species and *Laportea interrupta* (L.) Chew but differs by phenotypic characters such as cordate leaves, unbranched inflorescence, stem hairs types, linear cystoliths with varying shapes, and presence of 2 to 3 stolons arising from basal node of stems. The similarity with the allied species is due to similar habitats occupancy, but isolated geographically from each other. *Laportea stolonifera* is assessed as Endangered, and the population data, ecological parameters and associated taxa are also presented. A key to the genus *Laportea* is prepared for India has been presented in the text.

Introduction

The nettle family, Urticaceae Juss. (1789: 400), comprised of 1465 species and 54 genera in the world (TPL, 2016), includes herbs, shrubs, small trees, and vines distributed primarily in tropical, subtropical and temperate belts (Mabberley, 2008). While studying the floristic composition of Jammu and Kashmir (India) in Northwestern Himalaya, and during recent plant collection trips to Jasrota forest, samples from three interesting populations of tribe Urticeae were collected which upon critical investigation revealed to be a new species of the genus *Laportea* Gaudich.

The genus *Laportea* comprise of 25 species (TPL, 2016), chiefly distributed in Africa and Madagascar with a few pantropic species (Chew, 1989). In India it is represented by 4 species: *L. aestuans* (L.) Chew; *L. bulbifera* (Siebold & Zucc.) Wedd.; *L. interrupta* (L.) Chew and *L. stolonifera* B.L. Bhellum & B. Singh, *sp. nov.* Most of the *Laportea* species are confined to Northeast India, Northwest Himalaya and South India. *Laportea* sp. collected during recent field survey in Northwest India is morphologically similar to *L. ovalifolia* (Schumach. & Thonn.) Chew. After thorough scrutiny of the specimens, literature survey and geographic distribution, authors found that the unknown species is new to science. The species is named as *Laportea stolonifera* B.L. Bhellum & B. Singh, and is quite different from the species described in published literatures (Ramaswamy and Razi, 1973; Fyson, 1974; Bennet, 1987; Chew, 1989; Sharma, 2010). The specimen have been taxonomically described, photographed and illustrated. Data on populations, ecological parameters and associated taxa are also presented. The holotype voucher specimen is deposited at Janaki Ammal Herbarium (RRLH) at CSIR-Indian Institute of Integrative Medicine, Jammu (India).

[1]Corresponding author. Email: blbhellum@gmail.com / drbikarma@iiim.ac.in
[2]CSIR-Indian Institute of Integrative Medicine, Canal Road, Jammu-Tawi, India.

Materials and Methods

Critical morphological work out on the newly discovered specimen were done by comparison and assessment of voucher specimen of the new taxa with that of the allied holotypes, isotypes and specimens housed at the various herbaria, viz. ASSAM, CAL, BSD, BPL, CVH and RRLH. Herbarium acronyms followed were those of Thiers (2015). Specimens were also examined together with Google images from the Plant Illustration (http://www.plantillustrations.org), TROPICOS (http://www.tropicos.org), and Global Plants Initiative (http://gpi.myspecies.info). While describing the species, field data pertaining to habit, ecology, shape and size of floral elements, leaves, mild stinging hairs, fruits and seeds were incorporated from the field diary. Associated plants of this particular species were presented along with the photographs. Fresh flowering materials were examined under trinocular microscope at 10X and 40X for proper description and illustration.

Laportea stolonifera B.L. Bhellum & B. Singh, **sp. nov.** **(Figs 1&2)**.

Diagnosis: The new species, *Laportea stolonifera*, is highly similar to the African endemic, *L. ovalifolia*, and some characters related with *L. interrupta*, but differ from them by the presence of cordate leaves, unbranched inflorescence, three types of stem mild hairs, linear cystolith with varying shapes, and salient characters of 2 or 3 stolons arising from each basal node of stems.

Type: INDIA. North-West Himalaya: Jammu and Kashmir State, Jammu Province, Jasrota forest, shady places along the forest margin, altitude 294 m above sea level, 32°33'59.1"N 75°02'17.9"E, 2 September 2014, *BL Bhellum & B Singh* 21214 (*holotype* RRLH).

Annual herbs, 23–57 cm high; spreading along the ground by stolons and roots; rooting at the nodes, adventitious roots, thin, ca. 12 cm long. Stems woody at base, furrowed, half way glabrous, usually unbranched, greenish, clothed with three different types of mild stinging hairs, larger recurved beset with one, two or three stinging points, middle single with stinging hair spreading, base bulbous and shorter hair appressed, slightly curved. Petioles varies in size, depending on leaf blade size, 0.4–9.6 cm in length, presence of spreading stinging hairs; stipules interpetiolar, oblong, biclefted, segments linear, each segment prominently veined, hairy, sometimes exstipulate. Leaves simple, alternate, persistent, ovate-broadly ovate, varies in size, 1.0–8.4 cm long, 0.6–6.1 cm wide, matured ones cordate at base, young ones truncate at base, shortly acuminate at apex, membranous, upper surface packed with tiny sand-like cystoliths; margins crenate-dentate or coarsely serrated; stinging hairs on the upper surface short and shining, while those on the lower surface remotely hairy on lateral and basal nerves beneath; lateral nerves 5-6 pairs, prominent, basal nerves reaching more than half of the length of the blade; cystoliths linear, visible. Stolons arise from each node at the base of stem directly forming inflorescence, 2 or 3 in number. Inflorescence unisexual, arise from the nodes, racemes solitary, axillary spikes arranged in lax fascicles, 3–6 cm long, soon falling off after maturing; spikes armed with stinging hairs. Flower buds ellipsoidal, pale yellow or light green. Male flowers 1 mm long, tepals 4 or 5, subvalvate, depressed, usually inflexed in flower buds; stamens as many as tepals; rudimentary ovary clavate or subglobose; anthers bithecous, white, filaments inflexed in bud, pistillode short. Female flowers pedicellate, tepals 4, free or connate at base, lobes subequal, two large plus two opposite smaller, ciliolate along the margin, 1 mm long; ovary 1 mm long, oblique; stigmas deeply trilobed, usually filiform, central lobe longer than the other two lobes, papillae on one side, ovary with one erect ovule; staminodes absent. Achenes minute, 1.0-1.5 mm in diam., oblique, flattened, compressed, completely surrounded by a narrow membranous wing along the margin, obliquely stipitate, embraced by persistent hairy perianth; fruiting pedicels small, with stinging hairs.

Phenology: Flowering occurs from early August and continues till early October; fruiting usually seen in October to late November. The collection site is relatively seasonal; having almost seven months dry period and five months cold winter and semi dry.

Etymology: The specific epithet of *Laportea stolonifera* is named after the unique character, 'stolonifera', the presence of 2 or 3 stolons.

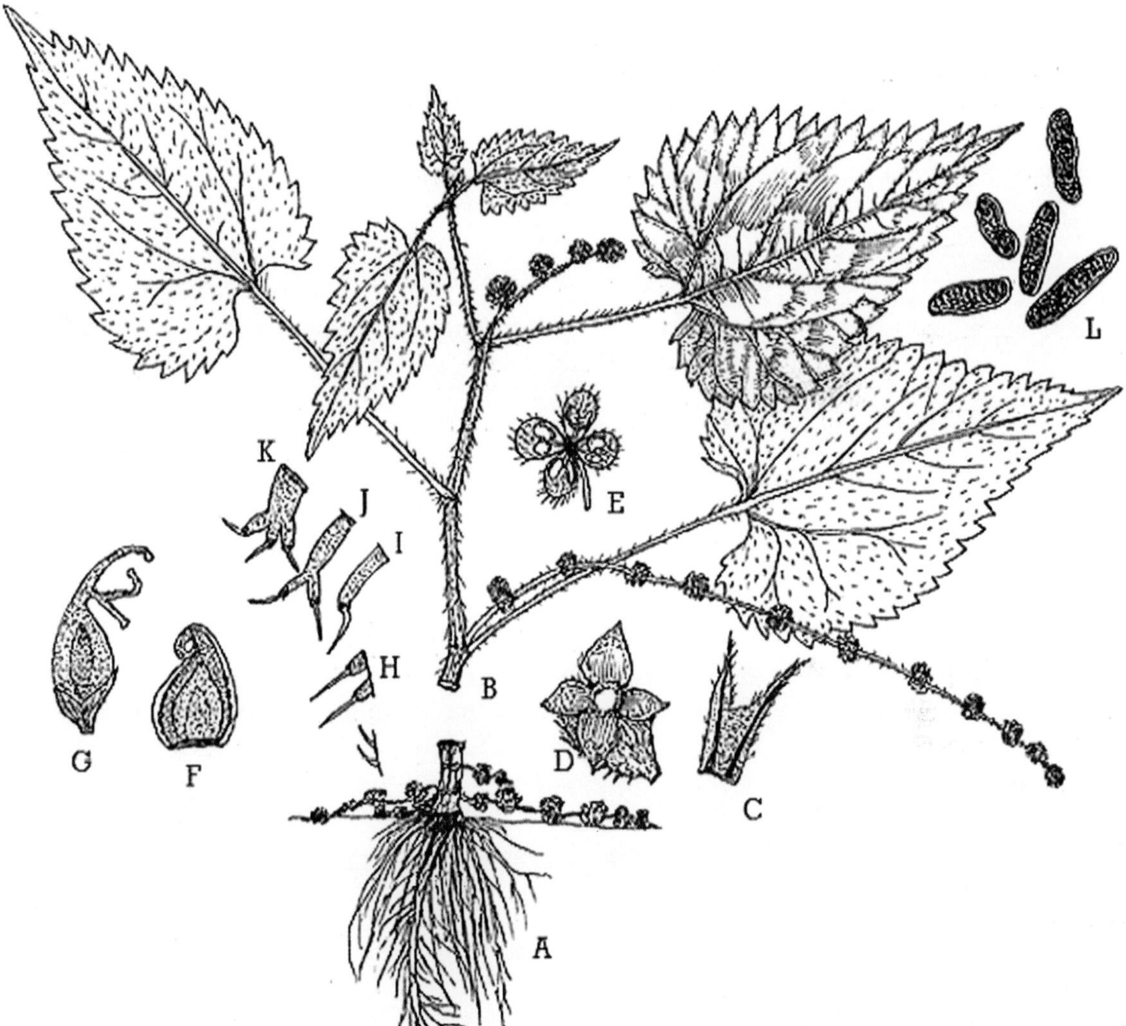

Fig. 1. Illustration of *Laportea stolonifera* B. L. Bhellum & B. Singh, **sp. nov.;** (A) basal portion of stem bearing roots and stoloniferous inflorescence, (B) upper portion of stem bearing inflorescence, (C) stipule, (D) female flower, (E) male flower, (F) fruit, (G) pistil, (H-K) different kind of hairs, and (L) cystoliths. Illustration based on living material and voucher specimens deposited at RRLH.

Distribution and *habitat ecology*: Lower subtropical forest areas of Jasrota in Jammu province. The plant species grows gregariously as annual herbs along the stream-sides under moderately dense forest canopy cover. The sites where it occurs are located at an elevation of 250 to 400 m above sea level (ASL). Plant prefers clayey to sandy types of edaphic habitat. Commonly associated species includes *Oxalis corniculata* L., *Scutellaria prostrata* Jacq. *ex* Benth., *Mazus pumilus* (Burm. f.) Steenis and *Urtica dioica* L.

Conservation status: Laportea stolonifera specimens were studied from three populations confined to the Jasrota forest. From these, 1196 matured individuals and 61 seedlings were recorded. In total 1257 individuals of plants were recorded in an area of less than 10 km^2. The assessment of the IUCN status of *L. stolonifera* were done following 2001 IUCN Red List Categories and Criteria version 3.1, and for this particular species IUCN Guidelines for Application of IUCN Red List Criteria at Regional and National Levels Criteria version 4.0 (IUCN, 2010) were followed. Data reveals that *L. stolonifera* should be treated as Endangered (EN) [A1(a,b,c,d).B2(a,b(i,ii,iv,v).C2a(i.ii)].

Fig. 2. *Laportea stolonifera* B.L. Bhellum & B. Singh, **sp. nov.;** (A) Habit, (B) Portion of stem showing hairs, (C) Inflorescence, (D-E) Basal portion of stems bearing roots and stoloniferous inflorescence.

Discussion

After detailed studies, *Laportea stolonifera* is found to be very close to *L. ovalifolia* morphologically, and few characters were also matching with that of *L. interrupta*. The altitudinal distributions of the allied species were also geographically isolated from each other. *L. stolonifera* is confined to Jasrota forest belts of Jammu province only, which occupies the innermost Northwest Himalayan belts of India, whereas the related species *L. ovalifolia* and *L. interrupta*, were reported from Northeast India (Indo-Myanmar Hotspot region) in Eastern Himalaya and African endemic, respectively. The occurrences of these two species do not over-lap as they are geographically isolated by deserts, hills and mountains.

While examining the new species, *Laportea stolonifera*, from national and international herbaria, and comparing the description of the species, the authors developed a key for identification of of *Laportea* species from India.

Key to the genus *Laportea* in India

1.	Often herbs with woody bulbils in leaf axils; pedicels of female flowers conspicuously laterally and symmetrically winged; achenes articulated on pedicel; stolons absent	*L. bulbifera*
-	Herbs without bulbils in leaf axils; pedicels of female flowers slightly dorsiventrally and asymmetrically winged, or not winged; achenes not articulated on pedicel; stolons present	2
2.	Inflorescence paniculate; stigma ligulated	*L. aestuans*
-	Inflorescence spicate; stigma 3-fid	3
3	Leaves broadly ovate to cordate, 1.0–8.4 cm long, 0.6–6.1 cm wide, upper surface packed with tiny sand-like cystoliths; inflorescence paniculate; 1-6 cm long, arise from base of stem directly from nodes; solons present, 2 or 3 in numbers	*L. stolonifera*
-	Leaves ovate, 5–8 cm long, 4.0–5.5 cm wide, both surfaces packed with cystoliths; inflorescence paniculate, 8-28 cm long, arise from upper half of the stem; solons absent	*L. interrupta*

Acknowledgements

Authors are grateful to Prof. A. K. Koul, Centre for Biodiversity, Baba Gulam Shah Badshah University (Rajouri), and Prof. Rani Magotra Department of Botany, University of Jammu (Jammu) for encouragement. Thanks are due to Director, Dr. Ram. A. Vishwakarma, CSIR-Indian Institute of Integrative Medicine for providing Herbarium facilities and keeping the records of voucher samples. Authors would like to thanks the reviewers for suggestions. Thanks are also due to Mr. Pyrus Bhellum for technical assistance. This article represents CSIR-IIIM institutional communication number IIIM/1899/2016.

References

Anonymous 2015. 200 new animal and plant species discovered in the Himalayas. http://www.wired.co.uk.

Bennet, S.S.R. 1987. Name changes in flowering plants of India and adjacent regions. Triseas Publishers, Dehradun, India.

Chew, W.L. 1989. Urticaceae. In: Flora of Australia. vol. **3.** Australia Government Publishing series, Canberra 190, pp. 68-93.

Fyson, P.F. 1974. The Flora of Nilgiri and Pulney Hilltops. Bishen Singh Mahendra Pal Sing, Dehra Dun, India. First ed. 1915.

IUCN. 2001. IUCN Red List Categories and Criteria. Version 3.1. http://www.iucnredlist.org/technical-documents/categories-and-criteria/2001-categories-criteria

IUCN 2010. Guidelines for using the IUCN Red List Categories and Criteria, Version 4.0. http://intranet.iucn.org/webfiles/doc/SSC/RedList/RedList Guidelines.pdf

Mabberley, D.J. 2008 Mabberley's Plant-book: a portable dictionary of plants, their classification and uses. Third Edition. Cambridgen University Press.

Ramaswamy, S.V. and Razi, B.A. 1973. Flora of Bangalore District. Parasaranga, University of Mysore, India.

Sharma, B.M. 2010. Illustrations of Jammu Plants, a supplement to flora of Jammu and plants of neighbourhood with general key and annotation. Bishen Singh Mahendra Pal Singh, Dehra Dun, India.

Thiers, B. 2015. Index Herbariorum: A global directory of public Herbaria and associated staff. New York Botanical Garden's Virtual Herbarium. Available from: http://sweetgum.nybg.org/ih/ (accessed 07 March 2016)

TPL.2016. The Plant List, a working list of all known plant species. Version 1.1, released in September 2013. Online: http://www.theplantlist.org.

Wilmot-Dear, C.M. 2009 Urticaceae for the non-specialist: Identification in the Flora Malesiana region, Indochina and Thailand. Blumea **54**(1-3): 233–241.

SYNECHOCOCCUS SALSUS SP. NOV. (CYANOBACTERIA): A NEW UNICELLULAR, COCCOID SPECIES FROM YUNCHENG SALT LAKE, NORTH CHINA

HONG-RUI LV, JIE WANG, JIA FENG, JUN-PING LV, QI LIU AND SHU-LIAN XIE[1]

School of Life Science, Shanxi University, Taiyuan 030006, China

Keywords: New species; China; *Synechococcus salsus*; DNA barcodes; Taxonomy.

Abstract

A new species of the genus *Synechococcus* C. Nägeli was described from extreme environment (high salinity) of the Yuncheng salt lake, North China. Morphological characteristics observed by light microscopy (LM) and transmission electron microscopy (TEM) were described. DNA barcodes (16S rRNA+ITS-1, *cpcBA*-IGS) were used to evaluate its taxonomic status. This species was identified as *Synechococcus salsus* H. Lv *et* S. Xie. It is characterized by unicellular, without common mucilage, cells with several dispersed or solitary polyhedral bodies, widely coccoid, sometimes curved or sigmoid, rounded at the ends, thylakoids localized along cells walls. Molecular analyses further support its systematic position as an independent branch. The new species *Synechococcus salsus* is closely allied to *S. elongatus,* C. Nägeli, but differs from it by having shorter cell with length 1.0–1.5 times of width.

Introduction

Synechococcus C. Nägeli (Synechococcaceae, Cyanobacteria) was first discovered in 1849 and is a botanical form-genus comprising rod-shaped to coccoid cyanobacteria with the diameter of 0.6–2.1 μm that divide in one plane. It is a group of ultra-structural photosynthetic prokaryote and has the close genetic relationship with *Prochlorococcus* (Johnson and Sieburth, 1979), and both of them are the most abundant phytoplankton in the world's oceans (Huang *et al.*, 2012). The genus is a taxon of unicellular cyanobacteria with an ongoing debate concerning its revision into several genera (Rippka *et al.*, 1979).

The genus *Synechococcus* is surely heterogeneous and its taxonomy needs to be solved by ultrastructural features and methods of molecular taxonomy (Komárek *et al.*, 2014). The genetic diversity and phylogeny of *Synechococcus* in freshwater were determined by nucleotide sequence such as 16S rRNA gene, 16S-23S rRNA gene internal transcribed spacer (ITS-1) and *cpcBA*-IGS (phycocyanin operon) (Jasser *et al.*, 2011). 16S rRNA gene is the most widely used gene sequence in the classification of cyanobacteria. Due to its highly conserved structure, it is only appropriate for evolutionary studies at the level of genera and above (Long and Chen, 2006). With the high variability and the rapid rate of change in molecular evolution, the internal transcribed spacer is often used to study the phylogenetic relationships at the taxa of genus, species, subspecies and so on (García-Martínez *et al.*, 1996). In addition, *cpc*BA-IGS, the phycocyanin operon, can be used as the classification basis in identifying the level of species and the following (Bolch *et al.*, 1996).
Yuncheng salt lake is located in southern suburbs of Yuncheng city where the algal distribution has its particularity. On the north of beach of Yuncheng salt lake, there were nearly 2×10^5 m^2

[1] Corresponding author: Email: xiesl@sxu.edu.cn

waters extensively appeared rosy during June, 2015. In order to find the reasons why the water turns red, we specifically took water samples and a coccoid strain with flickering red light of *Synechococcus* species was isolated.

In the present paper, we describe *Synechococcus* species isolated from Yuncheng salt lake as a new species *Synechococcus salsus* sp. nov. and discuss its taxonomic position based on morphological property observed by light microscopic and transmission electron microscopic, molecular data of 16S rRNA gene with the adjacent ITS, *cpc*BA-IGS and flanking regions were also used to support the entity as a new species.

Materials and Methods

Sample preparation and culture-dependent method

The unicellular, coccoid strain in this study was collected from Yuncheng salt lake, Shanxi Province, North China, in June 2015. It was preliminarily identified belonging to the genus *Synechococcus* through light microscopic examination (Olympus BX-51, Tokyo, Japan). We prepared BG-11 medium and then added it to a 24-well cell culture plate, each hole with 2 ml. Pasteur Micropipette was used to pick individual cells (Rippka, 1988). Cyanobacteria were isolated into unialgal culture and they were inoculated on the medium in the 24-well cell culture plate after washing eight times by sterile water. Then, it was cultured in a light incubator (BSG-300, Shanghai, China) at 25°C under a 12 h/12 h light/dark photoperiod. The light intensity was 25 $\mu mol \cdot m^{-2} \cdot s^{-1}$. Three weeks later, they reached exponential phase, then we examined them with microscope. In the case of growing well and axenic culture, the homogenates were transferred to conical flasks with 200 ml BG-11 liquid medium. Similarly, they were placed in the same culture condition for large scale culture. To make them fully absorb nutrients and equally distribute, we manually shook the conical flasks every two days. The unialgal strain was deposited in Algae Culture Collection at Shanxi University.

Morphological observations

Morphological characteristics were examined under a light microscope when the cyanobacteria in large scale culture reached stationary phase. Photographs of main characters were taken with a digital camera (CAMEDIA C5060WZ, Olympus) and a CCD camera (DP72, Olympus) mounted on the microscope. Moreover, transmission electron microscope (JEM-1011, JEOL, Tokyo, Japan) was used to observe the ultrastructure through the process of centrifuging, fixation, dehydration, embedding, sectioning and dye (Lim *et al.*, 2012).

Molecular analyses

Genomic DNA was extracted from 10 ml axenically exponential phase cultures according to the modified cetyltrimethyl ammonium bromide (CTAB) method (Doyle, 1987).PCR was carried out in 20 μl reaction volume containing 12.3 μl of double-distilled water, 2 μl of 10×Taq polymerase reaction buffer (Takara, Dalian, China), 0.2 μl of *EasyTaq* DNA polymerase (5 U/μl, Takara, China), 1.5 μl of each primer (10 μM), 2 μl dNTP mix (2.5 mM each; Takara) and 0.5 μl of undiluted genomic DNA. Part of the 16S ribosomal RNA gene (16S rRNA) and all of the 16S-23S rRNA internal transcribed spacer region (16S-23S ITS) were amplified using the primers P1 (5'-CTC TGT GTG CCT AGG TAT CC-3') and P2 (5'-GGG GAA TTT TCC GCA ATG GG-3') as described by Boyer *et al.* (2001). The *cpc*BA-IGS region of the phycocyanin operon were amplified with the cyanobacterium-specific primer pairs (Crosbie *et al.*, 2003a) *cpc*BF (5'-TAG TGT AAA ACG ACG GCC AGT TGY YTK CGC GAC ATG GA-3') and *cpc*AR (5'-TAG CAG GAA ACA GCT ATG ACG TGG TGT ARG GGA AYT T-3'). Polymerase chain reaction (PCR) was performed in a My Cycler thermal cycler (Bio-Rad, Hercules, CA, USA). The program consisted of the following condition: 5 min at 95 °C, 35 cycles of 45 s at 94 °C, 45 s at 55 °C, 1 min at 72 °C, and a final extensionstep of 10 min at 72 °C. Sequencing was carried out from purified PCR products by BGI (Beijing, China). Sequences generated from the research were deposited in GenBank (accessions KU925869 for 16S rRNA gene+ITS-1, KU925870 for phycocyanin operon).

Sequences were aligned using Bioedit and then manually adjusted. Phylogenetic trees were constructed from the aligned gene sequences using neighbor-joining (NJ), maximum likelihood (ML), and Bayesian (BI) methods. NJ, ML and BI analyses were performed through MEGA 5.0 (Tamura *et al.*, 2011), PhyML 3.0 (Guindon and Gascuel, 2003) and MrBayes version 3.1.2 (Ronquist and Huelsenbeck, 2003), respectively. Best fit models were selected under the Akaike information criterion (AIC) using Modeltest 3.7 (Posada and Crandall, 1998). For NJ analyses, evolutionary distances were computed using the Kimura 2-parameter method with 1000 bootstrap replicates. Treeview was used to view the phylogenetic trees. Final graphic refinement of all trees was done in Adobe Illustrator CS5 (Adobe Systems, San Jose, CA, USA).

Results and Discussion

The information and GenBank accession numbers for *Synechococcus salsus* sp. nov. and the other eighty-nine strains analyzed in this study were listed in Table 1. For the 16S rRNA+ITS-1 sequences, 44 strains of cyanobacteria were used and the aligned sequence dataset comprised 1321 nucleotides, of which 515 (38.99%) were variable sites and 455 (34.44%) were parsimony-informative sites. The aligned phycocyanin operon (*cpc*BA-IGS) dataset was 449 bp long. Forty-eight aligned *cpc*BA-IGS sequences had 307 (68.37%) variable sites and 281 (62.58%) parsimony-informative sites. The best fit model used for each locus was presented in Table 2.

For both the 16S rRNA+ITS-1 and the *cpc*BA-IGS data, tree topologies resulting from three methods had some similarities and differences. Based on 16S rRNA+ITS-1 sequences, the tree derived from the BI analysis is presented in Figure 1 to explain the relationships between *Synechococcus salsus* sp. nov. and the other accessions. *Leptolyngbya* sp. KIOST-1 was chosen as out group. *S. salsus* sp. nov. is grouped together with many *Synechococcus* strains. They are in a well-supported clade. Nevertheless, the bootstrap value by NJ method is less than 50%. The phylogenetic tree recovered by BI analysis of *cpc*BA-IGS is shown in Figure 2. *Cyanidium caldarium* was chosen as outgroup. In this topology, *S. salsus* sp. nov. and many other *Synechococcus* strains also constitute one cluster. The strain *Synechococcus* PCC9005 AF223465 was confirmed as closely related to *S. salsus* sp. nov., but the support value was low.

Synechococcus salsus H. R. Lv *et* S. L. Xie, **sp. nov.** **(Fig 3).**

Diagnosis: LM observations-Unicellular, cells solitary or agglomerated in groups (Figs A, B), but without common mucilage. Cells widely coccoid, sometimes curved or sigmoid, rounded at the ends, with a mean diameter of 2.0–3.5 μm and a length of 3–5 μm, usually pale blue-green, rarely olive-green, bright blue-green or pinkish, division always by binary fission, perpendicular to the longer axis of the cell, reproduction by solitary cells (Figs C, D). TEM observations-Cells with several dispersed or solitary polyhedral bodies, thylakoids localized along cells walls, photosynthetic pigments attached to it (Figs E, F).The new species is closely allied to *S. elongatus* C. Nägeli, but differs from it by having shorter cell with length 1.0–1.5 times of width, and cell length of *S. elongatus*is 1.5–3.0 times of width.

Type: China. Shanxi: Yuncheng City, Yuncheng salt lake, 21 June 2015, *Jie Wang & Chaoyan Gong*, SAS15yc (*Holotype*: SXU).

Table 1. Strains information and GenBank accession numbers for *Synechococcus salsus* H. Lv *et* S. Xie. and other taxa analyzed in phylogenetic comparison based on 16S rRNA+ITS-1 and Phycocyanin Operon (*cpc*BA-IGS).

Taxon	Strain/Isolate	16S rRNA + ITS-1	Phycocyanin Operon
Arthrospirafusiformis (Voronikhin) Komárek&J.W.G. Lund	AB2002/11	AY575929	
Arthrospiraindica Desikachary & N. Jeeji Bai	PD1997/ram	AY575931	
Arthrospira maxima Setchell & N.L. Gardner	FACHB-438	FJ826622	
Arthrospira platensis Gomont	FACHB-834	FJ826623	
Cyanidium caldarium (Tilden) Geitler			S77125
Cyanobium sp.	NS01		KF528824
	Sai001		GU935387
	Sai004		GU935390
	Sai005		GU935391
Leptolyngbya boryana (Gomont) Anagnostidis & Komárek	UTEX 'B 488'	EF429295	
Leptolyngbya corticola J.R. Johansen, Kovácik, Cassamata, Fucikova & Kastovsky	CCALA 085	EF429299	
Leptolyngbya saxicola (N.L. Gardner) Anagnostidis	BDU 91391		DQ829688
Leptolyngbya tenerrima (Hansgirg) Komárek	UTCC 77	EF429288	
Leptolyngbya valderiana (Gomont) Anagnostidis & Komárek	BDU 30501		AJ973264
	BDU 20041		AJ973263
	BDU 40231		DQ829696
	BDU 41001		DQ829697
	BDU 80221		DQ829698
	BDU 91712		DQ829699
	BDU 140441		AJ973265
Leptolyngbya sp.	KIOST-1	JX401929	
	HA4237-MV6	KJ939031	
Merismopedia sp.	AICB1014	KJ746509	
	AICB1015	KJ746510	
Microcoleus chthonoplastes Thuret *ex* Gomont	SAG 2209	EF654055	
Microcystis aeruginosa (Kützing) Kützing	2009-A		JN226767
	H107		KF840319
	PCC7806		AF195177
	UAM-MarB5		EU643824
	UAM-VMA-12		EU643811
Microcystis botrys Teiling	H179a		KF840323
Microcystis sp.	CYN06	EF634465	
	CYN10	EF634466	
	KLL-C018	KP726243	
	KLL-C005	KP726246	
	KLL-C019	KP726247	
	UWOCC Q		AF195179
Nostoc calcicola Brébisson *ex* Bornet & Flahault	Ind30		JF923546
Nostoc commune Vaucher *ex* Bornet et Flahault	NC3-K1	EU586723	
	NC5	EU586728	
	WY1KK1	EU586733	
Nostoc linckia Bornet ex Bornet & Flahault	PACC 5085		AY466120

Taxon	Strain/Isolate	16S rRNA + ITS-1	Phycocyanin Operon
Nostoc muscorum C. Agardh *ex* Bornet & Flahault	Ind33		JN402387
Nostoc punctiforme Hariot	NC6	EU586731	
Nostoc spongiaeforme C. Agardh *ex* Bornet & Flahault	Ind42		JF923547
Nostoc sp.	CAVN2		KJ511230
	CAVN10		KJ511236
	HA4355-MV2	HQ847576	
	HA4356-MV1	HQ847577	
	HK-01		JF740675
	KNUA003		JF740672
	MCC2741		KT166439
	Os-1	KP001508	
	PCC 6720		JF740673
	UAM 307	HM623782	
Pannus brasiliensis C.F.da Silva Malone *et al.*	CCIBt3594	KF668649	
Phormidium cf. *terebriformis*	AB2002/07	AY575933	
Spirulina subsalsa Oersted *ex* Gomont	FACHB-351	FJ826621	
	PD2002/gca	AY575935	
Synechococcus salsus H. Lv *et* S. Xie.	SAS15yc	KU925869	KU925870
Synechococcus sp.	ACT 0613		HQ859449
	ACT 0616		HQ859452
	ACT9701		GQ888569
	BE0807F	FJ763770	
	BE0807G	FJ763772	
	BE0807H		FJ763791
	BE0807L	FJ763778	
	CCY9201		EF513488
	EC-LC05PC		EU413956
	EC-LM05PC		EU420180
	MA0607E		FJ763807
	MI0608F		FJ763799
	MW97C4		AY151223
	NIBB 1026	AB610894	
	NIBB 1070	AB610895	
	NIBB 1071	AB610896	
	PCC 7918		AF223462
	PCC9005		AF223465
	PS673		AF223434
	Suigetsu-CG2	AB610891	
	WH8101		KF528825
Synechocystis sp.	AICB51	KJ746512	
	LSCB 01		KJ018112
	PUPCCC 62		KF573457
	Sai001	GU935367	
	Sai002	GU935368	GU935393
Uncultured *Synechococcus* sp.	MB11E09	AY033308	
Uncultured *Synechocystis* sp.	clone AF10-37-C9	JQ410261	
Uncultured marine bacterium	SPOTSOCT00 5m10	DQ009323	
	SPOTSOCT00 5m14	DQ009324	

Table 2. Model selection results for each phylogenetic dataset. Best fit models were selected under the Akaike information criterion (AIC) using Modeltest 3.7.

Gene	Model selected	Base frequencies	Substitution model
16S rRNA+ITS-1	GTR + I + G		R(a) [A-C] = 0.8415
	- lnL = 10568.6123	freqA = 0.2750	R(b) [A-G] = 2.0325
	K = 10	freqC = 0.2136	R(c) [A-T] = 1.1733
	AIC = 21157.2246	freqG = 0.2855	R(d) [C-G] = 0.5105
	(I) = 0.5199	freqT = 0.2259	R(e) [C-T] = 2.9532
	(G) = 0.7675		R(f) [G-T] = 1.0000
Phycocyanin Operon (*cpcBA*-IGS)	TIM+I+G		R(a) [A-C] = 1.0000
	- lnL = 5649.5356	freqA = 0.2653	R(b) [A-G] = 2.5283
	K = 8	freqC = 0.2977	R(c) [A-T] = 1.6029
	AIC = 11315.0713	freqG = 0.1944	R(d) [C-G] = 1.6029
	(I) = 0.2250	freqT = 0.2426	R(e) [C-T] = 3.8615
	(G) = 1.2670		R(f) [G-T] = 1.0000

Etymology: The specific epithet refers to its living environment with the high salinity.

Habitat and distribution: *Synechococcus salsus* was discovered in Yuncheng salt lake in Shanxi Province, North China, which is located at 34°48′ to 35°30′ N and 110°12′ to 111°41′ E. The lake is about 1.3×10^8 m^2 in area, with a salinity of 6.8%and pH 7.54 in June.

Note: The genus *Synechococcus* belongs to Cyanobacteria, Cyanophyceae, Synechococcophycideae, Synechococcales, Synechococcaceae. *Synechococcus* is polyphyletic and currently recognized as a distinct genus. The *Synechococcus* species are unicellular and rod-shaped to coccoid. Such tiny cells can be identified through the fluorescence of the main pigments phycoerythrin (PE) they contain, which is the evidence with respect to the phenomenon that Yuncheng salt lake appears rosy during June, 2015. According to Algae Base (http://www.algaebase.org/), 38 species have been flagged as currently accepted taxonomically. Zhao *et al.* (2010) studied spatial and temporal patterns of plankton assemblage structure of the saline lake Namuka Co in Northern Tibet and found *Synechococcus* strain. In the study of taxonomic and functional diversity in the saline Qinghai Lake, Huang *et al.* (2014) pointed out that *Synechococcus* was the dominant genus. Among the reports about cyanobacteria in Yuncheng Salt Lake region, only one *Synechococcus* species (*S. aeruginosus* Nägeli) was reported (Li and Xie, 2006). However, it is generally accepted belonging to the genus *Cyanothece* Komárek, namely *C. aeruginosa* (Nägeli) Komárek. *Synechococcus salsus* sp. nov. has no homology with it.

Synechococcus is an important type genus and species of this genus grow well in many different mediums. Most are freshwater species, and some grow within mats and colonies of other algae, or form fine colonies on wet substrates including mud, wood, stones, etc. A group of species is known from the metaphyton and periphyton of thermal and mineral springs (Dor, 1967), and few are described from aerophytic sites (Skuja, 1964). There are also some marine species (Jao, 1948; Komárek, 1956). Interesting picoplanktonic or planktonic species were found in oceans as well as in freshwater reservoirs (lakes) (Komárek and Anagnostidis, 1995).

In the past, the classification of cyanobacteria only relied on traditional morphological criteria. However, it is difficult to distinguish the members of *Synechococcus* because of their small differences, and its taxonomy should be resolved by help of ultrastructural procedures and molecular approaches. With the development of electron microscope technique and molecular phylogeny, the accuracy of describing the phylogenetic relationship about cyanobacteria has been greatly improved (Komárek, 2010).

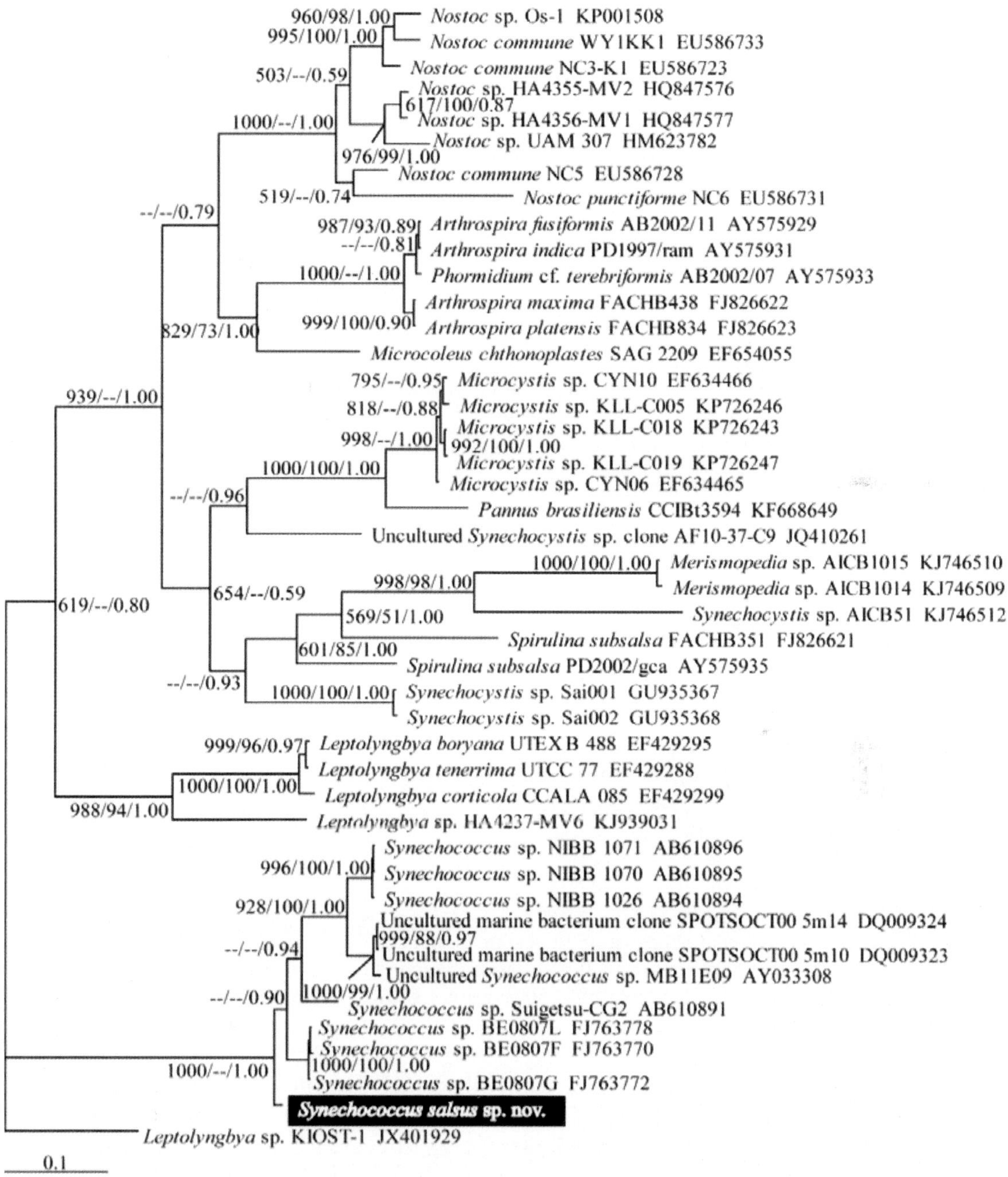

Fig. 1. Phylogenetic tree reconstructed using Bayesian inference from the aligned 16S rRNA+ITS-1 sequence. Numbers at nodes represent maximum likelihood bootstrap/neighbor-joining bootstrap/Bayesian posterior probabilities values. Support values less than 50% are not shown. Scale bar represents 0.1 substitutions per site.

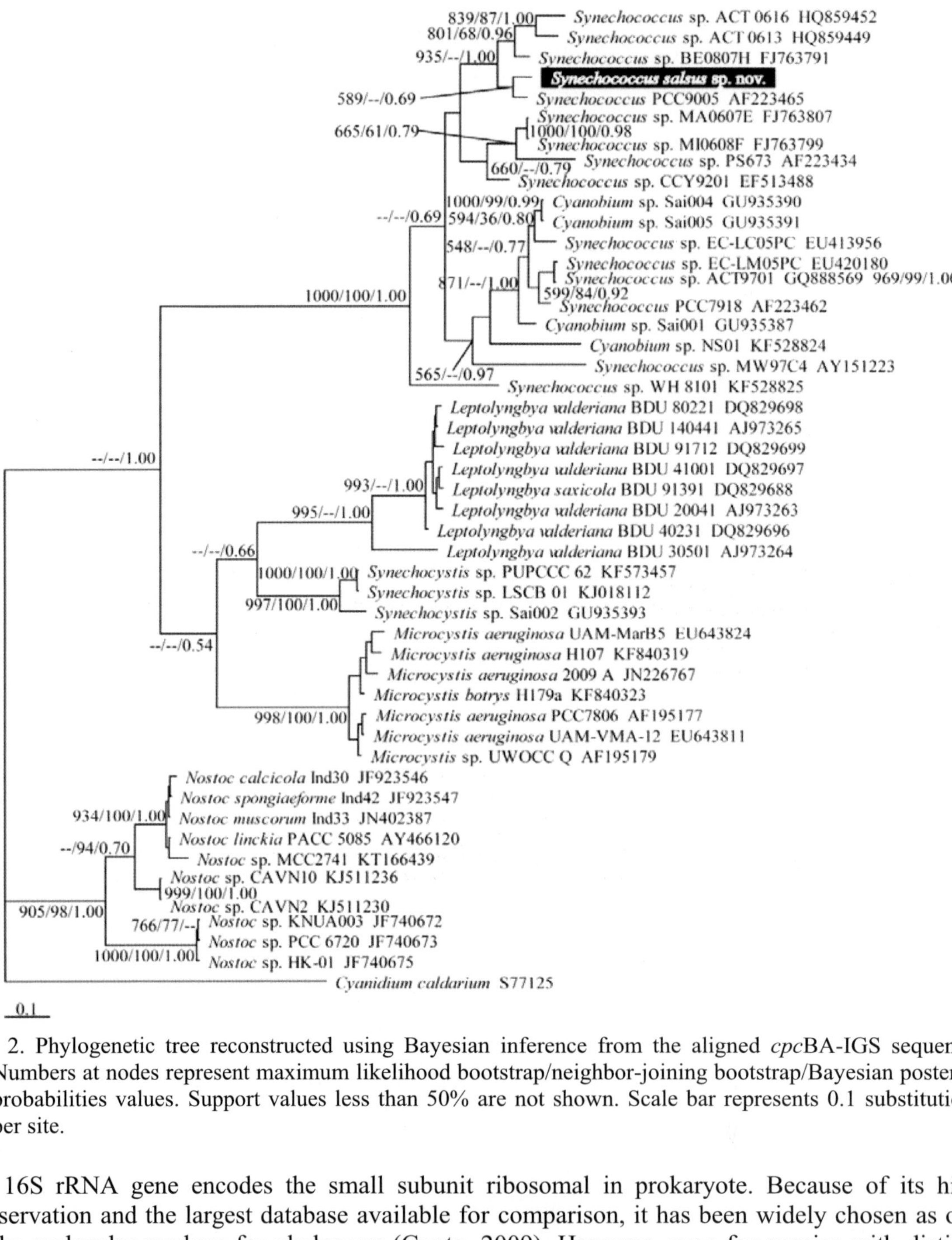

Fig. 2. Phylogenetic tree reconstructed using Bayesian inference from the aligned *cpc*BA-IGS sequence. Numbers at nodes represent maximum likelihood bootstrap/neighbor-joining bootstrap/Bayesian posterior probabilities values. Support values less than 50% are not shown. Scale bar represents 0.1 substitutions per site.

16S rRNA gene encodes the small subunit ribosomal in prokaryote. Because of its high conservation and the largest database available for comparison, it has been widely chosen as one of the molecular markers for phylogeny (Gupta, 2009). However, even for species with distinct physiological differences, 16S rRNA does not have enough distinguishing ability in solving the relationship problems (Jaspers and Overmann, 2004). ITS (internal transcribed sequences) with relatively high mutation rate, when it comes to some closely related species, is quite effective. Despite three pure cultures of *Synechococcus* strains and a *Prochlorococcus* strain was close to each other in the phylogenetic tree based on the 16S rRNA, but the length of their ITS sequences differ greatly (Laloui *et al.,* 2002). Rocap *et al.* (2002) also indicated that the length of the ITS sequences and G+C content vary greatly among different strains. It is always used to distinguish *Synechococcus* strains with highly similar sequences but different in physiological characteristics (e.g. different pigment composition, of light and nutrients of different growth response, etc.). The phycocyanin operon is composed of the two genes (*cpc*B and *cpc*A) that code for phycocyanin, a short and quite variable intergenic region (IGS) and three linker polypeptides (Belknap and Haselkorn, 1987). It can avoid many non-specific amplified products when used to study the diversity of *Synechococcus*.

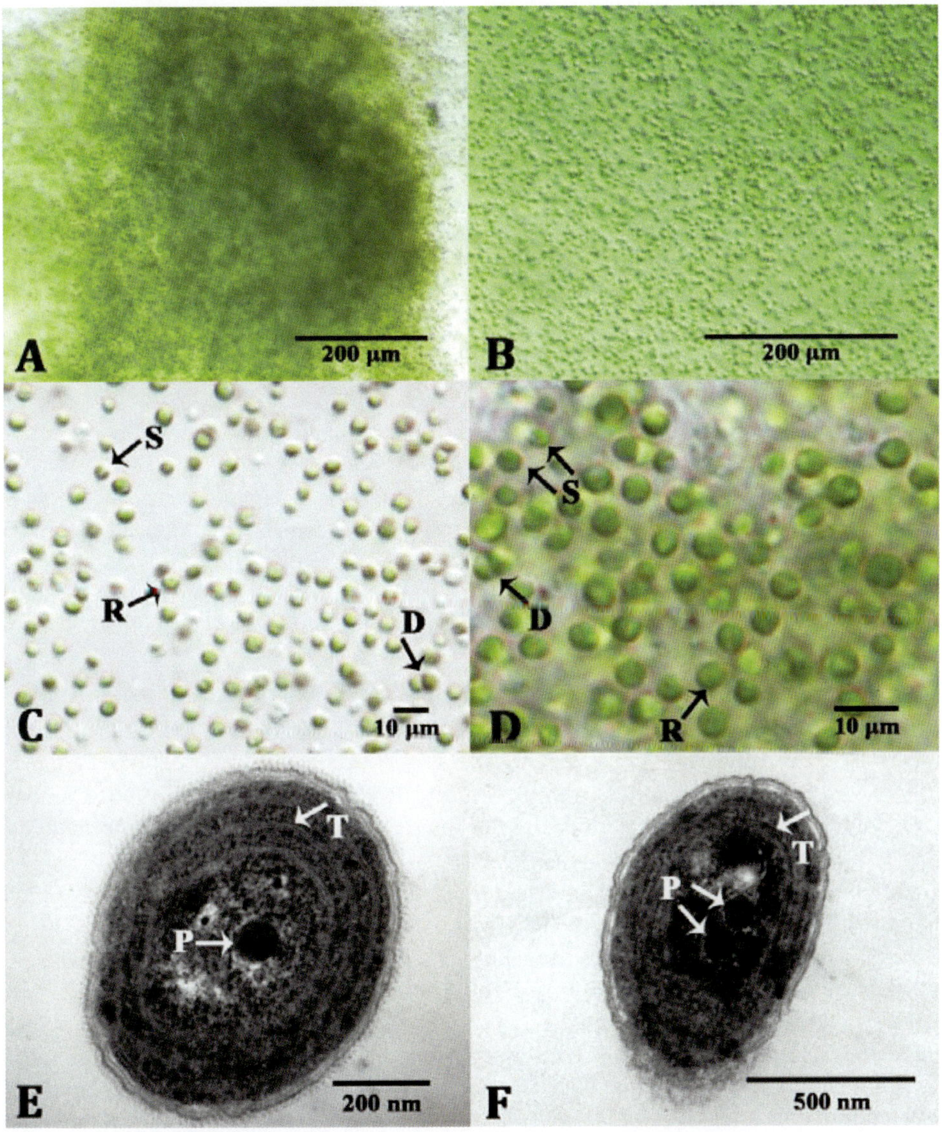

Fig. 3. LM and TEM of *Synechococcus salsus* H.R. Lv*et*S.L. Xie. A. Cells agglomerated in groups. B. Cells solitary. C-D. Cells usually pale blue-green, rarely olive-green, bright blue-green or pinkish, widely oval or coccoid, sometimes curved or sigmoid (S, arrowhead), rounded at the ends (R, arrowhead). Cell division (pinching or cleavage) always by binary fission (D, arrowhead). E-F. Thylakoids localized along cells walls (T, arrowhead). Photosynthetic pigments attached to it. Cells with dispersed or solitary polyhedral bodies (P, arrowhead).

This study relatively accurately evaluated the taxonomic status of *S. salsus* sp. nov. based on the 16S rRNA+ITS-1 gene sequence and phycocyanin operon (*cpc*BA-IGS). Based on 16S rRNA+ITS-1 sequences, the tree showed the relationships between *Synechococcus salsus* sp. nov. and the other accessions. The members of *Synechococcus* are gathered together into one cluster, which is in a well-supported rate. But in the *Synechococcus*-clade, it is quite obvious that *S. salsus* sp. nov. is separated from the other *Synechococcus* strains and alone in a small clade. On the phylogenetic tree recovered by BI analysis of *cpc*BA-IGS, the genus *Synechococcus* is not a monophyletic group. The strain *Synechococcus* PCC9005 AF223465 was confirmed as closely related to *S. salsus* sp. nov., but the support values were weak. From the above, it is reasonable to make it as a new *Synechococcus* species. Finally, salt lake (salinity > 0.5 g/L) is a special habitat. Although the salinity cannot be a criterion to justify a new species (Anagnostidis and Komarek, 2005), it may indeed contain many special organisms.

Acknowledgements

This study was supported by the National Natural Science Foundation of China (No.31170193 to Shulian Xie) and the Foundation of Plate form Construction Project of Infrastructure for Science and Technology of Shanxi (No. 2015091004-0102 to Shulian Xie).

References

Belknap, W.R. and Haselkorn, R. 1987. Cloning and light regulation of expression of the phycocyanin operon of the cyanobacterium *Anabaena*. EMBO J.**6**: 871–884.

Bolch, C.J.S., Blackburn, S.I., Neilan, B.A. and Grewe, P.M. 1996. Genetic characterisation of strains of cyanobacteria using PCR–RFLP of the *cpc*B Aintergenic spacer and flanking regions. J. Phycol. **32**: 445–451.

Boyer, S.L., Flechtner, V.R. and Johansen, J.R. 2001. Is the 16S–23S rRNA internal transcribed spacer region a good tool for use in molecular systematics and populations genetics? A case study in cyanobacteria. Mol. Biol.Evol. **18**: 1057–1069.

Dor, I. 1967.Algues des sources thermales de Tibèriade. Bull.Sea Fish. Res. Stat. Haifa **48**: 1–29.

Doyle, J.J. 1987. A rapid DNA isolation procedure for small quantities of fresh leaf tissue. Phytochem. Bull. **19**: 11–15.

García-Martínez, J., Martínez-Murcia, A., Antón, A.I. and Rodríguez-Valera, F. 1996. Comparison of the small 16S to 23S intergenic spacer region (ISR) of the rRNA operons of some *Escherichia coil* strains of the ECOR collection and *E. coil* K-12. J. Bacteriol. **178**: 6374–6377.

Guindon, S. and Gascuel, O. 2003. A simple, fast, and accurate algaorithm to estimate large phylogenies by maximum likelihood. Syst. Biol. **52**: 696–704.

Gupta, R.S. 2009. Protein signatures (molecular synapomorphies) that are distinctive characteristics of the major cyanobacterial clades. Internat. J. Syst. Evol. Micr. **59**: 2510–2526.

Huang, Q., Briggs, B.R., Dong, H., Jiang, H., Wu, G., Edwardson, C., De Vlaminck, L. and Quake, S. 2014. Taxonomic and functional diversity provides insight into microbial pathways and stress responses in the saline Qinghai Lake, China. PLoSOne **9**: 1–10.

Huang, S., Wilhelm, S.W., Harvey, H.R., Taylor, K., Jiao, N. and Chen, F. 2012. Novel lineages of *Prochlorococcus* and *Synechococcus* in the global oceans. Internat. Soc. Micr. Ecol. J. **6**: 285–297.

Jao, C.C. 1948. The marine Myxophyceae in the vicinity of Friday Harbor, Washington. Bot.Bull. Acad. Sin. **2**: 161–177.

Jaspers, E. and Overmann, J. 2004. Ecological significance of micro-diversity: identical 16S rRNA gene sequences can be found in bacteria with highly divergent genomes and ecophysiologies. Appl. Envir. Micr.**70**: 4831–4839.

Jasser, I., KrÓlicka, A. and Karnkowska-Ishikawa, A. 2011. A novel phylogenetic clade of picocyanobacteria from the Mazurian lakes (Poland) reflects the early ontogeny of glacial lakes. FEMS Microbio.Ecol. **75**: 89–98.

Johnson, P.W. and Sieburth, J.M. 1979.Chroococcoid cyanobacteria in the sea: A ubiquitous and diverse phototrophic biomass. Limnol. Oceanogr. **24**: 928–935.

Komárek, J. 1956. Some interesting blue-green algae from Bulgarian coast of Black Sea near Burgas. Acta Univ. Carol. Biol. **2**: 91–123.

Komárek, J. 2010. Recent changes (2008) in cyanobacteria taxonomy based on a combination of molecular background with phenotype and ecological consequences (genus and species concept). Hydrobiologia **639**: 245–259.

Komárek, J. and Anagnostidis, K. 1995. Nomenclatural novelties in chroococcaleancyanoprokaryotes. Preslia**67**: 15–23.

Komárek, J. and Anagnostidis, K.2005. Cyanoprokaryota: 2. Teil: Oscillatoriales. Elsevier GmbH, München, pp. 1–759.

Komárek, J., Ankaštovský, J., Mareš, J. and Johansen, J.R. 2014.Taxonomic classification of cyanoprokaryotes (cyanobacterial genera) 2014, using a polyphasic approach. Preslia **86**: 295–335.

Laloui, W., Palinska, K.A., Rippka, R., Partensky, F., Herdman, M. and Iteman, I. 2002. Genotyping of axenic and non-axenic isolates of the genus *Prochlorococcus* and the OMF-'*Synechococcus*' clade by size, sequence analysis or RFLP of the internal transcribed spacer of the ribosomal operon. Microbiology **148**: 453–465.

Li, Z. and Xie, S. 2006. Species compositions and distributions of cyanophyta in the salt lake region of Yuncheng County, Shanxi Province. ActaBotan. Bor.-Occid. Sin. **26**: 384–388.

Lim, D.H., Jang, J., Kim, S., Kang, T., Lee, K. and Choi, I.H. 2012. The effects of sub-lethal concentrations of silver nanoparticles on inflammatory and stress genes in human macrophages using cDNA microarray analysis. Biomaterials**33**: 4690–4699.

Long, W. and Chen, C. 2006. Application of 16SrRNA Sequence in bacteria indication. J. Beijing Technol. Business Univ. (Nat. Sci. Ed.) **24**: 10–12.

Posada, D. and Crandall, K.A. 1998. Modeltest: Testing the model of DNA substitution. Bioinformatics **14**: 817–818.

Rippka, R.1988. Recognition and identification of cyanobacteria. Methods Enzymol. **167**: 28–67.

Rippka, R., Deurelles, J., Waterbury, J.B., Herdman, M. and Stanier, R.Y. 1979. Generic assignments, strain histories and properties of pure cultures of cyanobacteria. J. Gen. Appl. Micr. **111**: 1–61.

Rocap, G., Distel, D.L., Waterbury, J.B. and Chisholm, S.W. 2002. Resolution of *Prochlorococcus* and *Synechococcus* ecotypes by using 16S-23S rDNA internal transcribed spacer sequences. Appl. Envir. Micr. **68**: 1180–1191.

Ronquist, F. and Huelsenbeck, J.P. 2003. MrBayes 3: bayesian phylogenetic inference under mixed models. Bioinformatics **19**: 1572–1574.

Skuja, H. 1964.Grundzüge der Algenflora und Algenvegetation der Fjeldgegenden um Abisko in Schwedisch-Lappland. Nova Acta Reg. Soc. Sci. Upsal. (Ser. 4) **18**(3): 1–465.

Tamura, K., Peterson, D., Peterson, N., Stecher, G., Nei, M. and Kumar, S. 2011. MEGA5: molecular evolutionary genetics analysis using maximum likelihood, 888 evolutionary distance, and maximum parsimony methods. Mol. Biol. .Evol. **28**: 2731–2739.

Zhao, W., Zhao, Y., Wang, Q., Zheng, M., Zhang, P. and He, Z. 2010. Spatial and temporal patterns of plankton assemblage structure in a high altitude saline lake, Namuka Co in northern Tibet, China. Clean-Soil, Air, Water **38**: 960–968.

BRACHYSTELMA SESHACHALAMENSE (APOCYNACEAE), A NEW SPECIES FROM ANDHRA PRADESH, INDIA

K. PRASAD[1] AND P.V. PRASANNA[2]

Botanical Survey of India, Deccan Regional Centre, Hyderabad 500048, India

Keywords: Brachystelma seshachalamense; New species; Asclepiadoideae; Ceropegieae; Seshachalam hills.

Abstract

Brachystelma seshachalamense, a new species belonging to the family Apocynaceae, is described from the Seshachalam hills of Kadapa District of Andhra Pradesh, India. The new species is closely allied to *Brachystelma beddomei* Hook. f., but differs in usually solitary flowers, corolla tube large with a ring of hairs around the corona, corolla lobes spreading with white hairy along the margins and pubescent inside, coronal cup lemon yellow, interstaminal corona shallowly lobed with hairs.

Introduction

Brachystelma Sims is the second largest genus of the tribe *Ceropegieae* with c. 160 species, distributed mainly in the Old World tropics, particularly in sub-Saharan Africa, India, Sri Lanka, South East Asia and Northern Australia (Prasad *et al*., 2016). In India *Brachystelma* is represented by 23 species and, 22 of them are endemic to the country (Prasad *et al*., 2016; Venu and Prasad, 2015). The genus is represented by 8 species in Eastern Ghats (Hooker, 1883; Gamble, 1921; Rao *et al*., 2011; Prasad and Rao, 2013; Rasingam *et al.,* 2013; Swamy *et al*., 2013). Two species are with twining habit, namely, *Brachystelma beddomei* Hook. f. & *B. volubile* Hook. f., which are endemic to Tamil Nadu and Andhra Pradesh, respectively. These species were recollected by Swamy *et al.* (2012) (*B. volubile*, Kadapa hills, Andhra Pradesh) and Vijayasankar *et al.* (2003) (*B. beddomei*, Thiruvannamalai district, Tamil Nadu).

In botanical explorations in the Kadapa district of the Seshachalam hills, the authors spotted a *Brachystelma* species in the rocky crevices among grasses in dry deciduous forests. A critical study of the specimens, literature (Hooker, 1883; Gamble, 1921; Karthikeyan *et al*., 2009; Venu and Prasad, 2015; Prasad *et al*., 2016) and a comparison with herbarium specimens at K, CAL, MH, BSID and SKU, revealed that the collected specimens did not belong to any of the known species of *Brachystelma* in India and elsewhere. Hence, we described here as a new species, *Brachystelma seshachalamense* **sp. nov.**

Brachystelma seshachalamense K. Prasad & Prasanna, **sp. nov.** (**Figs 1 & 2**).

Diagnosis: *Brachystelma seshachalamense* is closely allied to *B. beddomei,* but differs in having flowers usually solitary, rarely 2-flowered; corolla tube 4.5-5.0 mm long, with a ring of hairs around the corona; corolla lobes spreading, 1.4–1.6 cm long, 2-3 mm long white hairy along the margins and pubescent inside; coronal cup shallow, lemon yellow; interstaminal corona shallowly bilobed with hairs.

[1]Corresponding author. Email: prasad.orchids@gmail.com

[2]Central National Herbarium, Botanical Survey of India, Acharya Jagadish Chandra Bose Indian Botanic Garden, Howrah 711 103, India.

Type: India, Andhra Pradesh: Kadapa district, Seshachalam hills, Sanipaya forest range, 14°07'04.7"N, 78°58'09.4"E, 550 m, 27 December 2014, *K. Prasad 6454* (*Holotype:* CAL!, *Isotype:* BSID!)

Twining herbs, c. 2.5 m high. Tubers depressed-globose, 2.0–2.5 x 3.0–3.5 cm, dirty brownish without, white when cut open. Stem unbranched, slender, cylindric, glabrous. Leaves opposite-deccussate, petiolate; petiole 2–5 mm long, puberulous; lamina linear or linear-lanceolate, 4–7 x 0.2–0.6 cm, acute at apex, cuneate at base, margin hairy; midrib prominent below, puberulous, glabrous above. Inflorescence extra-axillary, reduced cymes, usually solitary, rarely 2-flowered; peduncle pinkish-green, terete, 2.6–3.0 mm long, pubescent. Flowers drooping; floral bract linear, c. 1 mm long, acute, glabrous; bracteoles 2, linear, less than 1 mm long, acute, glabrous; pedicels terete, 6–8 mm long, pubescent. Calyx pinkish, glabrous; lobes linear, 2.0–2.4 x 0.8–1.0 mm, acute. Corolla spreading; corolla tube campanulate, 4.5–5.0 mm long, a ring of hairs around the corona and pubescent inside, green with pink tinge outside, yellow between the lobes inside; corolla lobes greenish-yellow outside and dark pink inside, broad at base, narrowed progressively towards tip, 1.4–1.8 cm long, acute at apex, pubescent inside, densely white hairy along the margins; hairs 2–3 mm long. Corona yellow, biseriate, c. 4 mm in diameter, staminal and interstaminal parts fused to a shallow cup shaped structure; interstaminal corona cupular, 5 angled, c. 0.6 mm long, shallowly bilobed with few hairs; staminal corona lobes 5, incumbent, overlapping the anther lobes, lemon yellow, ovate-oblong, 1.5 mm long, glabrous. Pollinia yellow, obovate-oblong, c. 0.2 x 0.1 mm, with pellucid margin and basally attached light-brown, tubular caudicles to a brown corpusculum. Carpels 2, c. 1 mm long, narrow, free; style abscent; stigma head pentangular, discoid.

Flowering: December–January.

Habitat: Rarely distributed in red or black soils of grass dominated dry deciduous open forests, at an of 400–600 m altitude.

Distribution: The Seshachalam hills of Kadapa district of Andhra Pradesh.

Etymology: The new species is named after the type locality, the Seshachalam hills of Eastern Ghats, Andhra Pradesh.

Notes: *Brachystelma seshachalamense* is closely allied to *B. beddomei* but differs from the latter which are appended in Table 1.

Table 1. Comparison of characters of *Brachystelma seshachalamense* sp. nov. and *B. beddomei.*

Characters	*Brachystelma seshachalamense* **sp. nov.**	*B. beddomei*
Inflorescence	Usually solitary, rarely 2-flowered	Usually 3-4-flowered, rarely 1-flowered
Corolla tube	4.5-5.0 mm long, a ring of hairs around the corona and pubescent inside, green with pink tinge outside yellow between the lobes	c. 3 mm long, glabrous, greenish yellow without and within
Corolla lobe	Spreading, 1.4-1.8 cm long, with broad base, 2-3 mm long white hairs along the margins and pubescent inside, greenish-yellow outside and dark pinkish inside	Cohering at apex, 2-3.5.0 cm long, without broad base, glabrous, greenish-yellow outside and purple-brown inside
Coronal cup	Shallow	Deep
Interstaminal corona	Shallowly bilobed, with hairs	Deeply bilobed, glabrous
Staminal corona colour	Lemon yellow	Pinkish-cream

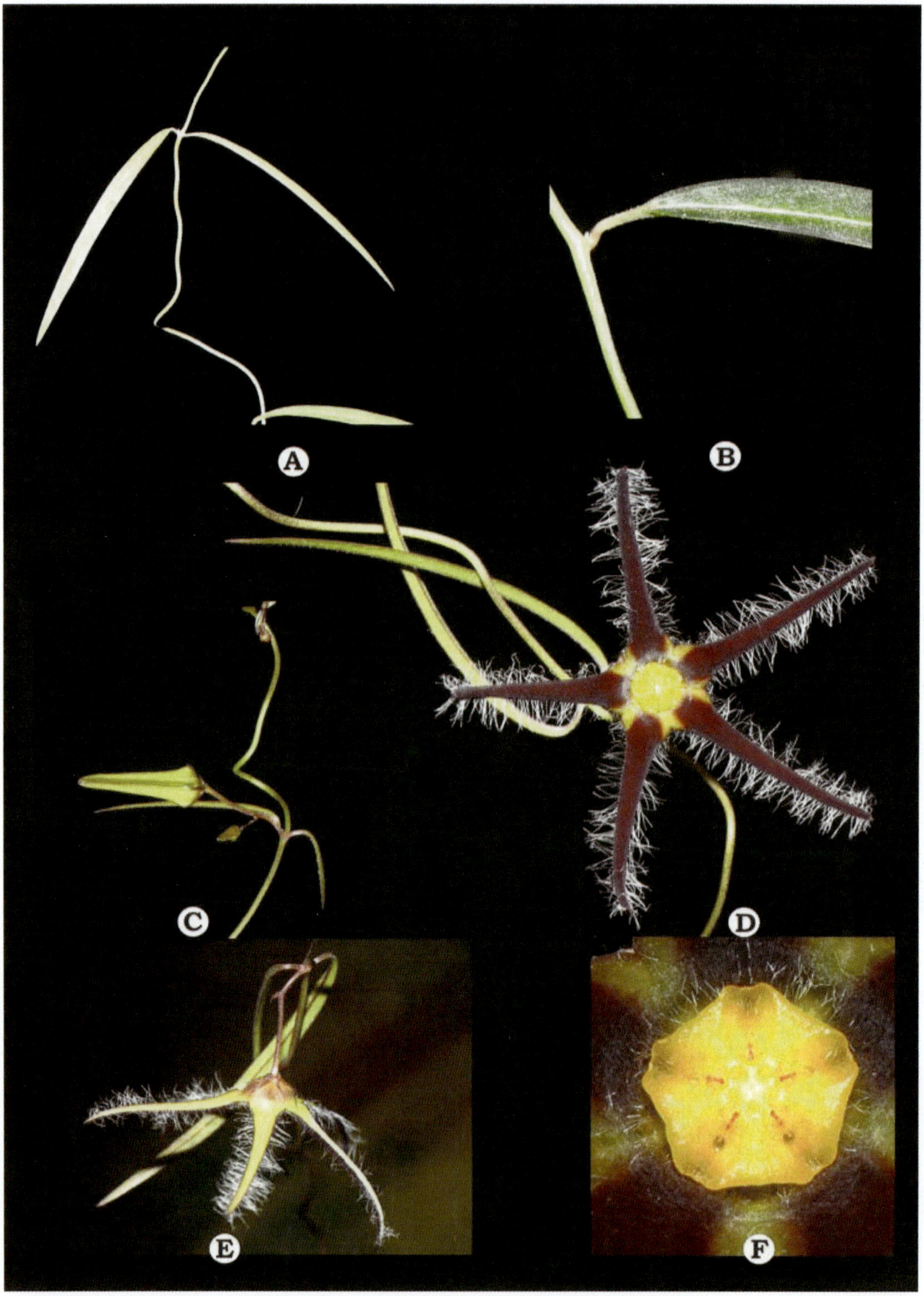

Fig. 1. **Brachystelma seshachalamense** K. Prasad & Prasanna, **sp. nov.** A. Leaves; B. Hairs on petiole and margins (close up); C. Flower bud; D & E. Flower; F. Corona (top view).

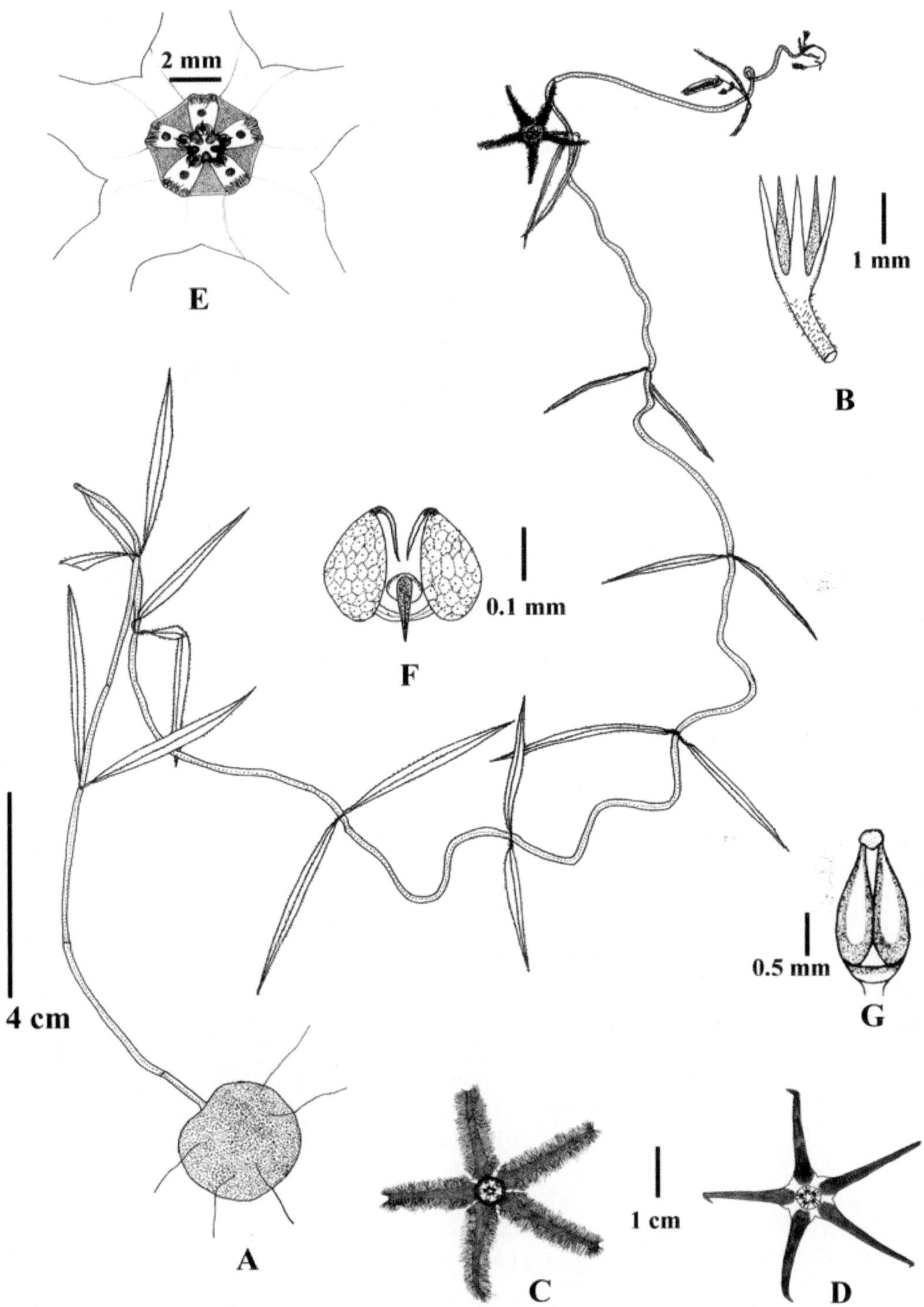

Fig. 2. **Brachystelma seshachalamense** K. Prasad & Prasanna, **sp. nov.** A. Habit; B. Calyx; C. Flower; D. Flower without hairs; E. Corona; F. Pollinia; G. Gynostegium.

Key to the twining *Brachystelma* species in India

1.	Corolla tube base quite flat, bulging limitedly with a constricted neck and in near urceolate shape, more than 1 cm long	***B. volubile***
-	Corolla tube campanulate, less than 5 mm long	2
2.	Corolla tube glabrous; lobes connate at apex, glabrous; interstaminal corona glabrous; staminal corona flesh coloured	***B. beddomei***
-	Corolla tube with a ring of hairs around the corona and pubescent inside; lobes spreading, hairy; interstaminal corona with hairs; staminal corona lemon yellow	***B. seshachalamense***

Acknowledgements

The first author gratefully acknowledges Department of science and technology (SERB-DST), New Delhi for financial assistance. Authors are thankful to Dr. P. Singh, Director, Botanical Survey of India (BSI), Kolkata and Dr. P. Venu, Scientist–F, Botanical Survey of India, Hyderabad for facilities and support. Authors gratefully acknowledge to all the forest officials and field staff of Rajampet forest division, Andhra Pradesh forest department.

References

Gamble, J.S. 1921. *Brachystelma. In:* Gamble, J.S., Flora of the Presidency of Madras, Vol. **2**, Adlard & Son, London, pp. 850–852.

Hooker, J.D. 1883. *Brachystelma. In*: Hooker, J.D., *Flora of British India,* Vol. **IV**. L Reeve & Co., London. pp. 65–66.

Karthikeyan, S., Sanjappa, M. and Moorthy, S. 2009. Flowering Plants of India: Dicotyledons.Vol. **1**. (Acanthaceae–Avicenniaceae). Botanical Survey of India, Kolkata. pp. 157–158.

Prasad, K., Prasanna, P.V., Meve, U., Rao, M.S., and Thulasaiah, T. 2016. *Brachystelma annamacharyae* sp. nov. (Apocynaceae) from the Seshachalam hills of Andhra Pradesh (India). Nord. J. Bot. Online publication on 29 March 2016.

Prasad, K. and Rao, B.R.P. 2013. *Brachystelma nallamalayanum sp. nov.* (Apocynaceae: Asclepiadoideae–Ceropegieae) from India. J. Threatened Taxa **5**(14): 4904–4906.

Rao, B.R.P., Prasad, K., Sadasivaiah, B., Basha, K.S., Suresh Babu, M.V. and Prasanna, P.V. 2011. A new species of *Brachystelma* R. Br. (Apocynaceae: Asclepiadoideae–Ceropegieae) from India. Taiwania **56**(3): 223–226.

Rasingam, L., Chorghe, A., Meve, U., Rao, M.S. and Prasanna, P.V. 2013. *Brachystelma penchalakonense* (Apocynaceae: Asclepiadoideae), a new species from Andhra Pradesh, India. Kew Bull. **68**: 663–667.

Swamy, R.K., Sandhya Rani, S. and Pullaiah T. 2013. *Brachystelma ciliatum* (Apocynaceae): A new record for Eastern Ghats of Andhra Pradesh, India. Nelumbo **55**: 191–195.

Swamy, R.K., Sandhyarani, S., Karuppusamy, S. and Pullaiah, T. 2012. The rediscovery of *Brachystelma volubile* (Apocynaceae–Asclepiadoideae). Rheedea **22**(2): 107–110.

Venu, P. and Prasad, K. 2015. The existential crisis in Indian *Brachystelmas* (Apocynaceae). Current Science **109**(4): 680–682.

Vijayasankar, R., Ravikumar, K. and Ravichandran, P. 2003. Endemic species, *Brachystelma brevitubulatum* (Bedd.) Gamble (Asclepiadaceae), relocated after a century. Phytotaxonomy **3**: 130–133.

22

DIMERIA KALERII (POACEAE: PANICOIDEAE), A NEW SPECIES FROM NORTHERN KERALA, INDIA

P. BIJU[1,3], E.J. JOSEKUTTY[2,3] AND AUGUSTINE JOMY[3]

Department of Botany, Government College, Kasaragod, Vidyanagar P.O., Kasaragod – 671123, Kerala, India

Keywords: Dimeria; Endemic species; Kerala; New species; India.

Abstract

Dimeria kalerii, a new species collected from the lateritic plateaus of Northern Kerala, India is described and illustrated. It is allied to *Dimeria gracilis* in robust habit, densely bearded nodes, lax racemes, long clavate pedicels with cupuliform apex but differs in ciliate apex of ligules, numerous racemes on the peduncles (7-32), hairy pedicels, smaller spikelets (3.8–4.2 mm), shorter cilia on the upper and lower glume, long bristly hairs at the apex of upper glume, hairy column of awns and smaller anthers (1.8–2.0 mm).

Introduction

Dimeria R. Br. (Brown, 1810) is a widely distributed genus in the tropical and subtropical regions of the world. Kiran Raj *et al.* (2015) recognized 65 species in their revised infrageneric classification of *Dimeria* R. Br. and classified them into four sections. The genus is represented by 40 species in the Peninsular India (Kiran Raj *et al.,* 2008). Recently a new species (Gosavi *et al.,* 2016) and a new subspecies (Kiran Raj *et al*., 2016) were added to the list. The genus can be easily distinguished by solitary, laterally compressed spikelets and flat or filiform rachis without joints (Sreekumar and Nair, 1991). The laterite plateau of Western Ghats shows luxuriant growth of *Dimeria* R. Br. populations during the post monsoon season and gives characteristic golden yellow colour to the plateau when they dry out.

During the floristic exploration of seasonal ponds in northern Kerala, India, the authors collected an interesting specimen of *Dimeria* with fascicled racemes. It grows densely inside two seasonal pools with approximately 1,000 individuals in the Karakkode lateritic plateau (2 sq. km. area) in Kasaragod district, Northern Kerala. The new grass showed extended distribution in Erikkulam lateritic plateau in Kasaragod district.The detailed study of specimens from three different collections (Biju & Jomy 1166, Biju & Jomy 1193, Biju & Jomy 2810) of three populations revealed that it is an interesting specimen of *Dimeria* R. Br. and differed from all other species of the genus known so far. Hence it is described here as a new species, *Dimeria kalerii*. The new species shares many characters of *Dimeria* sect. *Capillares* such as presence of numerous racemes, thin and wingless rachis and distantly arranged spikelets which are not readily disarticulating from the pedicel.

[1,3] Corresponding author. Email: bijuarimba@gmail.com

[2,3]Department of Botany, Government Brennan College Thalassery, Dharmadam P.O., Kannur 670106, Kerala, India.

[3] Department of Botany, St. Thomas College, Pala, Arunapuram P.O, Pala 686574, Kerala, India.

Results and Discussion

Dimeria kalerii P. Biju, E.J. Josekutty & Augustine Jomy, **sp. nov.** **(Figs 1-3)**.

Diagnosis: *Dimeria kalerii* is allied to *D. gracilis* in robust habit, densely bearded nodes, lax racemes, long clavate pedicels with cupuliform apex but differs in ciliate apex of ligules, numerous racemes (7–32), hairy pedicels, smaller spikelets (3.8–4.2 mm), shorter cilia on the upper glume and lower glume, long bristly hairs at the apex of upper glume, hairy column of awn and shorter anthers (1.8–2.0 mm).

Type: India, Kerala, Kasaragod district, Karakkode lateritic plateau, 12°36'18.9"N, 075°00'06.4"E, 75 m above sea level, 12 Dec 2014, *Biju & Jomy 1166* (*Holotype*: CAL; *Isotype*: MH).

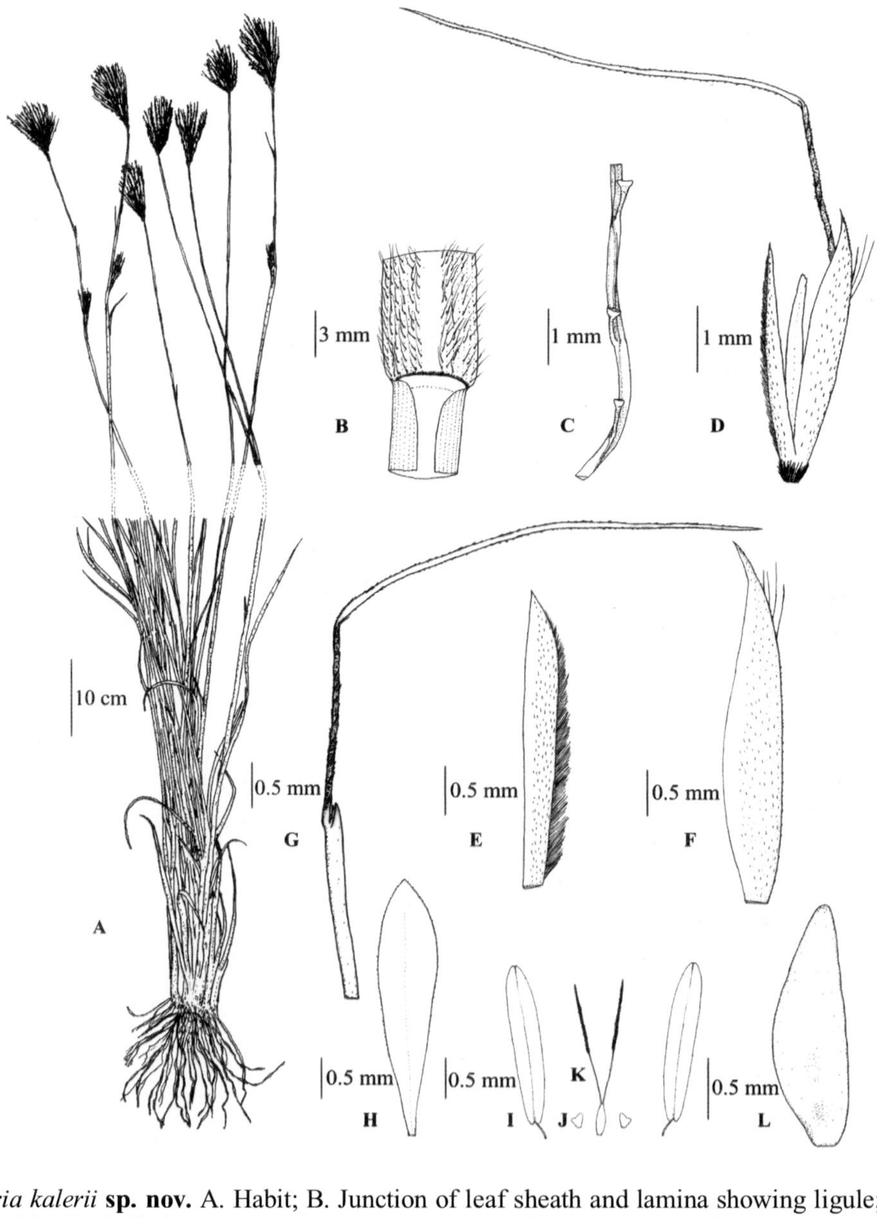

Fig. 1. *Dimeria kalerii* **sp. nov.** A. Habit; B. Junction of leaf sheath and lamina showing ligule; C. A portion of the rachis; D. Spikelet; E. Lower glume; F. Upper glume; G. Upper lemma; H. Lower lemma; I. Stamen; J. Lodicule; K. Carpel; L. Grain.

Tall, robust, tufted, marshy-terrestrial grass; culms 1.2–2.0 m high, branched or unbranched; internodes 10–25 cm long, 2–3 mm in diam., glabrous; nodes densely bearded, hairs white, up to 3 mm long. Leaves ascending; sheath 5–13 cm long, hairy above, basal sheaths extends to higher internode, upper sheaths shorter than internodes, laterally flattened, keeled; ligules membranous, 0.8–1.0 mm long, ciliate at apex; leaf blade linear-lanceolate, 15–45×0.4–0.5 cm, covered by long bulbous-based hairs on both sides, hairs 2–3 mm long; tip acuminate; base round-attenuate; margin minutely serrulate with distantly arranged bulbous based hairs, revolute in older leaves;

Fig. 2. *Dimeria kalerii* **sp. nov.** A. Habitat; B. Habit; C. Fascicled raceme.

midrib prominent, keeled on the midrib below. Racemes 7–32, usually 18–23, sub-digitate, 4–15 cm long with 34–55 spikelets; rachis thin, trigonous, tough and continuous, up to 0.4 mm wide, glabrous; spikelets oblanceolate, 3.8–4.2 mm long, distantly arranged, 0.5–5.0 mm apart, alternate, distichous, laterally compressed; pedicels 0.5–0.8 mm long, hairy, clavate, apex cupulate, sometimes pedicels curved bringing all the spikelets on the same side; callus short, 0.1–0.2 mm long, truncate at base, covered with golden hairs; hairs 0.5–0.6 mm long; lower glume 3.3–3.6x 0.4–0.5 mm, oblong-linear, single nerved, acute-acuminate, laterally flattened, one side convex, other side concave, coriaceous, hirsute on dorsal side, minutely keeled, long cilia along the keel, margin hyaline, minutely ciliate; upper glume 3.6–4.0×0.8–1.0 mm, elliptic, 2-nerved, coriaceous, round at back, acuminate, hirsute, few long bristly hairs towards apex, margin hyaline, ciliate. Florets two, lower floret reduced to lower lemma, oblanceolate, 2.8–3.2×0.5–0.6 mm, thin, hyaline, single-veined, margin minutely ciliate, acute, rarely emarginate; upper floret bisexual, upper lemma 2.0–2.2×0.4–0.6 mm, single-nerved, awned, awn 9.0–9.3 mm long; column 2.7–2.9 mm long, hairy; palea absent; lodicules 2, obovate, 0.1×0.15 mm, bilobed, hyaline; stamens 2; anthers yellow, 1.8–2.0×0.2–0.3 mm; filaments 0.8-1.0 mm long; ovary ovoid-elliptic, glabrous, 0.3–0.4×0.1–0.15 mm; style 0.5–0.7 mm long; stigma plumose, 0.8–0.9 mm long. Grain greyish brown, elliptic, 1.0–1.5×0. 4–0.5 mm.

Fig. 3. *Dimeria kalerii* **sp. nov.** A. Bearded node; B. A portion of leaf blade; C. A portion of rachis with pedicel; D. Spikelet; E. Lower glume; F. Upper glume.

Etymology: The new species is named to honour Mr. Om Prakash Kaler, a dedicated conservationist and former Chief Conservator of Forests, Kerala, India.

Additional specimens examined: India, Kerala, Kasaragod district, Karakkode lateritic plateau, 12°33'19.1"N, 075°04'19.7"E, 73 m asl, 20 Jan 2014, *Biju & Jomy 1193* (St. Thomas College Herbarium, Pala, Kerala, India); Kerala, Kasragod district, Erikkulam lateritic plateau, 12°34'17.4"N, 075°03'21.3"E, 110 m asl, 25 Nov 2016, *Biju & Jomy 2810* (St. Thomas College Herbarium, Pala, Kerala, India).

Related species: The new species *Dimeria kalerii* is allied to *D. gracilis* Nees *ex* Steud. (Steudel, 1854) but differs in many characters as given in Table 1.

Table 1. Distinguishing characters of *Dimeria kalerii* sp. nov. and *D. gracilis*.

Characters	*Dimeria kalerii* **sp. nov.**	*Dimeria gracilis*
Culm	Up to 200 cm long	Up to 90 cm long
Leaves	Up to 45 cm long, densely covered with long hairs; sheaths laterally flattened, hairy above; ligule ciliate at apex	Up to 25 cm long, sparsely covered with hairs; sheath terete, glabrous; ligule lacerate at apex
Racemes	Up to 32	Up to 11
Spikelets	3.8–4.2 mm long	5.0–5.5 mm long
Callus	Very short, up to 0.15 mm long	Up to 0.5 mm long
Pedicel	Hairy, laterally flattened-triangular	Glabrous, terete
Lower glume	Minutely keeled, densely ciliate along the keel; dorsal surface is covered with short white hairs; margin ciliate	Covered all over with long white hairs or glabrous; margin not ciliate
Upper glume	Covered with short white hairs, few long bristly hairs towards the apex; margin ciliate	Covered with long white hairs, long bristly hairs are absent at the apex, margin not ciliate.
Awn	Column up to 2.8 mm, hairy	Column up to 4 mm, glabrous

Distribution and conservation status: *Dimeria kalerii* is restricted to the marshy areas and seasonal ponds in the lateritic plateaus of Karakkode and Erikkulam in Kerala, India. The new species has so far been reported only from two localities, so further studies are needed to assess its distribution and conservation status. As per the currently available data the species should be assigned to the category 'Data Deficient' (DD) of IUCN (2012). This new species shows healthy populations in two seasonal pools in the type locality. The distribution of this rare, tall and robust endemic grass is limited. The increasing human activities and habitat destruction cause threat to the existence of this rare and interesting species.

Notes: The new species *Dimeria kalerii* is the largest species of *Dimeria* reported so far in the world. It is a perennial grass showing robust habit and growing up to 2 m high. The culms are 2-3 mm in diameter and the leaves are up to 45 cm long. The inflorescence fascicled with 7-32 racemes, which are up to 15 cm long. It can be easily distinguished from the allied species by its numerous racemes, hairy pedicels, smaller spikelets, long bristly hairs at the apex of upper glumes.

Acknowledgements

The authors are indebted to the Principal, St. Thomas College, Pala, Kerala for providing necessary laboratory facilities for the work. The first author is grateful to the Principal, Govt. College, Kasaragod, Kerala for providing necessary support for the work. The first and second authors are thankful to University Grants Commission, Government of India for providing financial support.

References

Brown, R. 1810. Prodromus Florae Novae Hollandiaeet Insulae van-Diemen Vol. **1**. J. Johnson, London, 204 pp.

Gosavi, K.V.C., Kamble, M.Y., Chandore, A.N. and Yadav, S.R. 2016. A new species of *Dimeria* (Poaceae) from Andaman and Nicobar Islands, India. Phytotaxa **270**(4): 295–300.

IUCN. 2012. IUCN Red List Categories and Criteria: version 3.1. 2nd edition. Gland, Switzerland and Cambridge, UK, iv+ 32 pp.

Kiran Raj, M.S. and Sivadasan, M. 2008. A new species of *Dimeria* R. Br. (Poaceae, Panicoideae, Andropogoneae) from Goa, India. Novon **18**(2): 183–186.

Kiran Raj, M.S., Sivadasan, M., Veldkamp, J.F., Alfarhan, A.H. and Amal Thamimi, A.S.M. 2015. A revised infrageneric classification of *Dimeria* R. Br. (Poaceae: Andropogoneae). Bangladesh J. Plant Taxon. **22**(1): 47–54.

Kiran Raj, M.S., Sivadasan, M., Dileep, P. and Alfarhan, A.H. 2016. A new subspecies of *Dimeria hohenackeri* Hochst. *ex* Miq. (Poaceae) from India. Bangladesh J. Plant Taxon. **23**(1): 27–31.

Sreekumar, P.V. and Nair, V.J. 1991. Flora of Kerala–Grasses. Botanical Survey of India, Calcutta, 81 pp.

Steudel, E.G.1854. Synopsis Plantarum Glumacearum, Vol. **1** .J.B. Metzler, Stuttgart, 413 pp.

23

A NEW SPECIES OF *ISCHAEMUM* L. (POACEAE) FROM KERALA, INDIA

C.N. Sunil, V. Nithya Madhanan, C.R. Remya Krishnan, V.V. Naveen kumar[1], M.S. Simi and K.J. Jyothi

Post Graduate & Research Department of Botany, S.N.M. College, Maliankara, Ernakulam District, Kerala, India

Key words: *Ischaemum;* new species; Poaceae; Kerala; India

Abstract

Ischaemum sreenarayanii, a new species of *Ischaemum* L. is described and illustrated with detailed notes on distribution, conservation status and phenology. The new species is similar to *I. santapaui* and *I. nairii* but differs in number of attributes having up to 1.3 cm long filiform tip of the leaf apex, villous pseudopetiole of upper leaves and 4-6 conspicuous annular transverse ridges and winged margins of the sessile spikelet.

Introduction

The genus *Ischaemum* L. belongs to the family Poaceae. All the species of *Ischaemum* are distributed in Warm and tropical regions of the world especially in Asia (Mabberley, 2008) with about 81 species (Plant list, 2013).

The genus *Ischaemum* was treated in different ways by various authors in India. Sur (2001) recognized 51 taxa in India. Later, Singh and Rao (2008) proposed changes in status for some species and new combinations. Thus in their study, only 36 taxa were reported including two new additions to the Indian flora. Srivastava and Nair (2010) worked on Indian *Ischaemum* and reported 56 taxa, indicating that they overlook the changes made by Sing and Rao (2008). In India, the genus is mostly represented in Southern Western Ghats, especially in Kerala. Sreekumar and Nair (1991) identified 28 species, 3 varieites and 2 sub varieties including 11 new species. Nayar *et al.* (2006) included 39 taxa of *Ischaemum* in 'Flowering plants of Kerala' and Srivastava and Nair (2010) reported 40 taxa from Kerala state. After the publication of 'Genus *Ischaemum* (Poaceae) in India' (Srivastava and Nair, 2010), two new species were described from India *viz.* I. *kasaragodensis* Dileep and G.G. Nair (Dileep and Geetha, 2015) and *I. sayajiraoi* Raole and R.J Desai (Vinay *et al.*, 2011). Hence the total number of *Ischaemum* in India becomes 58 and that of Kerala is 41.

During the floristic studies in central Kerala, authors collected interesting specimens of *Ischaemum* from wetland areas of Thrissur and Palakkad districts of Kerala. Critical studies using relevant literature and type materials of similar species revealed its novelty and distinctness from the hitherto known species and described here as new species *Ischaemum sreenarayanii.*

Ischaemum sreenarayanii Sunil, Nithya and Remya **sp.nov.** **(Figs 1 & 2)**.

Diagnosis: *Ischaemum sreenarayanii* is similar to *Ischaemum santapaui* Bor. in its tall and robust habit with stilt rooted lower nodes, glabrous glumes, narrowed and tapering base of lower leaves and rounded base of upper leaves but differs in having large bulbous-pilose leaves with

[1]Corresponding author. Email: naveenkumar2389@gmail.com

Fig. 1. *Ischaemum sreenarayanii* Sunil, Nithya and Remya: A. Habit; B. A portion of upper leaf base showing ligule & villous Pseudopetiole; C. Rachis.

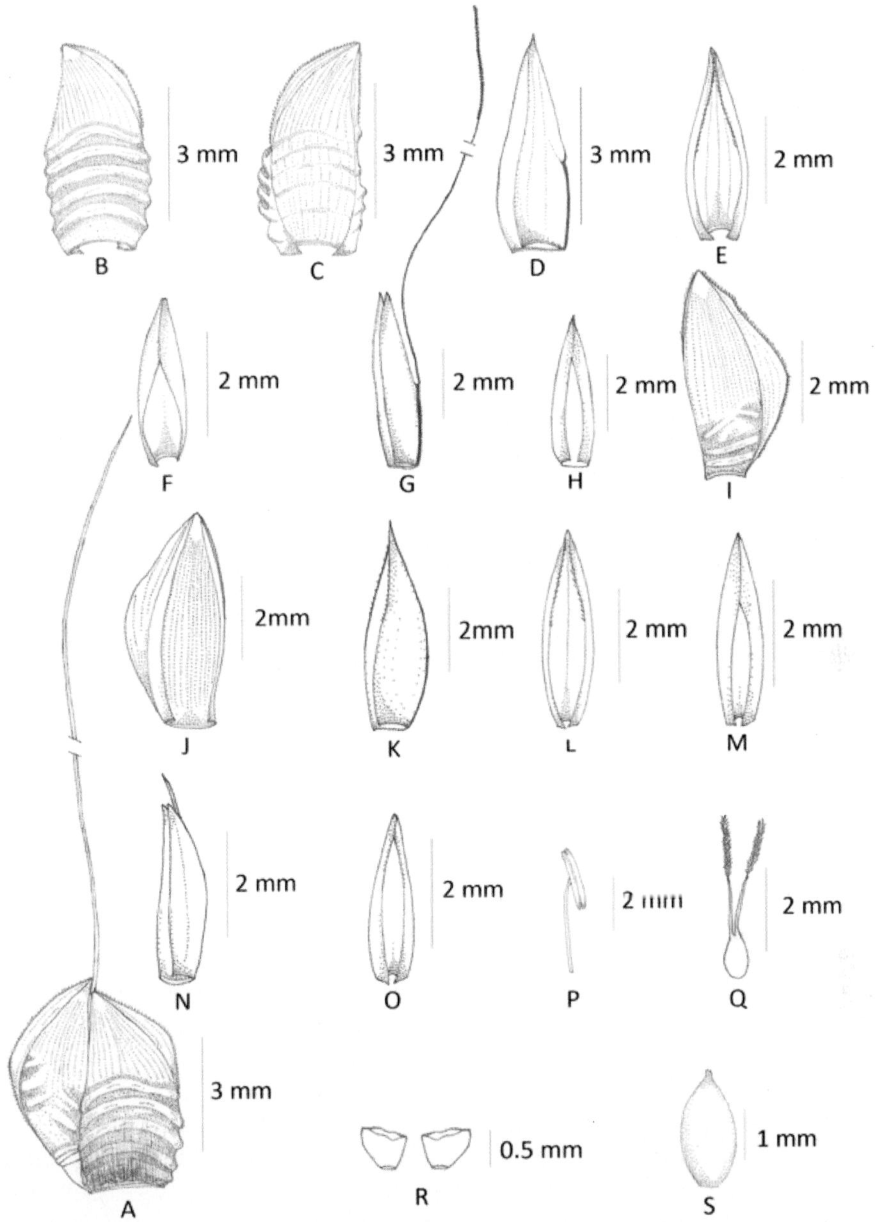

Fig. 2. *Ischaemum sreenarayanii* Sunil, Nithya and Remya: A. Paired spikelet; B-H Sessile spikelet: B & C. Dorsal & ventral view of Lower glume; D. Upper glume; E. Lower lemma; F. Lower palea; G. Upper lemma; H. Upper palea; I-O Pedicelled spikelet: I&J. Dorsal & ventral view of Lower glume; K. Upper glume; L. Lower lemma; M. Lower palea; N. Upper lemma; O. Upper palea; P. Stamen; Q. Pistil; R. Lodicules; S. Grain.

long filiform apex, villous pseudopetiole on upper leaves, long and appressed raceme, lower glume of the sessile spikelet with 4-6 conspicuous annular transverse ridges and winged margins, long awned upper lemma of sessile spikelet and well developed pedicelled spikelets. The new species is also shows some similarities with *Ischaemum nairii* Nair Sreekumar in its robust habit, tubercle based hairy leaves, glabrous glumes but differs in having stilt root, leaf apex with up to 1.3 cm long filiform tip, villous pseudopetiole on upper leaves, 4-6 conspicuous upwardly directed annular transverse ridges and winged margins on one side of lower glume of sessile spikelet, shorter lower palea of sessile spikelet, long upper lemma of sessile spikelet with 18- 24 mm long awn, pedicel 1.0-1.5 mm long and less than one-third length of lower glume of sessile spikelet.

Type: INDIA, Kerala: Thrissur district, Kombazha, ±11 m, 10^0 34′ 38.5″ N 76^0 23′ 50.0″ E, 8 December 2015, *Sunil* & *Nithya* 9007 (*Holotype*: MH!; *Isotypes*: CALI!, SNMH!).

Densely tufted and deep-rooted perennials. Culms robust, erect to decumbent, up to 160 cm long, 2.5-7.0 mm across and stilt rooted at the base, generally reddish-brown, glaucous below the node; nodes densely villous. Leaves all along the culm; blade 8-40 × 1.0-2.7 cm, linear-lanceolate to linear-elliptic, lower ones tapering into a short pseudopetiole, upper ones rounded with densely villous up to 3 mm long pseudopetiole, scaberulous on margins, acuminate with up to 1.3 cm long filiform apex, mid-vein thickly prominent below, canaliculate above, bulbous-pilose on both surfaces; ligule sub-coriaceous, 1-10 mm long, oblong, rounded to truncate at apex, sparsely pilose; sheath 8-20 cm long, rounded on the back, bulbous-pilose along margins. Inflorescence of two racemes, exserted, 8-14 cm long, appressed to give the appearance of a single raceme; joints of rachis 3- 4 × 1.5- 2.0 mm, triquetrous, clavate-turbinate to linear-clavate, crustaceous, fistular, strawcoloured, ciliate on the outer angle and glabrous otherwise. Sessile spikelet 5.0-5.5 × 2.0-2.2 mm, elliptic-oblong, awned, lower floret male, upper floret bisexual; callus ca. 0.5 mm long, long-villous with hair up to 2 mm long. Lower glume 4.5-5.5 × 2.0-2.2 mm, elliptic-oblong, acute at apex, crustaceous in the lower ½ - $^2/_3$ regions with 4-6 conspicuous upwardly directed annular transverse ridges, thickly coriaceous and smooth above, laterally2- keeled with incurved margins and the keel winged on one side towards apex, scaberulous on the margin of the wing, weakly 11-13-nerved. Upper glume 4.5- 5.5 × 1.8-2.0 mm, lanceolate, boat- shaped, acute at apex, keeled on back and humped below middle, shortly incurved on sparsely ciliate margins, coriaceous, straw-coloured, 5-7-nerved. Lower lemma 4-5 × 1.2-1.6 mm, elliptic- lanceolate, acute at apex, hyaline, 3- nerved, margins incurved and shortly ciliate on upper half. Lower palea 3-4 × 1.0-1.2 mm, elliptic-lanceolate, obtuse to truncate at apex, hyaline, 2-nerved, infolded along the nerves and minutely scaberulous along keel. Upper lemma 4.0-4.5 × 1.2-1.4 mm, elliptic, cleft to the middle with acuminate lobes, awned from the sinus, hyaline, glabrous, 3-nerved; awn 18-24 mm long, geniculate with a column of 8-10 mm long, scaberulous except on the chestnut brown column. Upper palea 2.5-3.0 × 1.0-1.2 mm, ovate-lanceolate, acuminate at apex, hyaline, 2-nerved, infolded along the nerves. Pedicelled spikelet 4.5-5.5 × 2.0-2.5 mm, oblong-elliptic, acute at apex, with rudimentary awn or awnless; lower floret male or empty, upper bisexual; pedicels 1.0-1.5 × ca. 1 mm, triquetrous, long ciliate on the outer angle and glabrous otherwise. Lower glume 4.3-5.4 × 2.0-2.5 mm, obliquely oblong-elliptic, acute at apex, thickly sub - crustaceous with 2 – 4 - annular transverse ridges or nodulose in the lower half, coriaceous-herbaceous above, laterally keeled with incurved margins and the keel broadly winged on one side and narrowly winged on the other, scaberulous on the margin of the wing, 11-13– nerved, glabrous. Upper glume 4-5 × 1.4-1.6 mm, lanceolate, boat- shaped, acute at apex, coriaceous, straw -coloured, minutely keeled on back, not humped, shortly incurved on margins with the margins sparsely ciliate or glabrous, 5-7-nerved. Lower lemma 3.5-4.5 × ca. 1 mm, elliptic-lanceolate, acute to acuminate at apex, hyaline, 3- nerved, incurved on the margins, minutely scabrid on keels and sparsely ciliate on the margins. Lower palea 3-4 × ca. 1 mm, elliptic-lanceolate, acute at apex, hyaline, 2- nerved, incurved along the nerves, minutely scabrid on keels towards apex. Upper lemma 2.7-3.4 × ca. 1 mm, ovate-lanceolate, shortly notched at apex and imperfectly awned with an awn up to 2 mm long from the sinus or entire and unawned, hyaline, glabrous, 3- nerved. Upper palea 2-3 × 1.0-1.2 mm, ovate-

Table 1. Distinguishing characters of *Ischaemum sreenarayanii* from *Ischaemum santapaui* Bor. and *Ischaemum nairii* V.J. Nair and P.V. Sreekumar.

Characters	*I. santapaui*	*I. sreenarayanii sp. nov.*	*I. nairii*
Stilt root	Present	Present	Absent
Leaves	0.4 – 1.0cm wide, glabrous, scabrid on both side, acute at apex, upper ones shallowly cordate at the base without a villous pseudopetiole	1.0- 2.7 cm wide, bulbous – pilose on both sides, acuminate with up to 1.3 cm long filiform tip at apex , upper ones rounded at the base with up to 3 mm long densely villous pseuodopetiole	1- 2 cm wide, densely villous with tubercle based hairs, acuminate, upper ones round or shallowly cordate without a villous pseudopetiole
Joints of rachis	Long hairy on one angle, smooth and glabrous on the other two	Ciliate on the outer angle and glabrous otherwise	Long villous along margins and dorsal angles
Raceme	In fascicles, upto 7.5 cm long, divergent	Not fascicled, 8-14 cm long, appressed	Not fascicled, 5-12 mm long, appressed
Lower glume of the sessile spikelet	Coriaceous, flat in the lower $^{3}/_{4}$ region, herbaceous above, margins not winged, 10-11- nerved	Crustaceous in the lower ½ - $^{2}/_{3}$ regions with 4-6 conspicuous upwardly directed annular transverse ridges, coriaceous above, margins winged on one side towards apex, 11-13- nerved	Crustaceous in the lower $^{1}/_{3}$ - $^{2}/_{3}$ regions with several side nodules joined by sharp ridges, scabrid above, not winged, 9-11- nerved
Upper glume of the sessile spikelet	Mucronate at apex, scabrid on dorsal surface, rounded on back below middle	Acute, not mucronate at apex, not scabrid on dorsal surface, humped on back below middle	Acute, tip minutely winged and scabrid, smooth, coriaceous,humped on back below middle
Lower palea of the sessile spikelet	Ca. 3.0 mm long	3-4 mm long	4.5-5.4 mm long
Upper lemma of Sessile spikelet	3.5-4.0 mm long; awn 10- 15 mm long, column ca. 6 mm long	4.0-4.5 mm long; awn 18 - 24 mm long, column 8-10 mm long	3.5-4.0 mm long; awn 14-16 mm long, column 6-7 mm long
Pedicel of pedicelled spikelet	Ca. 2 mm long, more than 1/3 rd length of lower glume of sessile spikelet	1-1.5 mm long, less than 1/3 rd length of lower glume of sessile spikelet	2-2.5 mm long, more than 1/3 rd length of lower glume of sessile spikelet
Pedicelled spikelets	Rudimentary, with rudimentary glumes without lemma or if present smaller in size	Well developed with well developed glumes and lemmas	Well developed with well developed glumes and lemmas
Lower glume of pedicelled spikelet	Glumes rudimentary often a small scale not winged	Broadly winged on one side and narrowly winged on the other	Narrowly winged on one margin

lanceolate, margins scabrid, truncate at apex, hyaline, 2- nerved infolded along the nerves. Lodicules 2, 0.5- 0.8 × 0.8-1.0 mm, obovate, obliquely truncate and wavy at apex, broadly cuneate at base,hyaline.Stamens 3; filaments 2.5-3.5 mm long; anthers 1.8- 2.2 mm long, oblong, yellow. Ovary ca. 1.0 × 0.4 mm, ovoid; style 1.0-1.4 mm long; stigmas 1.5-2.0 mm long, feathery, purple. Grain 2- 2.5 × 1.0-1.2 mm, oblong-ellipsoid, obtusely trigonous, smooth, straw-coloured.

Flowering and fruiting: November – March.

Habitat and associated species: Ischaemum sreenarayanii was found to grow in abandoned paddy fields and other marshy localities in association with *Hydrolea zeylanica* (L.) Vahl., *Isachne miliacea* Roth., *Ammania baccifera* L., *Fimbristylis littoralis* Gaudich., *Leersia hexandra* Sw., *Ludwigia hyssopifolia* (G. Don) Exell etc.

Conservation status: The new species was observed in two localities with very few populations. The extent of occurrence is estimated to be about 34 km and distributed as isolated patches. This was found to grow in the wetlands near the road side; hence the chance for getting extinction is more. The human encroachment and developmental activities leads to the demolition of the natural habitat of this species. By following IUCN criteria (CR B1 ab (i,ii,iv); 2ab (i,ii,iv); D) for assessing the status of Rare and Threatened plants, *Ischaemum sreenarayanii* is assessed as belonging to Critically Endangered (CR) category (IUCN 2014).

Etymology: The species is named in the honour of the great saint and social reformer, Sree Narayana Guru, in whose name a number of educational institutions have been established all over Kerala.

Additional specimens examined (*Paratype*): India, Kerala: Palakkad District, Alathur, Padur ±16 m 22 November 2014, *C.N. Sunil* 4928 (SNMH!); Thrissur district, Kombazha ±11, 27 January 2016, *Remya & Sunil* 8237 (CMPR!).

Acknowledgements

The authors are thankful to the Principal & Staff of the Department of Botany, S.N.M. College, Maliankara.

References

Dileep, P. and Geetha, G.N. 2015. *I. kasaragodensis* (Poaceae- Panicoideae), a new species from Western Ghats, India. Annals of Plant Sciences **4**(10): 1199–1201.

IUCN 2014. IUCN Red List Categories and Criteria, ver. 3.1. - IUCN Species Survival commission.

Mabberley, D.J. 2008. The Plant Book. Cambridge University Press.

Nayar, T.S., Rasiya Beegam, A., Mohanan, N. and Rajkumar, G. 2006. Flowering Pants of Kerala. Tropical Botanical Garden and Research Institute , Thiruvananthapuram, India

Singh R.K. and Rao, P.S.N. 2008.Genus *Ischaemum* L. (Poaceae) In India. J. Econ. Taxon. Bot. **32**(4): 797–835.

Sreekumar, P.V. and Nair, V.J. 1991. Flora of Kerala – Grasses. Botanical Survey of India, Calcutta.

Srivastava, S.K and Nair, V.J. 2010. Genus *Ischaemum* L. (Poaceae) In India. Nelumbo **52:** 63–92.

Sur, P.R. 2001. A Revision of the genus *Ischaemum* L. (Poaceae) in India. J. Econ. Taxon. Bot. **25**: 407–438.

The Plant List. 2013. Ver. 1.1. Published on the internet; http: //www.theplantlist.org/ accessed 28 July 2016.

Vinay, M.R., Rinku, J.D. and Veldcamp, J.F. 2011. *I. sayajiraoi,* a new species of Poaceae from Gujarat, India. Kew Bulletin **66**: 303–306.

24

A NEW SUBSPECIES OF *CREPIS PALAESTINA* (ASTERACEAE) FROM TURKEY

HUSEYIN INCEER[1] AND NURSEN AKSU KALMUK

Karadeniz Technical University, Faculty of Science, Department of Biology, 61080 Trabzon, Turkey

Keywords: Crepis; Asteraceae; *babcockii* subsp. nov.; Chromosome number; Turkey.

Abstract

Crepis palaestina subsp. *babcockii* Inceer & Aksu Kalmuk subsp. nov. (Asteraceae, Cichorieae) is described and illustrated. It grows in shady places and red pine forest in southwest Anatolia, Turkey. The chromosome number of the new subspecies is $2n = 2x = 8$. The diagnostic morphological characters that distinguish *C. palaestina* subsp. *babcockii* from morphologically similar taxa *C. palaestina* subsp. *palaestina* and *C. pulchra* are discussed, and a conservation status for the new taxon is suggested.

Introduction

Crepis L. is a large, critical and taxonomically difficult genus in the tribe Cichorieae of the family Asteraceae. It comprises over 200 species (Bremer, 1994), mainly distributed throughout the northern hemisphere and Africa (Enke, 2009). Ekim (2012) listed 42 Turkish taxa of *Crepis*, but our recent taxonomic data obtained from revision of *Crepis* in Turkey indicate that the genus together with the inclusion of the new subspecies described here has 40 taxa in Turkey, of which 8 are endemic. Among the species occurring in Turkey, *C. palaestina* (Boiss.) Bornm. is one of the rare species and is found only in Manavgat of Antalya province in southwest Anatolia (Lamond, 1975). According to a recent taxonomic review of *Crepis*, this species belongs to the section *Intybellioides* Froel. (Enke, 2009).

C. palaestina was reported with a brief description from a single locality, which had no completely mature achenes, in the *Flora of Turkey and the East Aegean Islands* (Lamond, 1975). Lamond (*l.c.*) pointed out that the specimens of *C. palaestina* in Turkey differed from East Mediterranean specimens in having glandular-pubescent basal leaves. Hence, a detailed taxonomic treatment of this species was necessary.

During our field work for the taxonomic revision of *Crepis* in Turkey, we collected some intriguing specimens of *C. palaestina* from the Antalya Province. After studying the morphological characters, examining the specimens deposited in the herbaria ANK, BULU, EGE, GAZI, HUB, IST and VANF, and consulting relevant floras and literature (Post and Dinsmore, 1933; Babcock, 1947; Lamond, 1975; Mouterde, 1983), we concluded that the specimens represented an undescribed subspecies of *C. palaestina*.

Material and Methods

Plant material

The materials were collected in the field from native populations in the Antalya Province, Turkey. Vouchers were deposited in the herbarium at the Karadeniz Technical University, Department of the Biology (KTUB).

[1]Corresponding author. Email: inceer@ktu.edu.tr

Chromosome counts

Root tips obtained from the germinated achenes were pre-treated with 0.05% aqueous colchicine solution for 3–5 h at room temperature and then fixed in absolute ethanol-glacial acetic acid (3:1) for at least 24 h at 4°C (Inceer and Hayirlioglu-Ayaz, 2007). They were hydrolyzed in 1N HCl at 60°C for 12–15 min. Staining was carried out in 1% lacto-propionic orcein for 12–18 h at room temperature and squash preparations were made in 45% acetic acid (Inceer *et al.*, 2016). Five well-spread metaphase plates were used for chromosome counts.

Results and Discussion

Crepis palaestina subsp. **babcockii** Inceer & Aksu Kalmuk, **subsp. nov.** (**Fig. 1**).

Diagnosis: *Crepis palaestina* subsp. *babcockii* closely resembles *C. palaestina* subsp. *palaestina*, but differs in having glandular-pubescent basal and cauline leaves (not eglandular-pubescent), small ligule teeth (0.2–0.3 mm long, vs 0.4–1.0 mm), corolla tubes (3.25–3.5 mm long, vs. 5.0–6.5 mm), style branches (1.0–1.2 mm long, vs. 1.4–2.25 mm) and achene morphology (not biform).

Type: Turkey, C3 Antalya: Manavgat, 10 m, 24 April 2015, *Inceer* 1142 (*Holotype:* KTUB!; *Isotype*: ANK!).

Fig. 1. *Crepis palaestina* subsp. *babcockii*, **subsp. nov.** A. Habit; B. Branch with synflorescences; C. Cauline leaves; D. Capitulum from side; E. Capitulum from top side.

Caulescent annual, 65–120 cm long with short caudex. Stem erect, slender or robust, terete, striate, glandular pubescent above, ± pubescent near base. Basal leaves 15–23×2.0–3.5 cm, soon withering, lyrate-pinnatifid, terminal segment large, reniform, glandular-pubescent, lateral lobes triangular, glandular-pubescent. Cauline leaves numerous, 2.0–8.5×0.5–4.0 cm, lowest similar to the basal leaves, middle ones ovate-lanceolate, acute, sessile, auriculate, glandular-pubescent, upper ones ± bract-like. Peduncles 2.0–7.5 cm long, strict or arcuate, glandular-pubescent below, glabrous above, swollen near base of fruiting heads. Capitula 30–40 flowered. Involucre cylindrical, 11.5–15.0×8.0–9.5 mm, dark green, becoming straminous and indurate in fruit, glabrous or the bracts ± pubescent with pale glandless hairs. Outer bracts 6–8, minute, 2.5–4.0×0.5–1.5 mm, ovate, acute, pale-margined; inner bracts 10–13, 12–14×2.0–2.5 mm, lanceolate, acute, appressed-pubescent on inner face, sometimes pale-margined, becoming very prominently carinate dorsally and pale spongy-thickened confluent with base. Receptacle areolate, glabrous. Ligules yellow, 16–20×2.0–2.5 mm, teeth 0.2–0.3 mm, corolla tube 3.2–3.5×0.2–0.3 mm, densely pubescent. Anthers 5, coherent along most of their length and forming a tube 3.0–3.3 mm long around style, yellow, tinged green at summit; filament 0.7–0.8 mm long, appendages 0.4–0.5 mm long, lanceolate, acute. Style 7–8×0.1–0.2 mm, dark green below, yellow above, branches 1.0–1.2×0.05-0.1 mm, dark green, slightly expanded at tip. Achenes triform, straminous, 15–20 striate, outermost (marginal) achene 8.8–9.1×0.8-0.9 mm, ± obcompressed, and lateraly alate, narrowly summit, intermediate achene 8.0–8.4×0.5–0.6 mm, densely spiculate, gradually attenuate upward, with slightly expanded pappus disk, ± dilated at the hollow base, innermost achene 8.1–8.5×0.5–0.6 mm, striate, gradually attenuate upward, with slightly expanded pappus disk, ± dilated at the hollow base. Pappus white, 4.1–5.5 mm, multiseriate, fine, soft, flexuous, ± persistent, included in involucre.

Phenology: April to May.

Etymology: This new subspecies is named after Professor Ernest Brown Babcock, who contributed very much to the taxonomy and genetics of *Crepis*.

Distribution and habitat: *C. palaestina* subsp. *babcockii* is known only from type locality in southwest Anatolia. It grows in shady places and *Pinus brutia* (red pine) forest at an altitude of 10 m a.s.l.

Conservation status: CR: B1ab (i, ii, iii)+2ab (i, ii, iii). The population of *C. palaestina* subsp. *babcockii* in the type locality seems to be small and scattered. It should therefore be regarded as Critically Endangered CR (IUCN, 2014) because of its local distribution and small population size.

Additional specimen examined: Turkey, C3 Antalya, Manavgat, 10 m, 29 May 2014, *Inceer* 1086 (KTUB!).

Taxonomic and cytological notes: *C. palaestina* subsp. *babcockii* is also closely related to *C. pulchra* which is distributed in other regions of Antalya, and thus they are sympatric on this region. *C. palaestina* subsp. *babcockii* can be easily distinguished from *C. pulchra* by the shape of fruits (achene-cypsela). *C. palaestina* subsp. *babcockii* has lyrate basal leaves with a large terminal lobe, whereas *C. pulchra* has the basal leaves denticulate to runcinate-pinnatifid (Table 1).

The present study reveals that *C. palaestina* subsp. *babcockii* is a diploid taxon with $2n = 2x = 8$ chromosomes (Fig. 2). This taxon has the same chromosome number with the members of the section *Intybellioides* such as *C. palaestina* subsp. *palaestina*, *C. reuteriana*, *C. pulchra*, *C. stojanovii* and *C. pterothecoides* (Babcock, 1947).

Table 1. Comparison of the diagnostic characters of *Crepis palaestina* subsp. *babcockii* subsp. nov., *C. palaestina* subsp. *palaestina* and *C. pulchra*.

Characters	*C. palaestina* subsp. *babcockii* **subsp. nov.**	*C. palaestina* subsp. *palaestina*	*C. pulchra*
Basal leaves	Lyrate-pinnatifid, terminal segment large, reniform, glandular-pubescent	Oblanceolate, obtuse or subacute, lyrate-pinnatifid, terminal segment large, oblong-cordate to reniform, eglandular-pubescent	Oblanceolate or obovate, denticulate to runcinately dentate or pinnatifid, on both sides pubescent
Cauline leaves (middle ones)	Ovate-lanceolate, auriculate	Mostly lanceolate, runcinate-pinnatifid, broadly auriculate	Lanceolate, denticulate to subpinnatifid, sub-amplexicaul, on both sides pubescent, pale glandular hairs
Ligule teeth	0.2–0.3 mm long	0.4–1.0 mm long	0.1–0.2 mm long
Corolla tubes	3.25–3.5 mm long	5.0–6.5 mm long	4.0–4.5 mm long
Style branches	1.0–1.2 mm long	1.4–2.3 mm long	0.8–1.1 mm long
Achene forms	Triform	Biform	Biform
Outermost achenes	± Obcompressed, laterally alate, narrow summit	± Obcompressed, laterally broadly alate, narrow summit	± Obcompressed, ± attenuate, spiculate, slightly expanded pappus disk
Intermediate achenes	Densely spiculate, gradually attenuate upward, with slightly expanded pappus disk, ± dilated at the hollow base	Absent	Absent
Innermost achenes	Striate, gradually attenuate upward, ± dilated at the hollow base	Striate, gradually attenuate upward, conspicously dilated at the hollow base	Striate, ± attenuate

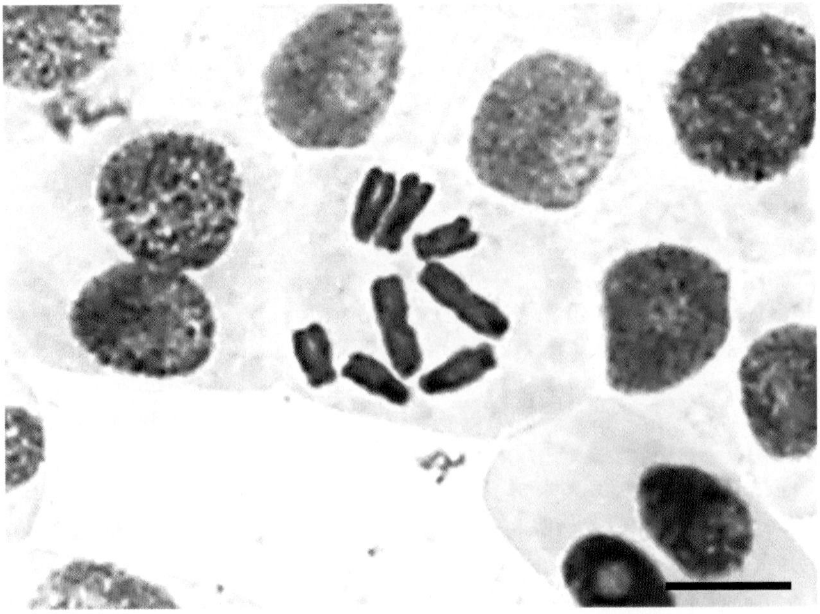

Fig. 2. Somatic metaphase chromosomes of *C. palaestina* subsp. *babcockii*. (Scale bar: 10 μm).

Acknowledgements

The authors thank the Scientific and Technological Research Council of Turkey (TUBITAK Project No. 112T132) for financial support.

References

Babcock, E.B. 1947. The Genus *Crepis*. Part Two: Systematic Treatment. University of California Press, Berkeley and Los Angeles.

Bremer, K. 1994. Asteraceae: Cladistics and Classification. Timber Press, Portland, Oregon.

Ekim, T. 2012. *Crepis*. *In*: Guner, A., Aslan, S., Ekim, T., Vural, M. and Babac, M.T. (Eds), Türkiye Bitkileri Listesi (Damarlı Bitkiler), Nezahat Gökyigit Botanik Bahçesi ve Flora Araştırmaları Derneği Yayını. Istanbul, pp. 150–154 (in Turkish).

Enke, N. 2009. Contributions towards a revised infrageneric classification of *Crepis* (Cichorieae, Compositae). Willdenowia **39**: 229–245.

Inceer, H. and Hayirlioglu-Ayaz, S. 2007. Chromosome numbers in the tribe Anthemideae (Asteraceae) from Turkey. Bot. J. Linn. Soc. **153**: 203–211.

Inceer, H., Aksu Kalmuk, N., Imamoglu, V.K., Duman, O., Hayirlioglu-Ayaz, S. and Arslan, G. 2016. Micromorphological, anatomical and cytogenetical studies in endemic *Crepis macropus* Boiss. & Heldr. (Asteraceae) from Turkey. Acta Bot. Croat. **75**(2): 173–178.

IUCN 2014. 2014 IUCN Red List of Threatened Species. <www.iucnredlist.org>. IUCN Red List Unit, Cambridge, U.K.

Lamond, J.M. 1975. *Crepis* L. *In*: Davis, P.H. (Ed.), Flora of Turkey and the East Aegean Islands. Vol. **5**. Edinburgh University Press. Edinburgh, pp. 814–841.

Mouterde, P. 1983. Nouvelle Flore du Liban et de la Syrie. Tome **III**. Dar El-Machreq Sarl., Beyrouth, Liban, pp. 536–544.

Post, G.E. and Dinsmore, J.E. 1933. Flora of Syria, Palaestine and Sinai: A Handbook of the Flowering Plants and Ferns, native and naturalized from the Taurus to Ras Muhammad and from the Mediterranean Sea to the Syrian Desert 2. American Press, Beirut, pp. 152–157.

DIDYMODON MONGOLICUS (BRYOPHYTA, POTTIACEAE), A NEW SPECIES FROM MONGOLIAN PLATEAU

DONG-PING ZHAO[1], TONG-RUI ZHANG, XUE-LIANG BAI AND DONG-MEI REN

Department of Biology, School of life Science, Inner Mongolia University, West College Road 235, Hohhot 010021, China

Keywords: China; *Didymodon*; Mongolia; Moss; Taxonomy.

Abstract

Didymodon mongolicus D. P. Zhao &T. R. Zhang is described as a new species from northern China and Mongolia. The new taxon is distinguished from others species of the genus by its small, ovate leaves, weakly recurved margins, smooth laminal cells and costa ending several cells below the apex, with 1 layer of ventral stereids in the lower part of the leaf. Drawings and light microscope photographs of the main characters are given, and its relationships with some closely related species within the genus are discussed.

Introduction

The genus *Didymodon* Hedw. (Pottiaceae) currently consists of 126 species with distribution nearly throughout the world (Zander, 1993; 2007). The genus concept of *Didymodon* has been controversial, especially the morphological delimitation of *Didymodon* versus *Barbula* Hedw. The taxonomic differences between the two genera are mainly on gametophytic characteristics: *Barbula* has entirely hyaline axillary hairs, while in *Didymodon* the hairs have one or two brown basal cells. The leaves of *Barbula* are ovate to long-elliptical, the basal cells are hyaline, elongate and well differentiated, while those of *Didymodon* are mostly lanceolate to long-lanceolate, the basal cells are usually green, shortly rectangular and little differentiated (Saito, 1975; Zander, 1993, 2007; Jiménez *et al.*, 2005; Werner *et al.*, 2005; Jiménez, 2006). Recently, Jiménez *et al.* (2012) described *Andina* J.A. Jiménez & M.J. Cano as a new genus, which contained several species earlier recognized within *Pseudocrossidium* R.S. Williams and *Didymodon*. *Andina* and *Didymodon* both have brown basal cells in axillary hairs; however, *Andina* can be easily distinguished from the latter by the cucullate leaf apices, strongly recurved to revolute leaves with margins infolded in the apex, ventral costal outgrowths differentiated as pads of bulging and papillose cells, filaments or lamellae, and upper laminal cells usually bulging on both surfaces.

Mongolian Plateau is divided politically and geographically by Mongolia in the north and the Inner Mongolia Autonomous Region of China in the south. *Didymodon* is the most species-rich genus in the Pottiaceae in this region and includes 30 species (Bai, 1997; Tsegmed, 2010; Zhao *et al.*, 2014, 2015). Over the last several years we have been studying *Didymodon* of the Mongolian Plateau (e.g., Zhao 2010; Zhao *et al.* 2013, 2014, 2015). In the course of this work we found that several specimens fit into the generic concept of *Didymodon* based on the presence of stem sclerodermis, axillary hairs with one brownish basal cells, costa with ventral and dorsal stereids, upper laminal cells rounded, subquadrate or oval, and basal cells little differentiated. Nevertheless, they were not referable to any known *Didymodon* species. After studying the types deposited in H and JE herbaria and after comparison between the specimens with published information (Saito, 1975; Zander, 1993; Zander and Ochyra, 2001; Li *et al.* 2001; Jiménez *et al.*, 2005; Jiménez,

[1] Corresponding author. Email: topalizdp@aliyun.com

2006; Zander, 2007; Kučera and Ignatov, 2015), we conclude that our samples belong to an undescribed species of *Didymodon*, which is described, illustrated and compared with the most closely related species in the study.

Didymodon mongolicus D. P. Zhao &T. R. Zhang, **sp. nov.** **(Figs 1 & 2).**

Diagnosis: *Didymodon mongolicus* is differentiated from other species in the genus by the following combination of characters: small sized, ovate leaves, weakly recurved margins up to 3/4 of the apex, smooth laminal cells, costa ending below the apex and cross-section of the costa with1 layer of ventral stereids in the lower part of the leaf.

Type: China. Inner Mongolia Autonomous Region, Helan Mountain National Nature Reserve, Xuelingzi valley, 38°39'54" N, 105°47'25" E,1900 m, on soil in *Juniperus communis* L. forests, 1 Aug. 2012, *X. L. Bai1459* (Holotype: HIMC; Isotypes: KUN).

Plants small, 0.5-0.9 cm high, green above, brownish below, in dense turfs. Stems erect, simple or branched, transverse section rounded, sclerodermis present, without hyalodermis, central strand differentiated. Rhizoidal tubers absent. Leaves monomorphic, appressed when dry, erect-patent to spreading when moist, ovate, not keeled, 0.5-1.0 × 0.3-0.5 mm; lamina unistratose, yellowish green with KOH; apex acuminate, not deciduous; margins entire, weakly recurved from base to 3/4 of the leaf, unistratose. Costa single, ending several cells below the apex; ventral cells of the costa, in the upper middle of the leaf, rectangular to subquadrate, dorsal cells of the costa, in the upper middle of the leaf, rounded or subquadrate; 5-7 rows of cells across ventrally at midleaf; costal transverse section elliptical at leaf base, with 4 guide cells in 1 layer, 1 layer of ventral stereids, 1-2 layers of dorsal stereids, without hydroids, ventral epidermis differentiated, not bulging, smooth, dorsal epidermis differentiated, papillose or smooth. Upper and middle laminal cells rounded, subquadrate or oval, 7.8-10.4 × 7.2-9.6 μm, smooth, lightly thick-walled; basal cells quadrate or shortly rectangular,10.4-15.6 × 10.4-13.0 μm, not differentiated, not hyaline, smooth, thick-walled, not pitted. Gemmae absent. Sexual condition unknown. Sporophyte unknown.

Etymology: Named after the type locality, Mongolian Plateau.

Habitat and Distribution: *Didymodon mongolicus* grows on soil and rocks in *Juniperus communis* forests. It forms small and mostly monospecific patches. Besides the type locality, we discovered four additional populations from Inner Mongolia and ones from Mongolia. The new species seems to be restricted to the Mongolian Plateau, but it is likely to be found elsewhere in this large plateau with progress in field studies.

Paratypes: China. Inner Mongolia Autonomous Region: Helan Mountain National Nature Reserve, Xuelingzi valley, 1 Aug. 2012, *X. L. Bai1452* (HIMC); Baotou City, Bayan-obo Mining District, 5 Aug. 2013, *D. M. Ren201308085* (HIMC); Ordos City, Jungar Banner, 1 July 2005, *X. L. Bais.n.* (HIMC); Mausoleum of Genghis Khan, 27 Oct. 2008, *X. L. Bai0810001* (HIMC); Xingan League, Bailang Town, 31 July 1988, *X. L. Bai1636* (HIMC); Mongolia. Bayankhongor Province: 13 June 1971, *N. S. Golubkova* and *U. Tsogt 5* (HIMC, UBA); Töv Province: 29 June 1974, *N. S. Golubkova* and *U. Tsogt 111* (HIMC, UBA); Khovd Province: 24 July 1984, *TS. Tsegmed 9045* (HIMC, UBA).

Note: *Didymodon mongolicus* is most closely related to *D. tectorum* (Müll. Hal.) K. Saito and *D. cordatus* Jur., sharing such charateristics as the shape of the leaves, yellowish green colour of the lamina with KOH reaction, rectangular to subquadrate ventral cells of costa above midleaf, and growing in dry soil and rock. In fact the small forms of *D. tectorum*, which lack gemmae cannot be easily distinguished from the new species. *Didymodon mongolicus* can be separated readily from *D. tectorum* by its leaf margins weakly recurved from base to 3/4 of the leaf (Fig. 2: A, F), cells smooth above midleaf (Fig. 2: C, D, H, I), and transverse section of costa with 1 layer of ventral stereids (Fig. 2: E, J). Although the type of *D. tectorum* does not contain gemmae,

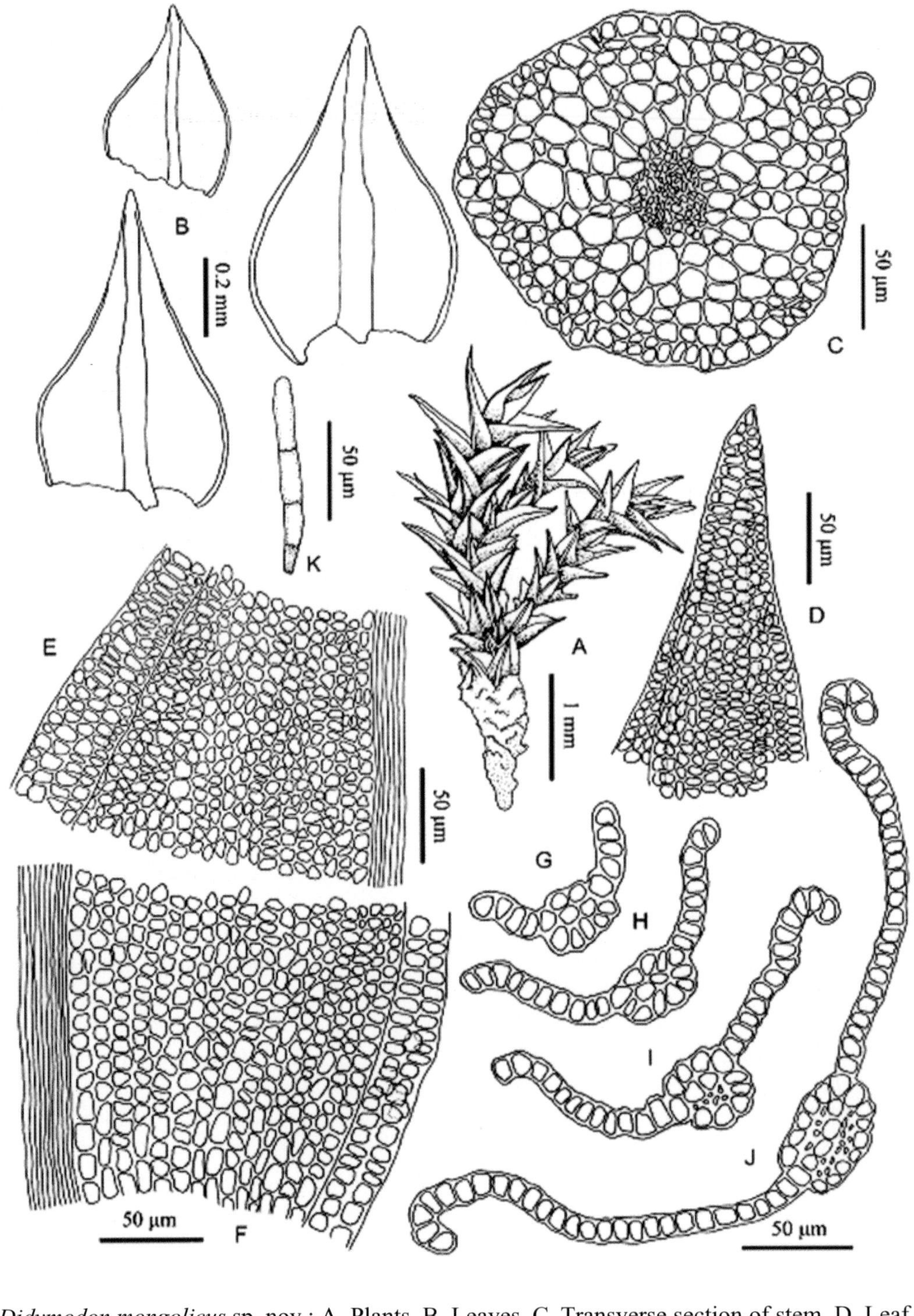

Fig. 1. *Didymodon mongolicus* sp. nov.: A. Plants, B. Leaves, C. Transverse section of stem, D. Leaf apex, E. Middle part of leaves,F. Basal part of leaves, G-J. Transverse section of leaves from apex to base, K. Axillary hair. (All drawn from the holotype).

Fig. 2. A-E. *Didymodon mongolicus* sp. nov. (holotype, *X. L. Bai 1459*, HIMC), F-J. *Didymodon tectorum* (isotype, *Rev. Jos. Giraldi 849*, H), K-O. *Didymodon cordatus* (isolectotype, *C. Schliephacke s.n.*, JE), A, F, K. Leaves. B, G, L. Leaf apex. C, H, M. Transverse section at upper leaves. D, I, N. Transverse section at midleaves. E, J, O. Transverse section near leaf bases.

numerous specimens of *D. tectorum* from China and Mongolia have copious gemmae. *D. cordatus* differs from *D. mongolicus* in having a stout costa that is shortly excurrent, ending in a wide mucro (Fig. 2: L) and with 2-5 layers of ventral steroids (Fig. 2: N, O), and margins revolute up to the apex.

Small forms of *Didymodon acutus* (Brid.) K. Saito may be confused with *D. mongolicus*, as both species have appressed leaves when dry, smooth laminal cells, slightly recurved margins, and absence of gemmae in the axils of the leaves, but *D. mongolicus* differs in its smaller size, more ovate leaf shape and never having lanceolate leaves, and the costa ending before the apex. *D. validus* Limpr., a species known from Central Asia and Europe (Jiménez, 2006), resembles *D. mongolicus* in the smooth laminal cells. It differs from the new species by its recurved or incurved leaves when dry, shape of leaves ovate-lanceolate or linear-lanceolate, costa long-excurrent and transverse section of the costa with 1-3 layers of ventral stereids while *D. mongolicus* has appressed leaves when dry, ovate leaves, costa ending several cells below the apex and transverse section of the costa with 1 layer of ventral stereids.

Acknowledgements

The first author thanks Dr. TS. Tsegmed for her assistance during our field work in Mongolia. We are grateful to the curators of herbaria H, JE and UBA for the loan specimens. This research was conducted with financial support from the National Natural Science Foundation of China (nos. 31260046, 31360142) and Open Major Basic Research Projects in Inner Mongolia (nos. 201503001).

References

Bai, X.L. 1997. Flora Bryophytarum Intramongolicarum. Inner Mongolia University Press, Hohhot, pp. 1-541.

Jiménez, J.A. 2006. Taxonomic revision of the genus *Didymodon* Hedw, (Pottiaceae, Bryophyta) in Europe, North Africa and Southwest and Central Asia. J. Hattori Bot. Lab. **100**: 211-292.

Jiménez, J.A., Cano, M.J., and Jiménez, J.F. 2012. Taxonomy and phylogeny of *Andina* (Pottiaceae, Bryophyta): a new moss genus from the tropical Andes. Syst. Bot. **37**(2): 293-306.

Jiménez, J.A., Ros, R.M., Cano, M.J. and Guerra, J. 2005. A Revision of *Didymodon* Section *Fallaces* (Musci, Pottiaceae) in Europe, North Africa, Macaronesia, and Southwest and Central Asia. Ann. Missouri Bot. Gard. **92**: 225-247.

Kučera, J. and Ignatov, M.S. 2015. Revision of phylogenetic relationships of *Didymodon* sect. *Rufiduli* (Pottiaceae, Musci). Arctoa **24**: 79-97.

Li, X.J., He, S. and Iwatsuki, Z. 2001. Pottiaceae. *In*: Li, X. J. and Crosby, M. R. (eds), Moss flora of China. Vol. **2**. Science Press, Beijing, Missouri Botanical Garden Press, St. Louis, pp. 114-249.

Saito, K. 1975. A monograph of Japanese Pottiaceae (Musci). J. Hattori Bot. Lab. **39**: 373-537.

Tsegmed, T.S. 2010. Moss flora of Mongolia. Izdatel'stvo Sel'khozakademii. Moscow, pp. 1-634.

Werner, O., Jiménez, J.A., Ros, R.M., Cano, M.J. and Guerra, J. 2005. Preliminary investigation of the systematics of *Didymodon* (Pottiaceae, Musci) based on nrITS sequence data. Syst. Bot. **30** (3): 461-470.

Zander, R.H. 1993. Genera of the Pottiaceae: mosses of harsh environments. Bull. Buffalo Soc. Nat. Sci. **32**: 1-378.

Zander, R.H. 2007. Pottiaceae. *In*: Flora of North America Editorial Committee (Ed.) Flora of North America North of Mexico. Vol. **27**. Oxford University Press, New York, pp. 476-642.

Zander, R.H. and Ochyra, R. 2001. *Didymodon tectorum* and *Didymodon brachyphyllus* (Musci, Pottiaceae) in North America. Bryologist **104**: 372-377.

Zhao, D.P. 2010. Pottiaceae. *In*: Bai, X. L. (Ed.), Bryophyte Flora of Helan Mountain. Ningxia People's Press, Yinchuan, pp. 61-128.

Zhao, D.P., Bai, X.L., Wang J.N. and Liu, Y. 2013. *Didymodon cordatus* Jur. (Pottiaceae), new to the moss flora of China. Bangladesh J. Plant Taxon. **20**(2): 259-261.

Zhao, D.P., Tsegmed, T.S. and Bai, X.L. 2015. *Didymodon rigidulus* var. *subulatus* (Thér. & Bartram ex E.B. Bartram) R.H. Zander, new to the moss flora of Mongolia and Asia. Bangladesh J. Plant Taxon. **22**(1): 63-66.

Zhao, D.P., Wang, J.N. and Zhao, X.D. 2014. *Didymodon baii* (Pottiaceae), a new moss species from China. Ann. Bot. Fennici **51**: 185-188.

26

A NEW SUBSPECIES OF *DIMERIA HOHENACKERI* HOCHST. *EX* MIQ. (POACEAE) FROM INDIA

M.S. KIRAN RAJ, M. SIVADASAN[1]*, P. DILEEP[2] AND A.H. ALFARHAN[1]

Department of Botany, Sree Narayana College, Cherthala, S. N. Puram P.O., Alappuzha–688 582, Kerala, India

Keywords: Dimeria hohenackeri subsp. *kodaguensis*; Endemic taxon; Peninsular India; Poaceae.

Abstract

A new subspecies, *viz. Dimeria hohenackeri* Hochst. *ex* Miq. subsp. *kodaguensis* Kiran Raj, Sivad. & Dileep from South Karnataka of the Southern Western Ghats in Peninsular India is described and illustrated. It mainly differs from *D. hohenackeri* Hochst. *ex* Miq. subsp. *hohenackeri* in having mat-forming habit with leaves crowded at culm-base, spikelets 3–4 mm long and glumes not widely divergent during anthesis.

Introduction

The palaeotropical genus *Dimeria* R. Br. with about 65 species is mainly adapted to humid or semi-arid habitat and distributed from Indian subcontinent to Malesiana, northern Australia and Madagascar (Bor, 1960; Clayton *et al*., 2006; Kiran Raj and Sivadasan, 2008; Kiran Raj *et al*., 2013, 2015a; Teerawatananon *et al.,* 2014). The genus is most peculiar in the tribe Andropogoneae by its solitary, laterally compressed spikelets and flat or filiform rachis without joints. Majority of the taxa (ca. 34 spp., 3 subspp. and one variety) are confined to Peninsular India with 26 endemics (Kiran Raj *et al*., 2015b). Moreover, the monotypic genus *Nanooravia* Kiran Raj & Sivad. belonging to the bitypic subtribe Dimeriinae, is also endemic to this region, indicating it to be the centre of diversity of the subtribe (Kiran Raj, 2008; Kiran Raj *et al*., 2015b).The infra-tribal classification of Andropogoneae and especially the systematic position of Dimeriinae are enigmatic and still controversial (Clayton and Renvoize, 1986; Simon, 2007) due to the lack of sufficient data from the representative taxa.

Interesting specimens of *Dimeria* were collected during botanical explorations in Southern Western Ghats in Karnataka, India, and detailed study showed them to be representing new taxon related to *Dimeria hohenackeri*. It is described herein as a new subspecies of *Dimeria hohenackeri* Hochst. *ex* Miq. The species belongs to *Dimeria* sect. *Capillares*, with distribution extending from lowland grassy slopes of Western Ghats in Maharashtra through Karnataka to Kerala in Peninsular India. *Dimeria* sect. *Capillares* is distinguished by capillary raceme-rachis with distantly arranged spikelets, and is endemic to Indian subcontinent (Kiran Raj *et al*., 2015b). Detailed description of the new subspecies with other relevant information including illustration is provided for easy identification of taxon in field. A key to the subspecies of *D. hohenackeri* in India is also provided.

---

*Corresponding author. Email: drmsivadasan@gmail.com

[1]Department of Botany & Microbiology, College of Science, King Saud University, P.O. Box 2455, Riyadh–11451, Kingdom of Saudi Arabia.

[2]Department of Botany, Regional Institute of Education (RIE), Mysore-570 006, Karnataka, India.

Dimeria hohenackeri Hochst. *ex* Miq. subsp. **kodaguensis** Kiran Raj, Sivad. & Dileep, **subsp. nov.** **(Fig. 1)**.

Diagnosis: *Dimeria hohenackeri* subsp. *kodaguensis* closely resembles subsp. *hohenackeri*, but differs in having mat-forming culms and branching at the extreme lower nodes, leaves crowded at culm-base, racemes 2–3 in number, spikelets 3–4 mm long and glumes not widely divergent during anthesis.

Type: India, Karnataka: Kodagu Dist., Chettalli, on the way to Madikeri from Siddapura, 3 Dec 2002, *Kiran Raj CU 92982* (*Holotype*: CALI!; *Isotype* : KFRI!).

Paratypes: India, Karnataka: Shimoga Dist., Jersopa, + 250 m, 28 Dec 2003, *Kiran Raj CU 92880* (CALI!); Shimoga Dist., Jersopa, 13 Dec 2013, *P. Dileep* 9311 (MH!).

Annuals. Culms procumbent, 12–25 cm long, crowded at base and mat-forming, nodes bearded at upper half of culm. Leaves confined to base of culm; sheath much shorter than internodes, keeled in the upper half, rounded below, striate, smooth, shining and glabrous in the lower two thirds, pilose in the upper third with bulbous-based hairs; ligules membranous, up to c. 1 mm long, truncate and fimbriate at apex; leaf blade linear-acuminate, 3–5 × 0.2–0.4 cm, abaxially keeled on the midrib with the keel continuous with that of the sheath, acuminate at apex, rounded at base, coarsely scabrid on the nerves on both surfaces and margins, bulbous-based hairs on upper surface and margins. Racemes 2–3, sub-digitate, 4–6 cm long, peduncle long exserted from spatheole; rachis terete or angled, c. 0.2 mm wide, glabrous, tough, with alternately arranged pedicelled spikelets; pedicels c. 1 mm long, lip cupuliform, terete below, glabrous; each raceme with 15–20 spikelets. Spikelets linear-oblong, 3–4 mm long; callus very short, c. 0.3 mm long, densely bearded, often spreading at right angles to the rachis; lower glume coriaceous, oblong-acuminate, 3.0–3.5 mm long, abaxially rounded and keeled towards apex, not winged, sparsely hairy along the keel towards apex, margins hyaline; upper glume coriaceous, linear-acuminate, 3.5–4.0 mm long and 1.0–1.5 mm wide, strongly compressed, straight on back, keeled at apical one-third, wingless, with few long hairs along the keel at apex; glumes not widely divergent during anthesis. Lower floret empty; lower lemma very hyaline, oblanceolate, c. 1.5 mm long, 1-nerved, margin ciliate towards apex. Upper floret bisexual; upper lemma elliptic, 2.0–2.5 mm long, bifid at apex with acute lobes, sub-hyaline, awned from the sinus, awn up to 12 mm long with a dark twisted c. 3.5 mm long column; palea lanceolate, c. 0.5 mm long, hyaline; lodicules 2, small, truncate, apically toothed. Stamens 2; anthers 1.2–1.5 mm long, yellowish brown. Ovary oblong; styles 2; stigmas plumose. Grain 1.5–1.8 mm long, oblong-elliptic, slightly compressed, hilum basal, linear-punctiform; embryo about one-third the length of grain, without epiblast and with a scutellate tail.

Flowering and Fruiting: October–December.

Habitat: Grassy slopes along the forest margins, alt. 250–300m.

Distribution and Conservation status: India, Southern Western Ghats in Karnataka. Known only from the type locality. In the absence of data on distribution and abundance, it is currently categorized as belonging to Data Deficient (DD) (IUCN, 2012; IUCN, 2014).

Etymology: The infraspecific epithet is indicative of name of the type locality – Kodagu, in Karnataka State, India.

Notes: *Dimeria hohenackeri* is quite distinct from all other species of the genus by its characteristic 3–5 slender racemes, capillary rachis, distantly placed, stalked spikelets and slender awns. *Dimeria hohenackeri* subsp. *kodaguensis* mainly differs from *D. hohenackeri* subsp. *hohenackeri* in having mat-forming culms with leaves crowded at culm-base and racemes 2–3 in number. Distinguishing morphological features of the two subspecies are provided in Table 1, and a key to the subspecies of *Dimeria hohenackeri* is provided below.

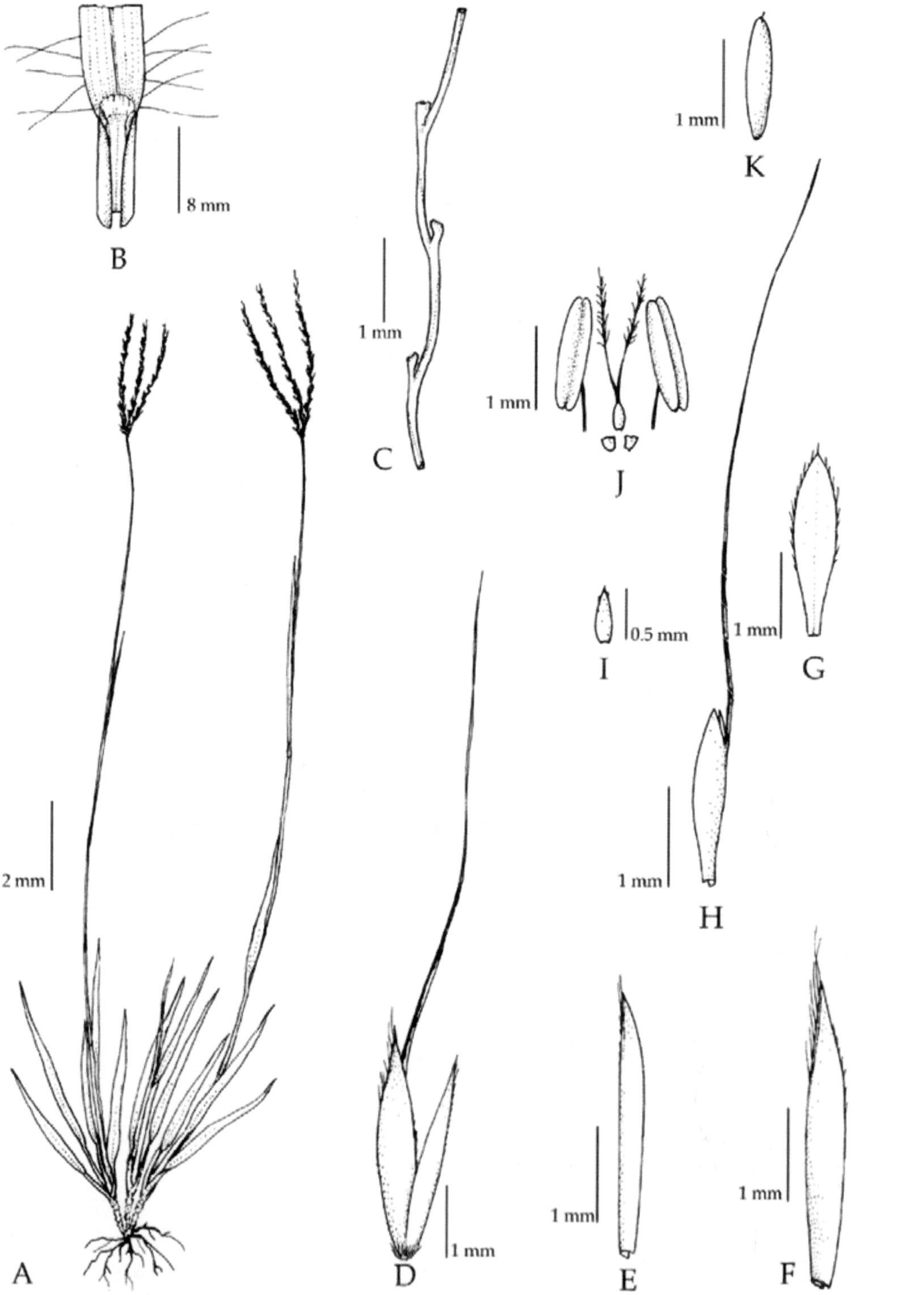

Fig. 1. *Dimeria hohenackeri* Hochst. *ex* Miq. subsp. *kodaguensis* Kiran Raj, Sivad. *et* Dileep, **subsp. nov.**: A. Habit with inflorescence; B. Junction of leaf-sheath and lamina showing ligule; C. A portion of rachis with pedicels; D. Spikelet; E. Lower glume; F. Upper glume; G. Lower lemma; H. Upper lemma with awn; I. Palea; J. Lodicules, Stamens & Pistil; K. Grain (Drawings by Kiran Raj from holotype).

Table 1. Distinguishing morphological features of the subspecies of *Dimeria hohenackeri.*

Characters	*D. hohenackeri* subsp. *hohenackeri*	*D. hohenackeri* subsp. *kodaguensis*
Habit	Culm up to 60 cm long, slender, erect, not tufted.	Culm up to 25 cm long, crowded, procumbent, mat-forming.
Leaves	Scattered all along the culm	Crowded at base of the culm
Pedicel	c. 1.5 mm long	c. 1 mm long
Glumes	Widely divergent during anthesis, lower glume 4.0–4.5 mm long	Not widely divergent during anthesis, lower glume 3.0–3.5 mm long

Key to the subspecies of *Dimeria hohenackeri*

1. Culm 35–60 cm long, slender, erect; leaves scattered along the culm; racemes 3–5; spikelets 30–40 in each raceme; glumes widely divergent during anthesis, lower glume 4.0–4.5 mm long. ***D. hohenackeri*** subsp. ***hohenackeri***

– Culm 12–25 cm long, profusely crowded, procumbent; leaves crowded at base of culm; racemes 2–3; spikelets 15–20 in each raceme; glumes not widely divergent during anthesis, lower glume 3.0–3.5 mm long. ***D. hohenackeri*** subsp. ***kodaguensis***

Acknowledgements

Authors are grateful to Dr. J. F. Veldkamp, Naturalis Biodiversity Center, Leiden, the Netherlands for providing copies of relevant literature. The first author is indebted to the University Grants Commission (UGC), New Delhi for granting a Minor Research Project in 2013 and to Dr. K. Anirudhan, Principal, Sree Narayana College, Cherthala for constant encouragement. The second and fourth authors gratefully acknowledge encouragement and support by the Deanship of Scientific Research, King Saud University, through the research group project No. RGP-135. The third author is thankful to the Head, Department of Studies in Botany, University of Mysore, India and also to the Principal, Regional Institute of Education, Mysore, Karnataka for providing necessary facilities.

References

Bor, N.L. 1960.The grasses of Burma, Ceylon, India and Pakistan. Pergamon Press, London, 137 pp.

Clayton, W.D. and Renvoize, S.A. 1986. Genera Graminium: Grasses of the World. Kew Bull. Additional Ser. **13**: 1–389.

Clayton, W.D., Vorontsova, M.S., Harman, K.T. and Williamson, H. 2006 (onwards). Grass Base- The Online World Grass Flora. http://www.kew.org/data/grasses-db.html. Accessed on 23August 2015.

IUCN.2012. IUCN Red List Categories and Criteria: Version 3.1. Second edition. IUCN, Gland, Switzerland and Cambridge, UK, iv+32 pp.

IUCN. 2014. IUCN Standards and Petitions Subcommittee (2014). Guidelines for using the IUCN Red List Categories and Criteria. Version 11. Prepared by the Standards and Petitions Subcommittee, 87 pp. Available from:http://www.iucnredlist.org/documents/RedListGuidelines/ (accessed 7 September 2015)

Kiran Raj, M.S. 2008.Taxonomic revision of the subtribe Dimeriinae Hack. of Andropogoneae (Poaceae-Panicoideae) in Peninsular India. Ph.D. Thesis (Unpublished). University of Calicut, India, 409 pp.

Kiran Raj, M.S. and Sivadasan, M. 2008. A new species of *Dimeria* R. Br. (Poaceae-Panicoideae-Andropogoneae) from Goa, India. Novon **18**(2): 183–186.

Kiran Raj, M.S., Sivadasan, M., Alfarhan, A.H. and Veldkamp, J. F. 2015a. *Dimeria raviana* (Poaceae-Panicoideae), a new species from South Western Ghats, India. Phytotaxa **195**: 193–196.

Kiran Raj, M.S., Sivadasan, M., Veldkamp, J.F., Alfarhan, A.H. and Amal Tamimi, A.S.M. 2015b. A revised infrageneric classification of *Dimeria* R. Br. (Poaceae-Andropogoneae). Bangladesh J. Plant Taxon.**22**(1): 47–54.

Kiran Raj, M.S., Sivadasan, M., Veldkamp, J.F., Alfarhan, A.H. and Thomas, J. 2013.*Nanooravia gen. nov.,* subtribe Dimeriinae (Poaceae-Panicoideae-Andropogoneae) from India. Nord. J. Bot. **31**: 161–165.

Simon, B.K. 2007. Grass Phylogeny and Classification: Conflict of Morphology and Molecules. Aliso **23**(1): 259–266.

Teerawatananon, A., Boontia, V., Chantarasuwan, B., Hodkinson, T.R. and Sungkaew, S. 2014. A taxonomic revision of the genus *Dimeria* (Poaceae: Panicoidea) in Thailand. Phytotaxa **186**: 137–147.

27

NEW RECORDS OF THREE SPECIES AND A GENUS OF THE EUPHORBIACEAE FOR BANGLADESH

M. Nazim Uddin[1], M. Oliur Rahman[2] and M. Atiqur Rahman

Department of Botany, University of Chittagong, Chittagong 4331, Bangladesh

Keywords: *Cleidiocarpon laurinum* Airy Shaw; *Euphorbia repens* K. Koch; *Euphorbia pycnostegia* Boiss.; First records; Euphorbiaceae; Bangladesh.

Abstract

Three species belonging to two genera of the family Euphorbiaceae are being reported here as new records for Bangladesh, viz., *Cleidiocarpon laurinum* Airy Shaw, *Euphorbia pycnostegia* Boiss. and *Euphorbia repens* K. Koch. The genus *Cleidiocarpon* Airy Shaw is also a new generic record for Bangladesh. One cultivated species, *Acalypha wilkesiana* Muell.-Arg., is reported as addition to the account of the Euphorbiaceae of Bangladesh. Description and photographs of these species are provided.

Introduction

The Euphorbiaceae, one of the largest, most complex and diverse families of angiosperms, is consisted of 334 genera (Webster, 1994) and over 8,000 species (Radcliffe-Smith, 2001). Wurdack *et al.* (2004) opined that as a pantropical family Euphorbiaceae (*s.l.*) is composed of 340 genera and approximately 8,000-9,000 species. The family is known from its latest treatment in the Encyclopedia of Flora and Fauna of Bangladesh to be represented in the flora of Bangladesh by 47 genera and 141 species (Ahmed *et al.*, 2008).While examining the collected specimens of the family Euphorbiaceae from different districts of the country and specimens preserved at DACB, DUSH, BFRIH, BCSIRH and HCU we came across some unidentified specimens, and after critical examination they have been identified as *Cleidiocarpon laurinum* Airy Shaw, collected from Chittagong, Cox's Bazar and Rangamati; *Euphorbia pycnostegia* Boiss., collected from Khulna, Pabna and Satkhira; and *Euphorbia repens* K. Koch, collected from Barisal and Khulna districts. These species were neither previously recorded from any part of Bangladesh, nor appeared in any previous relevant literature (Roxburgh, 1814; Hooker, 1886-1888; Prain, 1903; Heinig, 1925; Raizada, 1941; Datta and Mitra, 1953; Sinclair, 1956; Khan and Afza, 1968; Khan and Banu, 1972; Khan, 1972-1987; Alam, 1988; Khan and Rahman, 1989-2002; Khan *et al.*, 1994; Rahman and Hassan, 1995, 2017; Rahman and Uddin, 1997; Uddin and Rahman, 1999; Rashid *et al.*, 2000; Khan and Huq, 2001; Rahman *et al.*, 2001, 2012; Rahman and Khan, 2008; Tutul *et al.*, 2010; Uddin and Hassan, 2010; Arefin *et al.*, 2011; Rahman *et al.*, 2012, 2013; Rahman and Alam, 2013; Rashid and Chowdhury, 2013; Rahman *et al.*, 2015; Arefin *et al.*, 2017). Hence, these species are reported here as new records for Bangladesh.

Materials and Methods

We collected some interesting specimens belonging to the family Euphorbiaceae from Barisal, Chittagong, Cox's Bazar and Rangamati districts. All collected specimens and some other unidentified specimens of the family Euphorbiaceae preserved at Bangladesh National Herbarium (DACB), Dhaka University Salarkhan Herbarium (DUSH), Herbaria of Chittagong University

---

[1]Department of Botany, Rangamati Government College, Rangamati, Bangladesh.

[2]Department of Botany, University of Dhaka, Dhaka 1000, Bangladesh. Corresponding author. Email: prof.oliurrahman@gmail.com

(HCU), Bangladesh Forest Research Institute (BFRIH), Bangladesh Council for Scientific and Industrial Research (BCSIRH) were critically studied. Examination, identification and description of the specimens were made following standard taxonomic methods and consulting relevant literature (Hooker, 1886-1888; Prain, 1903; Hara *et al.*, 1982; Haridasan and Rao, 1987; Sharma *et al.*, 1993), which resulted in three new angiosperm records for the country. Each of these new records is cited with relevant taxonomic data on current name with synonyms, description, ecology, occurrence in Bangladesh, global distribution and specimens examined. Photograph of each newly recorded species is provided based on herbarium specimen.

Results and Discussion

The study resulted in recognition of occurrence of one more genus, *Cleidiocarpon* Airy Shaw with one species *C. laurinum* Airy Shaw, and two more species belong to the genus *Euphorbia* L., *viz.*, *E. repens* K. Koch and *E. pycnostegia* Boiss. in the flora of Bangladesh. The occurrence of these species from elsewhere in Bangladesh were not reported in any previous taxonomic treatment. The genus *Euphorbia* L. with 20 species were previously recorded from Bangladesh (Ahmed *et al.*, 2008). However, *Cleidiocarpon laurinum* Airy Shaw, *Euphorbia repens* K. Koch and *E. pycnostegia* Boiss. are reported here as addition to the account of the Euphorbiaceae of Bangladesh.

Cleidiocarpon laurinum Airy Shaw in Kew Bull. 19(2): 313 (1965). *Cleidion bishnui* Chakrab. & M. Gangop., J. Econ. Taxon. Bot. 12(2): 473 (1988 publ.1989). **(Fig. 1)**.

A large shrub or small tree. Leaves alternate, lanceolate, acuminate at apex, cuneate at base, margin entire, coriaceous, discoid gland 4-6, near base of lamina; petiole swollen, geniculate.

Fig. 1. *Cleidiocarpon laurinum* Airy Shaw

Flowers small, monoecious or dioecious. Male flowers: in long axillary racemes; calyx globose, sepals 3-4, valvate, refluxed, glabrous; stamens numerous; anthers 2-celled. Female flowers: solitary, axillary; sepals 3-5, imbricate, thick, ovate; ovary 1-celled; styles 2, very long, each again divided into 2 filiform arms. Fruits capsule, deeply 2-lobed.

Flowering and fruiting: December to April.

Ecology: Rain forests and forest margins.

Occurrence in Bangladesh: Chittagong, Cox's Bazar and Rangamati districts.

Global distribution: Myanmar and Bangladesh.

Specimens examined: Chittagong: Dhopachari, Chamachari, 16.9.1998, Rahman *et al.* 3393 (HCU); Chandanaish, Dhopachari, 14.6.2013, Uddin *et al.*10336 (HCU). Cox's Bazar: Panerchara, Tulabagan, 15.10.1998, Rahman *et al.* 3828 (HCU). Rangamati: Kaptai, Rampahar, 5.11.1999, Rahman 5945 (HCU).

Note: The genus *Cleidiocarpon* Airy Shaw was not known to be represented in the flora of Bangladesh and this is the new generic record for the country.

Euphorbia pycnostegia Boiss., Cent. Euphorb.: 9 (1860). *E. pycnostegia* Boiss. var. *zornioides* (Boiss.) Santapau, Bull. Bot. Soc. Bengal 8: 11 (1955). *Chamaesyce zornioides* (Boiss.) Soják, Cas. Nár. Mus., Odd. Prír. 140: 170 (1972). (**Fig. 2**).

An annual, erect, glabrous herb. Leaves opposite, leaf blade cordately linear-oblong, obtuse or mucronate, serrulate. Flowers distichously imbracting, broadly ovate, cordate. Involucres sub-solitary, glabrous, lobes fimbriate, limbs of glands large, entire, rosy, cocci, obtusely angled, glabrous. Seeds smooth or papillose.

Fig. 2. Habit of *Euphorbia pycnostegia* Boiss.

Flowering and fruiting: January to April.

Ecology: Sandy places.

Occurrence in Bangladesh: Khulna, Pabna and Satkhira districts.

Global distribution: India and Bangladesh.

Status of occurrence: Critically Endangered (CR).

Threat to the species: Habitat destruction.

Specimens examined: Khulna: Khulna, 4.1.1966, Tauhid 18 (DACB). Pabna: Pabna, 2.1.1966, Ali 27 (DACB). Satkhira: Satkhira, 27.3.1966, Salahuddin 81 (DACB).

Euphorbia repens K. Koch, Linnaea 21: 728 (1849). *Tithymalus repens* (K. Koch) Klotzsch & Garcke, Abh. Königl. Akad. Wiss. Berlin 1859: 96 (1860). **(Fig. 3)**.

A small, prostrate annual herb with branching stems. Leaves opposite, small, ovate, margin entire. Flower axillary, single on long stalks. Cyathia single, axillary, lobes of the involucre white and red, campanulate to turbinate, shortly pilose outside, marginal lobes 5, ovate, glands 4. Male flowers: few, slightly exserted from involucre. Female flowers: pedicel long; ovary shortly pubescent; style free; stigma 2-lobed. Fruits capsule, ovoid, smooth, shortly pubescent. Seeds ovoid.

Fig. 3. Habit of *Euphorbia repens* K. Koch

Flowering and fruiting: June to September.

Ecology: Mainly a ruderal weed, preferring rough open ground.

Occurrence in Bangladesh: Barisal and Khulna districts.

Global distribution: Africa and Bangladesh.

Status of occurrence: Critically Endangered (CR).

Threat to the species: Habitat destruction.

Specimens examined: Barisal: Bogra, Shahid Zia Smreeti road, Barisal City Corporation area, 18.9.2014, Rahman & Uddin 11207 (HCU). Khulna: Near Newsprint Mill area, 29.6.1973, Huq 1034 (DACB).

Note: It is reported here as a new record. The first collection of this species was made by A.M. Huq from Khulna in 1973 which is available in DACB. The last collection was made by Rahman and Uddin from Barisal City Corporation area in 2014 which is accessible in HCU.

In addition to above-mentioned three species reported here as new records for Bangladesh, one cultivated species, *Acalypha wilkesiana* Muell.-Arg., is also recorded as a new addition to the account of the Euphorbiaceae of Bangladesh.

Acalypha wilkesiana Muell.-Arg., Prodr. 15(2): 817 (1866).

A densely branched shrub, monoecious. Leaf blades elliptic-ovate to broadly ovate, obtusely acuminate at apex, cuneate or rounded at base, margin crenate-dentate, many nerved from the base, main veins on both surfaces at first glabrous, later glabrescent, green, coppery or bronze coloured and variously variegated with purple, red, pink, cream-coloured or white or else pink-cream or white margined, lateral nerves in pairs; stipules narrowly lanceolate, acute to acuminate. Inflorescence axillary, usually solitary, spicate, unisexual. Male flowers: sessile or so, buds tetragonal, reddish; anthers yellowish. Female flowers: sessile; sepals ovate; ovary sub-globose, tomentose; style united at the base, deeply laciniate, red. Fruits 3-lobed, pubescent.

Flowering and fruiting: December to April.

Ecology: Cultivated in gardens and along road sides.

Occurrence in Bangladesh: Barisal, Chittagong and Dhaka districts.

Global distribution: Worldwide in (sub-) tropical climates where it is often introduced as an ornamental.

Status of occurrence: Cultivated.

Medicinal uses: In Britain, leaves are used to treat diarrhoea. Likewise, fresh or heated leaves are externally applied to relieve rheumatic pain and inflammation.

Specimens examined: Barisal: Bogra, Shahid Zia Smreeti road, Barisal City Corporation area, 18.9.2014, Rahman & Uddin 11208 (HCU). Chittagong: Mirsarai forest range, 18.11.1996, Islam 5898 (HCU). Dhaka: Balda Garden, 23.1.1980, Begum 143 (DACB).

Acknowledgements

The authors are grateful to the authorities of DACB, DUSH, HCU, BFRIH and BCSIRH for giving permission to study their preserved specimens of the Euphorbiaceae.

References

Ahmed, Z.U., Hassan, M.A., Begum, Z.N.T., Khandker, M., Kabir, S.M.H., Ahmed, M., Ahmed, A.T.A., Rahman, A.K.T. and Haque, E.U. (Eds). 2008. Encyclopedia of Flora and Fauna of Bangladesh. Vol. **7**. Asiatic Society of Bangladesh, Dhaka, pp. 376–488.

Alam, M.K. 1988. Annotated Checklist of the Woody Flora of Sylhet Forest. Bulletin **5**. Plant Taxonomy Series, Bangladesh Forest Research Institute, Chittagong, pp. 35–44.

Arefin, M.K., Rahman, M.M., Uddin, M.Z. and Hassan, M.A. 2011. Angiospermic flora of Satchari National Park, Habiganj, Bangladesh. Bangladesh J. Plant Taxon. **18**(2): 117–140.

Arefin, M.S., Hossain, M.K. and Hossain, M.A. 2017. Plant diversity of Sonadia island – An ecologically critical area of south-east Bangladesh. Bangladesh J. Plant Taxon. **24**(1): 107–116.

Datta, R.M. and Mitra, J.N. 1953. Common plants in and around Dacca. Bull. Bot. Soc. Bengal. **7**(1&2): 25–30.

Hara, H., Chater, A.O. and Williams, L.H.J. 1982. An Enumeration of the Flowering Plants of Nepal. British Museum (Natural History), London **3**: 139–199.

Haridasan, K. and Rao, R.R. 1987. Forest flora of Meghalaya. Bishen Singh Mahendra Pal Singh, Dehra Dun, India, **2**: 479–480.

Heinig, R.L. 1925. List of Plants of Chittagong Collectorate and Hill Tracts. The Bengal Government Press, Darjeeling, India, pp. 55–59.

Hooker, J.D. 1886–1888. Flora of British India, Vol. **5**. L. Reeve and Co. Ltd., Kent, England, pp. 239–477.

Khan, M.S. (Ed.). 1972-1987. Flora of Bangladesh. Fasc. **1**–**59**. Bangladesh National Herbarium, Bangladesh Agricultural Research Council, Dhaka.

Khan, M.S. and Afza, S.K. 1968. A taxonomic report on the angiospermic flora of Teknaf and St. Martin's Island. Dhaka Univ. Studies, Part B. **16**: 35–37.

Khan, M.S. and Banu, F. 1972. A taxonomic report on the angiospermic flora of Chittagong Hill Tracts-2 (Dicotyledons). J. Asiat. Soc. Bang. **17**(2): 59–88.

Khan, M.S. and Huq, A.M. 2001. The vascular flora of Chunati Wildlife Sanctuary in south Chittagong, Bangladesh. Bangladesh J. Plant Taxon. **8**(1): 47–64.

Khan, M.S. and Rahman, M.M. (Eds). 1989-2002. Flora of Bangladesh. Fasc. **40**–**53**. Bangladesh National Herbarium, Dhaka.

Khan, M.S., Rahman, M.M., Huq, A.M., Mia, M.M.K. and Hassan, M.A.1994. Assessment of biodiversity of Teknaf Game Reserve in Bangladesh focusing on economically and ecologically important plant species. Bangladesh J. Plant Taxon. **2**(1&2): 47–79.

Prain, D. 1903. Bengal Plants, Vol. **2**. Bishen Singh Mahendra Pal Singh, Dhera Dun, India, pp. 694–716.

Radcliffe-Smith, A. 2001. Genera Euphorbiacearum. Royal Botanic Gardens, Kew, London, 464 pp.

Rahman, M.A. and Uddin, S.B. 1997. Assessment of plant diversity of Sitakunda in Chittagong. Bangladesh J. Plant Taxon. **4**(1): 17–36.

Rahman, M.M., Rashid, M.H. and Rashid, S.H. 2001. Assessment of plant biodiversity of sand dune ecosystem along the Cox's Bazar to Teknaf coast. Bangladesh J. Plant Taxon. **8**(1): 27–45.

Rahman, M.A., Uddin, M.N., Rashid, M.E. and Islam, M.M. 2012. Floristic diversity in Rampahar reserve forest of Kaptai, Rangamati. Biodiversity Bull. **6**: 1–31.

Rahman, M.O. and Alam, M.T. 2013. A taxonomic study on the angiosperm flora of Trishal upazila, Mymensingh. Dhaka Univ. J. Biol. Sci. **22**(1): 63–74.

Rahman, M.O. and Hassan, M.A. 1995. Angiospermic flora of Bhawal National Park, Gazipur (Bangladesh). Bangladesh J. Plant Taxon. **2**(1&2): 47–79.

Rahman, M.O. and Hassan, M.A. 2017. New angiospermic taxa for the flora of Bangladesh. Bangladesh J. Plant Taxon. **24**(2): 165–171.

Rahman, M.O. and Khan, B. 2008. Euphorbiaceae. *In*: Ahmed, Z.U., Hassan, M.A., Begum, Z.N.T., Khandker, M., Kabir, S.M.H., Ahmed, M., Ahmed, A.T.A., Rahman, A.K.T. and Haque, E.U. (Eds), Encyclopedia of Flora and Fauna of Bangladesh, Vol. **7**. Asiatic Society of Bangladesh, Dhaka, pp. 376–488.

Rahman, M.O., Begum, M. and Ullah, M.W. 2013. Angiosperm flora of Sadar upazila of Munshigonj district, Bangladesh.Bangladesh J. Plant Taxon. **20**(2): 213–231.

Rahman, M.O., Antara, R.T., Begum, M. and Hassan, M.A. 2012. Floristic diversity of Dhamrai upazila of Dhaka, Bangladesh with emphasis on medicinal plants. Bangladesh J. Bot. **41**(1): 71–85.

Rahman, M.S., Hossain, G.M., Khan, S.A. and Uddin, S.N. 2015. An annotated checklist of the vascular plants of Sundarban mangrove forest of Bangladesh. Bangladesh J. Plant Taxon. **22**(1): 17–41.

Rashid, M.H., Rahman, E. and Rahman M.A. 2000. Additions to the angiospermic flora of Moheskhali Island. Bangladesh J. Plant Taxon. **7**(1): 43–63.

Rashid, M.H. and Chowdhury, M.A.I. 2013. Additions to the angiosperm flora in the Sitapahar Reserve Forest of Kaptai, Rangamati, Bangladesh. Bangladesh J. Plant Taxon. **20**(2): 255–257.

Raizada, M.B. 1941. On the flora of Chittagong. Indian Forester **67**(5): 245–254.

Roxburgh, W. 1814. Hortus Bengalensis (*num. nud.*). Boerhave Press, Leiden (Holland),105 pp.

Sharma, B.D., Balakrishnan, N.P., Rao, R.R. and Hajra, P.K. 1993. Flora of India, Botanical Survey of India, Calcutta **1**: 1–467.

Sinclair, J. 1956. The Flora of Cox's Bazar, East Pakistan. Bull. Bot. Soc. Beng. **9**(2): 105–106.

Tutul, E., Uddin, M.Z., Rahman, M.O. and Hassan, M.A. 2010. Angiospermic flora of Runctia Sal forest, Bangladesh. II. Magnoliopsida (Dicots). Bangladesh J. Plant Taxon. **17**(1) 33–53.

Uddin, M.Z. and Hassan, M.A. 2010. Angiosperm diversity of Lawachara National Park (Bangladesh): A preliminary assessment. Bangladesh J. Plant Taxon. **17**(1): 9–22.

Uddin, S.B. and Rahman, M.A. 1999. Angiospermic flora of Himchari National Park, Cox's Bazar. Bangladesh J. Plant Taxon. **6**(1): 31–68.

Webster, G.L. 1994. Synopsis of the genera and suprageneric taxa of Euphorbiaceae. Ann. Miss. Bot. Gard. **81**: 33–144.

Wurdack, K.J., Hoffmann, P., Samuel, R., Bruijn, A., van der Bank, M. and Chase, M.W. 2004. Molecular phylogenetic analysis of Phyllanthaceae (Phyllanthoideae *pro parte*. Euphorbiaceae *s.l.*) using plastid *rbc*L DNA sequences. Am. J. Bot. **91**: 1882–1900.

CHEILOLEJEUNEA VITTATA (STEPH. *EX* G. HOFFM.) R.M. SCHUST. & KACHROO (LEJEUNEACEAE: MARCHANTIOPHYTA) – A NEWLY RECORDED SPECIES FROM INDIA

SHASHI KUMAR AND SUSHIL KUMAR SINGH[1]

Botanical Survey of India, Eastern Regional Centre, Shillong-793003, India

Keywords: Cheilolejeunea vittata; Lejeuneaceae; New record; India.

Abstract

Cheilolejeunea vittata (Steph. *ex* G. Hoffm.) R.M. Schust. & Kachroo is reported and described for the first time in Indian bryoflora from Manipur.

Introduction

Lejeuneaceae is the largest family of the liverworts with more than 1000 species in the world belonging to 68 genera. The genus *Cheilolejeunea* (Spruce) Steph. is one of the largest genera under the family Lejeuneaceae, contains c. 80-100 species, pantropical in distribution (Thiers, 1997; Ye *et al.*, 2015) belonging to nine sections, namely *Anomalolejeunea* (Schiffn.) W.Ye, Gradst. & R.L. Zhu, *Cheilolejeunea, Cyrtolejeunea* W.Ye, Gradst. & R.L. Zhu, *Euosmolejeunea* W.Ye, Gradst. & R.L. Zhu, *Leucolejeunea* W.Ye, Gradst. & R.L. Zhu, *Paroicae* W.Ye, Gradst. & R.L. Zhu, *Omphalanthus* W.Ye, Gradst. & R.L. Zhu, *Strepsilejeunea* W.Ye, Gradst. & R. L . Zhu and *Xenolejeunea* B. Thiers. However, in recently published "World checklist of hornworts and liverworts", the number varies to 169 species (incl. 10 uncertain species) belonging to four subgenera viz., subg. *Cheilolejeunea*, *Euosmolejeunea* (Spruce) Kachroo, *Renilejeunea* R.M.Schust. and *Xenolejeunea* Kachroo & R. M. Schust. (Söderström *et al.,* 2016).

The genus is characterized by 1) pale green to yellowish green colour, 2) thin stems with a 2(-4) cells wide ventral merophyte and enlarged epidermis cells, 3) leaf lobules with 1(-2) teeth and a hyaline papilla present distal side of the second tooth, 4) leaf cells with 1 – 3(-5) usually large, coarsely segmented oil-bodies, 5) underleaves usually bifid rarely entire, 6) gynoecia without or with 1 – 2 lejeuneoid or pycnolejeuneoid innovations, 7) perianth with 3 – 5 smooth keels, rarely pluriplicate or without plicae (Zhu *et al.*, 2002; Ye *et al.*, 2015).

In India, the genus represented by 21 species (Singh *et al.*, 2016) with prevalence in eastern Himalayan bryo-geographical territory (15 species) followed by Western Ghats (12 species). Andaman and Nicobar territory is represented by five species and Central India by 2 species, while four bryo-geographical territories have no representations of the genus. Following Söderström *et al.* (2016) subgeneric concept, the Indian species belong to three subgenera *Cheilolejeunea* (6 species), *Euosmolejeunea* (11 species), and *Xenolejeunea* (4 species). Three species namely, *Cheilolejeunea ghatensis* G. Asthana, S.C. Srivast. & A.K. Asthana, *Cheilolejeunea orientalis* (Gottsche) Mizut. and *Cheilolejeunea udarii* G. Asthana, S.C. Srivast. & A.K. Asthana are endemic to India.

Materials and Methods

Plant materials were collected from the forest along Pung-Pung river of Chandel district, Manipur, North-east India. The plants were growing as epiphyte in moist places. Critical morpho-

[1] Corresponding author: Email: sksbsinc@rediffmail.com

taxonomic investigation of the specimens was performed under Microscopes (Binocular Olympus SZ 51 Stereo zoom microscope & Trinocular Olympus CX 41 biological microscope). The specimens were identified based on the literature. A brief taxonomic description and line drawing illustration is provided. Nomenclature follows Söderström *et al.* (2016).

Table 1. Morphological and anatomical comparison of *Cheilolejeunea trapezia Cheilolejeunea ceylanica* and *Cheilolejeunea vittata.*

Characters	*Cheilolejeunea trapezia* (Dey & Singh, 2012)	*Cheilolejeunea ceylanica* (Zhu & So, 2001; Mizutani, 1980)	*Cheilolejeunea vittata* (our plant)
Plant length & width	18-25 mm long & 1.5-1.9 mm wide	5-15 mm long& 0.7-1.3 mm wide	8-12 mm long; 0.9 – 1.4 mm wide
No of cortical & medullary cells in stem	cort. 9-12 & med. 15-18	cort. 7 & med. 9-13	cort. 7 (-8) & med. 9-11
Ventral merophyte	2-4 cells wide	2 cells wide	2 cells wide
Leaf shape & Size	ovate, 0.8-1.1 mm long & 0.6- 0.75 mm wide	triangular-ovate , 0.4-0.6 mm long & 0.3-0.4 mm wide	ovate, 0.5- 0.7 mm long, 0.4 -0.5 mm wide
Vitta in leaf lobe	absent	often distinct, 3-7 cells long & 3-5 cells wide	distinct, 12 -16 cells long & 6-11 cells wide
Leaf lobule	3/5-2/3 of leaf length	3/5-2/3 of leaf length	(2/5-) 1/2 of leaf length
Second tooth of leaf lobule	1-4 (-5) cells long, erect or slightly curved, as long as the apical portion of leaf lobule near lobe attachment	4-7 cells long, usually curved at apex of leaf, as long as the apical portion of leaf lobule near lobe attachment	1-4 cells long, erect or occasionally slightly curved, usually shorter than the apical portion of leaf lobule near lobe attachment

Taxonomic description

Cheilolejeunea vittata (Steph. *ex* G. Hoffm.) R.M. Schust. & Kachroo, J. Linn. Soc. Bot. 56: 509. 1961; Mizutani, J. Hattori Bot. Lab. 47: 321. 1980. *Pycnolejeunea vittata* Steph. *ex* Hoffm., Ann. Bryol. 8: 115. 1935.

Plants small, pale green; shoots 8 – 12 mm long, 0.9 – 1.4 mm wide; branching irregulary, *Lejeunea*-type; cross-section of stem subglobose – oval in outline, 78.6 – 105.0 × 66.5 – 80.0 μm, 5 – 6 cells across the diameter; cortical cells in a layer of 7 (–8) cells, subquadrate – rectangular, 15.0 – 32.5 × 10.0 – 20.0 μm, thick-walled, medullary cells 9 – 11, polygonal, 10.0 – 22.5 × 7.5 – 12.5 μm, thick-walled; ventral merophytes 2 cells wide. Leaves imbricate, widely spreading; leaf lobes ovate, 0.51 – 0.70 mm long, 0.41 – 0.52 mm wide, antical margin arched, postical margin straight or sometimes curved, apex rounded, margin entire; marginal leaf cells towards apex small, subquadrate – quadrate, 7.5 – 12.5 × 5.0 – 7.5 μm; median leaf cells pentagonal – hexagonal, 15.0 – 30.0 × 10.0 – 22.5 μm; basal leaf cells elongated hexagonal – polygonal, 17.5 – 45.0 × 10.0 – 30.0 μm; cells thin or slightly thick-walled, with nodular trigones and intermediate thickenings prominent in median and basal leaf cells; vitta about 3/4 the length and 2/5 – 3/5 the

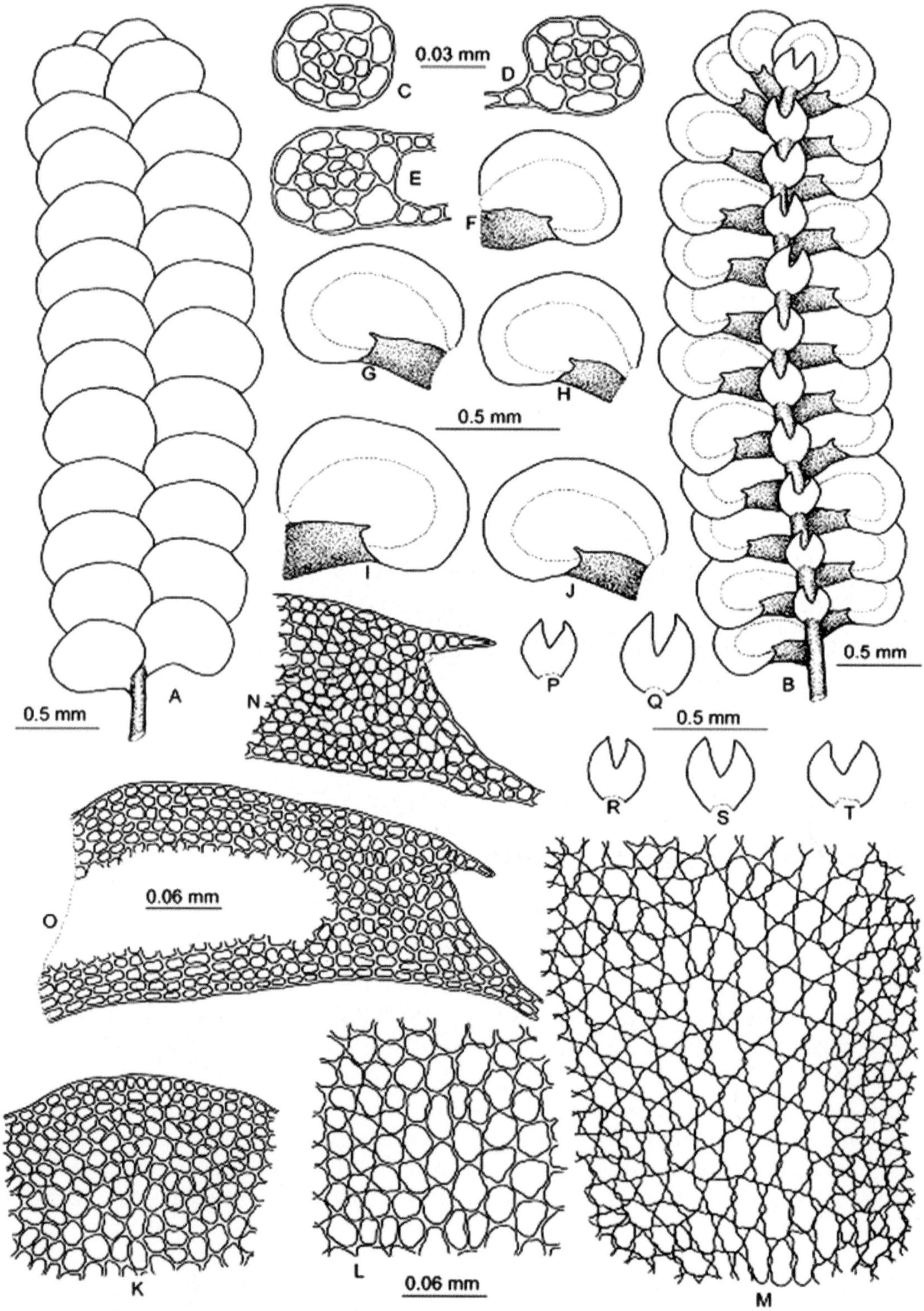

Fig. 1. **Cheilolejeunea vittata** (Steph. *ex* G. Hoffm.) R.M. Schust. & Kachroo: A. A portion of plant in dorsal view; B. The same in ventral view; C – E. Cross-sections of stem; F – J. Leaves; K. Marginal leaf cells; L. Median leaf cells; M. Basal leaf cells; N, O. Leaf lobules; P – T. Underleaves.

width of the leaf lobe, 12 – 16 cells long, 6 – 11 cells wide; cuticle slightly mammillose; oil-bodies not observed; leaf lobules inflated, rectangular, 1/2 (-2/5) as long as the lobe, 0.25 – 0.35 mm long, 0.15 – 0.18 mm wide, apex truncate, first tooth obsolete, second tooth elongated, 1 (-2) – 4 cells long in a row, hyaline papilla small, present at the distal side of second tooth. Under leaves distant, suborbicular, 2 – 3 times as wide as the stem, 0.20 – 0.33 mm long, 0.18 – 0.31 mm wide, bilobed to 3/5 of its length, lobes triangular, apex acute, sinus "V"-shaped, margin entire. Androecial and gynoecial branches not observed.

Habitat: Epiphytic, growing in moist and shady places.

Distribution: India [Manipur – present study], China (Zhu & So, 1999; Zhu *et al.,* 2002), Indonesia (Söderström *et al.,* 2010), Malaysia (Chuah-Petiot, 2011), Papua New Guinea (Grolle & Piippo 1984), Sri Lanka (Rubasinghe & Long, 2014), Thailand (Lai *et al.*, 2008; Pócs & Podani, 2015), Philippines (Mizutani, 1980), Australia (Thiers, 1992, 1997).

Specimen examined: Manipur, Chandel district, Pung-Pung river, 24°15′36.3′′N, 94° 17′ 37.2′′E, 220 m, 22.06.2014, *Shashi Kumar*, TSLI – 194, 195, 215 (ASSAM).

Notes: Cheilolejeunea vittata (belongs to subg. *Xenolejeunea)* is characterized by its leaf lobe cells forming a vitta (about 3/4 the length and 2/5 – 3/5 the width of leaf lobe, 12 – 16 cells long, 6 – 11 cells wide) and with sub-nodular – nodular or bulging trigones and few, small – prominent intermediate thickenings; leaf lobules rectangular, (2/5-) 1/2 as long as lobe with elongated 1 – 4 cells long, erect or slightly curved apical tooth which usually shorter than the apical portion of leaf lobule. However, it shows affinity with *Cheilolejeunea ceylanica* in having similar habit of plants, similar arrangement of leaves, robust leaf lobule with elongated apical tooth and under leaf structure, but the latter differs from the former in having comparatively larger leaf lobule 2/3 as long of the leaf lobe length with very long apical tooth which is 5 – 7 cells long and curved toward apex, usually as long as the apical portion of leaf lobule, leaf lobes usually triangular-ovate with elongate median and basal cells often forming a distinct (or indistinct) vitta (extending to the middle portion) vitta 3 – 7 cells long (Zhu & So, 2001). It also shows similarity with *Cheilolejeunea trapezia* in large leaf lobule, and lobular tooth (1 – 4 cells long, erect or occasionally slightly curved), but latter differ from the former in its leaf cells which is not forming vita and ventral merophyte 2 – 4 cells wide (Zhu & So, 2001; see also table 1).

Acknowledgements

The authors are thankful to Director, Botanical Survey of India, Kolkata and Head of the Office, BSI, Eastern Regional Centre, Shillong for facilities and encouragement; to the officials of Forest Department, Manipur for permission and help rendered during field exploration. One of us (Shashi Kumar) is also grateful to the Director, BSI, for financial assistance under 'Flora of India' project.

References

Chuah-Petiot, M. S. 2011. A checklist of Hepaticae and Anthocerotae of Malaysia. Polish Bot. J. **56**(1): 1–44.

Dey, M. and Singh, D. K. 2012. Epiphyllous Liverwort of Eastern Himalaya. BSI, Kolkata.

Grolle, R. & Piippo, S. 1984. Annotated catalogue of Western Melanesian bryophyte. I: Hepaticae and Anthoccrotue. Acta Bot. Fenn. **125**: 1–86

Lai, M. J., Zhu, R.L. and Chantanaorrapint, S. 2008. Liverworts and hornworts of Thailand: an updated checklist and bryofloristic accounts. Ann. Bot. Fenn. **45**: 321–341.

Mizutani, M. 1980. Lejeuneaceae from the Philippines. J. Hattori Bot. Lab. **43**: 127–136.

Pócs, T. & Podani, J. 2015. Southern Thailand and Bryophytes II: Epiphylls from the Phang-Nga area. Acta Bot. Hung. **57**(1-2): 183–193.

Rubasinghe, S. C. K. & Long, D.G. 2014. Liverworts and Hornworts of Sri Lanka: a revised checklist. Ceylon J. Sci. (Bio. Sci.) **43**(1): 1–36.

Söderström, L., Hagborg, A., von Konrat, M., Bartholomew-Began, S., Bell, D., Laura Briscoe, L., Brown, E., Cargill, D.C., Costa D.P., Crandall-Stotler B.J., Cooper E.D., Dauphin, G., Engel, J.J., Feldberg, K., Glenny, D., Gradstein, S.R., He, X.-L., Heinrichs, J., Hentschel, J., Ilkiu-Borges, A.L. Katagiri, T., Konstantinova, N.A., Larraín, J., Long, D.G., Nebel, M., Pócs, T., Puche, F., Reiner-Drehwald, E., Renner, M.A.M., Sass-Gyarmati, A., Schäfer-Verwimp, A., Moragues, J.G.S., Stotler, R.E., Sukkharak, P., Thiers, B.M., Uribe, J.M., Váňa, J., Villarreal, J. C., Wigginton, M., Zhang, L. & Zhu, R.-L. 2016. World checklist of hornworts and liverworts. Phyto Keys **59**: 1–828.

Söderström, L., Gradstein, S.R. and Hagborg, A. 2010. Checklist of the hornworts and liverworts of Java. Phytotaxa **9**: 53–149.

Thiers, B.M. 1992: A re-evaluation of *Cheilolejeunea* subgenus *Xenolejeunea*. Trop. Bryol. **5:** 10–21.

Thiers, B.M. 1997. *Cheilolejeunea* in Australia: description of new taxa and key. J. Hattori Bot. Lab. **82**: 321–328.

Ye, W., Gradstein, S.R., Shaw, A. J., Shaw, B., Ho, B.-C., Schäfer-Verwimp, A., Pócs, T., Heinrichs, J. and Zhu, R.-L. 2015. Phylogeny and classification of Lejeuneaceae subtribe Cheilolejeuneinae (Marchantiophyta) based on nuclear and plastid molecular markers. Cryptog. Bryol. **36**: 313–333.

Zhu, R.L. and So, M. L. 1999. Additions of Lejeuneaceae taxa to the hepatic flora of Yunnan, China. Ann. Bot. Fenn. **36:** 219–229.

Zhu, R.L. and So, M.L. 2001. Epiphyllous liverworts of China. Beih. Nova Hedwigia **121**: 1–418.

Zhu, R.L., So, M.L. and Wang, Y.F. 2002. The genus *Cheilolejeunea* (Hepaticae, Lejeuneaceae) in China. Nova Hedwigia **75:** 387–408.

IXORA PREDEEPII, A NEW SPECIES OF RUBIACEAE FROM SOUTHERN WESTERN GHATS, INDIA

Anoop P. Balan[1] and Harikrishnan Shanmugam

Indian Cardamom Research Institute, Spices Board, Myladumpara, Kailasanadu P.O., Idukki, Kerala 685553, India

Keywords: *Ixora*; New species; Rubiaceae; Southern Western Ghats.

Abstract

Ixora predeepii, a new species of the family Rubiaceae from Southern Western Ghats, India is described and illustrated. It is allied to *Ixora elongata* Heyne *ex* D. Don in general appearance, but differs from the latter by its small habit, short peduncled congested inflorescence, small purplish white flowers with glabrous corolla and scarlet berry.

Introduction

Ixora L. is the third largest genus of Rubiaceae with about 500 species (Mouly *et al.*, 2009) includes shrubs and small trees distributed in tropical and subtropical regions of the world. In India, the genus is represented by 46 species (Husain and Paul, 1989), of which 22 species are reported from Kerala state (Sasidharan, 2004, 2013; Nayar *et al.*, 2006). During a floristic exploration in the Cardamom Hills of Southern Western Ghats, India, the authors collected some interesting specimens of *Ixora,* resemble *I. elongata* in general appearance. On critical studies it turned out to be very distinct from *I. elongata* and upon perusal of herbarium specimens and literature (Hooker, 1880; Gamble, 1921; Nayar *et al.*, 2014), it was found to be an undescribed taxon which is described here as a new species, *Ixora predeepii*. A detailed description, illustration, photographs and relevant notes on ecology and phenology of the new species are provided to facilitate its easy identification in the field.

Ixora predeepii Anoop *et* Harikrishnan, **sp. nov**. **(Figs 1 & 2**).

Diagnosis: *Ixora predeepii* **sp. nov.** is allied to *I. elongata* in its general appearance, many–flowered, congested inflorescence with pubescent peduncle and calyx, but clearly distinct from the latter by its small habit, comparatively small leaves, short peduncled congested inflorescence with small flowers, long calyx lobes, glabrous corolla and scarlet berry (Table 1).

Type: INDIA, Kerala, Idukki District, Udumbanchola, Myladumpara, 9° 5″ N, 77° 9′ E, at 1100 m, 10 November 2013, *Anoop P. Balan* 20471 *(Holotype*: MH; *Isotype*: CAL).

Paratype: INDIA, Kerala, Idukki District, Chathurangappara, 1100 m, 15 December 2013, *Anoop P. Balan & Harikrishnan* 20490 (MH).

Shrubs, 1.0–1.5 m high; branches slender, glabrous, internodes 3–5 cm long. Stipules 6–8 × 2–3 mm, triangular, long cuspidate, glabrous. Leaves 8–13 × 2.5–5.0 cm, ovate–elliptic, acute at apex, cuneate at base, chartaceous, glabrous, margin weavy; petiole 6–8 mm long, glabrous; lateral

[1]Corresponding author. Email: anooppb01@gmail.com

veins 8–12 pairs, prominent beneath. Bracts supporting inflorescence foliaceous, 3.0–4.5 × 2.0–2.5 cm, ovate, acute at apex, rounded at base, persistent. Inflorescence terminal, 6–10 cm long, trichotomously branched corymbose cymes; peduncle branched at 1.0–3.5 cm from the base; branches 4–6 cm long, patently pubescent; bracts supporting flowers 2 mm long, linear, pubescent. Flowers sessile, bracteolate; bracteoles 1.5 mm long, linear, pubescent. Calyx tube 1 mm long, sparsely pubescent; lobes 4, c. 1.5 mm long, lanceolate, ciliate. Corolla deep red in bud, purplish white when open, glabrous; tube 10 mm long, broadened towards apex; lobes 4 × 2 mm, ovate–oblong, obtuse to subacute at apex, reflexed. Stamens 4, attached at mouth, alternating with corolla lobes; filaments c. 1mm long; anthers sub–basifixed, 3 mm long, tip pointed, sagitate at base. Ovary 2–celled, ovule 1 in each cell; style slender, 1.2 cm long, tip fusiform, longitudinally cleft forming two stigmatic lobes of c. 2 mm long. Berry didymous, 6–8 × 5–6 mm, glabrous, crowned by the permanent calyx, scarlet when ripe. Seeds 2, 5–6 × 5–6 mm, plano–convex, brown, testa membraneous.

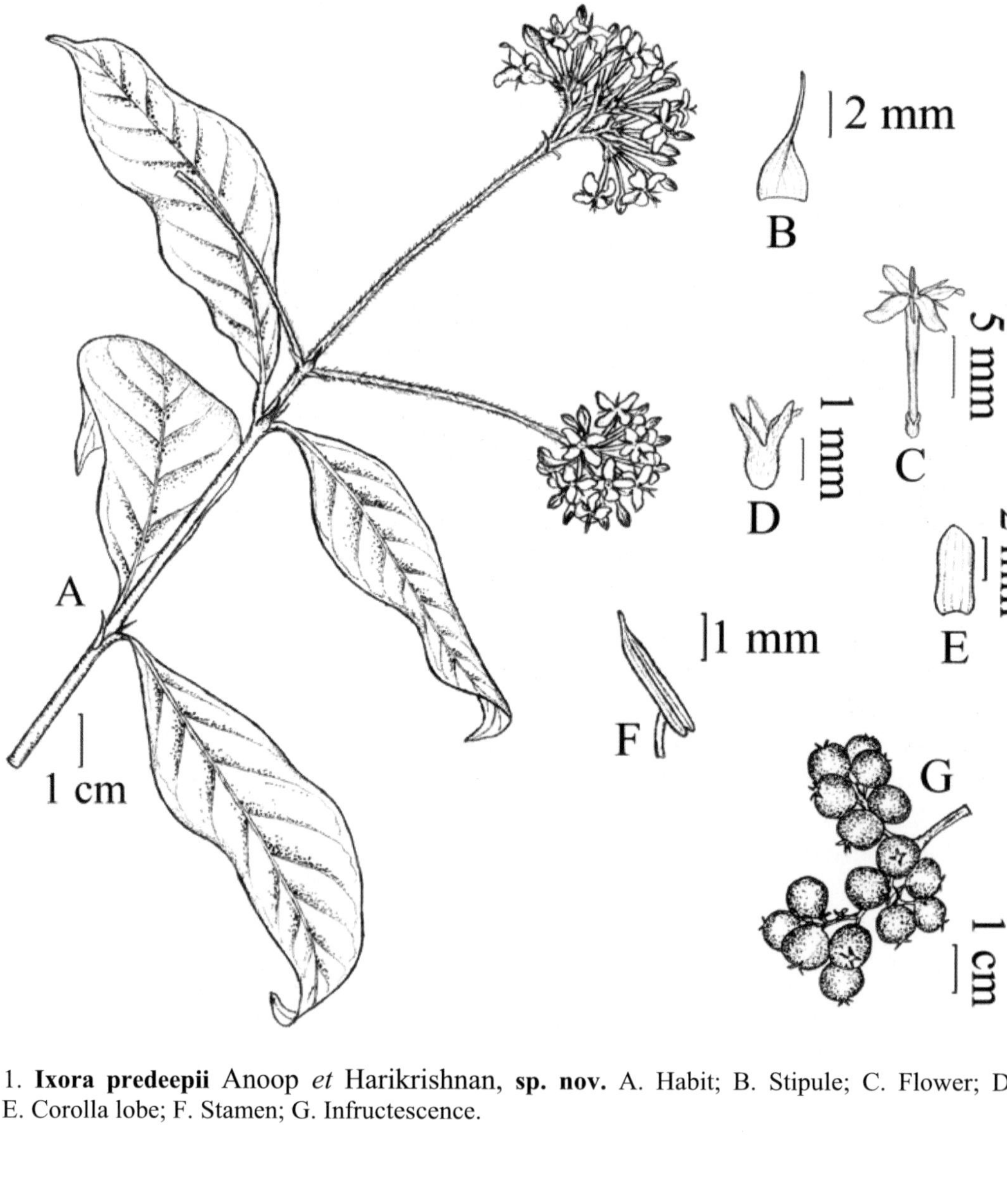

Fig. 1. **Ixora predeepii** Anoop *et* Harikrishnan, **sp. nov.** A. Habit; B. Stipule; C. Flower; D. Calyx; E. Corolla lobe; F. Stamen; G. Infructescence.

Fig. 2. **Ixora predeepii** Anoop *et* Harikrishnan, **sp. nov.** A. Habit; B. Nodal portion; C & D. Inflorescence; E. Flower; F & G. Infructescence; H. Seeds.

Phenology: Flowering from November to January; fruiting from February to April

Habitat: Cardamom plantations and the margins of evergreen forests at 1100 m elevation.

Distribution and ecology: *Ixora predeepii* is so far known only from the type locality, Cardamom hills of Southern Western Ghats, Kerala. It is an undergrowth shrub in Cardamom plantations and the margins of evergreen forests often found in association with *Dichapetalum gelonioides* (Roxb.) Engl., *Nothapodytes nimmoniana* (Graham) Mabb., *Ligustrum robustum* subsp. *walkeri* (Decne.) P. S. Green, *Psychotria nilgiriensis* Deb & Gangop. and *Lepisanthes erecta* (Thw.) Leenh.

Etymology: Specific epithet of the new taxon is in honor of Dr. S.V. Predeep, Department of Botany, SVR NSS College, Vazhoor for his valuable contributions to the field of Angiosperm taxonomy.

Table 1. Comparison between *Ixora predeepii* sp. nov. and its related species *I. elongata*.

Characters	*Ixora predeepii* **sp. nov.**	*Ixora elongata*
Habit	Small shrubs, 1.0–1.5 m high	Large shrubs, 3–5 m high
Petiole	0.6–0.8 cm long	1.0–1.5 cm long
Lamina	8–13 × 2.5–5.0 cm	12–22 × 3.5–10 cm
Lateral nerves	8–12 pairs	12–15 pairs
Inflorescence	6–10 cm long	16–25 cm long
Calyx	Lobes longer than tube	Lobes equal to tube
Corolla	Lobes glabrous	Lobes with spreading hairs
Fruit	Scarlet when ripe	Black when ripe

Note: The only known locality of this novel taxon is the Cardamom hills and adjacent evergreen forest patches of Idukki district of Kerala state. The natural flora of Cardamom hills are under severe threat due to human activities related to agriculture. Except large trees, all other plants are undergoing weeding process every year which may cause severe destruction of the population of *Ixora predeepii* also. A detailed population survey is yet to be undertaken to assess the present status of this taxon in order to ascertain a proper IUCN status to ensure the conservation of this species.

Acknowledgements

The authors are grateful to the curators of MH and CAL for allowing consultation of herbarium specimens. We are also thankful to the Director (Research) and our colleagues in Indian Cardamom Research Institute, Myladumpara for their constant support and encouragement and Dr. A.J. Robi, BAM College, Thuruthicadu, for his valuable comments on the identity of this species.

References

Gamble, J.S. 1921. The Flora of the Presidency of Madras, Vol. **2**. Adlard and Son Ltd., London, 445 pp.

Hooker, J.D. 1880. The Flora of British India, vol. 3. Reeve and Co., London, pp. 137-149.

Husain, T. and Paul, S.R. 1989. Taxonomic studies on Indian species of genus *Ixora* L. (Rubiaceae). J. Econ. Tax. Bot. Additional series **6**. Scientific publishers, Jodhpur, India.

Mouly, A., Razafimandimbison, S.G., Khodabandch, A. and Bremer, B. 2009. Phylogeny and Classification of the species rich pantropical showy genus *Ixora* (Rubiaceae–Ixoreae) with indications of geographical monophyletic units and hybrids. Am. J. Bot. **96**: 686–706.

Nayar, T.S., Beegum, A.R., Mohanan, N. and Rajkumar, G. 2006. Flowering Plants of Kerala - A handbook. JNTBGRI, Thiruvananthapuram, pp. 282–287.

Nayar, T.S., Beegum, A.R. and Sibi, M. 2014. Flowering Plants of the western Ghats, India, Vol. **1.** JNTBGRI, Thiruvananthapuram, 386 pp.

Sasidharan, N. 2004. Biodiversity documentation for Kerala. Part 6: Flowering Plants. Kerala Forest Research Institute, Peechi, Kerala, pp.232.

Sasidharan, N. 2013. A digital database of Flowering Plants of Kerala: CD-ROM ver. 2. Kerala Forest Research Institute, Peechi, Kerala.

30

PRIMULINA WENII (GESNERIACEAE), A NEW SPECIES FROM CHINA

JIAN LI, XIN CHEN, SHU LI[1] AND LI-JIAO YAN[2]

Institute of Ecology, College of Life Sciences, Zhejiang University, Hangzhou, Zhejiang province, China, CN-315008, P.R. China

Keywords: East of China; Flora of Fujian; Gesneriaceae; New taxon; *Primulina juliae*

Abstract

Primulina wenii (Gesneriaceae) is described and illustrated here as a new species. Morphologically, this new species resembles *P. juliae* (Hance) Mich. Möller & A. Weber, but it is distinct from the latter by some obvious features in leaf blade, bract, indumentum, corolla, filament, stigma and pistil. Nuclear ribosomal internal transcribed spacer (ITS) region and plastid *trn*L-F intronspacer (*trn*L-F) DNA sequence data from the new species and its 22 relatives are used to determine the systematic position of the new species within *Primulina*. Molecular evidence suggests that *P. wenii* is strongly supported as a sister to a clade in which *P. juliae* is included. The conservation status of *P. wenii* is assessed as "Critically Endangered" (CR) according to IUCN Red List categories and criteria.

Introduction

A great number of new species of *Primulina* have been described since 2011 at present the number has rapidly reached more than 150 species in China (Möller *et al.*, 2016). Undoubtedly, the limestone areas of South-west and South China and North Vietnam are considered to be of the highest biodiversity and differentiation for *Primulina* (Gesneriaceae) (Wei *et al.*, 2010; Wen *et al.*, 2015). It is easily understood that very few species of *Primulina* in Fujian province have been recorded. Before 1992, there were two species and one variety, namely *P. pinnatifida* (Hand.-Mazz.) Y.Z. Wang [former *Chirita pinnatifida* (Hand.-Mazz.) Burtt], *Primulina fimbrisepala* (Hand.-Mazz.) Y.Z. Wang (former *Chirita fimbrisepala* Hand.-Mazz.) and *Chirita gueilinensis* W.T. Wang var. *brachycarpa* W.T. Wang (Zhang, 1993). However, *Chirita gueilinensis* var. *brachycarpa* was brought into *Chirita juliae* Hance (Wang *et al.*, 1998; Li and Wang, 2004) based on the similar morphological characters of the two congeners. But soon afterwards, *Chirita* along with several other associated genera, namely almost all the species and *Chirita* sect. *Gibbosaccus* C.B. Clarke, two species of *Wentsaiboea* D. Fang & D.H. Qin (*W. renifolia* D. Fang & D.H. Qin (2004) and *W. luochengensis* Yan Liu & W.B. Xu (2010)) and all the species of *Chiritopsis* W.T. Wang (1981) were shifted to *Primulina* Hance (Li and Wang, 2007; Wang *et al.*, 2011; Weber *et al.*, 2011). It was originally the monotypic genus (Hance, 1883). Subsequently, a new species of *Primulina, viz., P. xiuningensis* (X.L. Liu & X.H. Guo) Mich. Möller & A. Weber (former *Chiritopsis xiuningensis* X.L. Liu & X.H. Guo) (Liu and Guo, 1989; Weber *et al.*, 2011), from Fujian, China was discovered and reported formally in recent years (Geng *et al.*, 2014). To sum it up, Fujian province in East China cannot still be considered to have high biodiversity of *Primulina* because only three species were found in this area.

---

[1]Gesneriaceae Conservation Center of China (GCCC), Guilin Botanical Garden, Guangxi Inst. of Botany, Guangxi Zhuang Autonomous Region and Chinese Academy of Sciences, CN-541006 Guilin, P.R. China.
[2]Corresponding author.Email: yanljzju@163.com

In 2012, one of the authors found a population of the peculiar Gesneriaceae plants growing on a rocky slope near a stream in a limestone gorge close to Fuzhou city, Fujian province, China. Firstly, we considered it to be *Prmulina juliae* because it is recorded in Ninghua County and Songxi County of Fujian. The next year we re-visited the same location to collect specimens with flowers, and we found that the two similar congeners are actually different. We sent some specimens with flowers and fruits to IBK (Dr. Fang Wen, FW) for identification. He thought they should be an unknown species and have never been seen before. After consulting relevant literature (Wang, 1985; Wang *et al.*, 1990, 1998; Li and Wang, 2004; Wei *et al.*, 2010; Wang *et al.*, 2011; Weber *et al.*, 2011; Möller *et al.*, 2011, 2016;Xu *et al.*, 2012; Möller and Clark, 2013) and checking herbarium specimens of *Primulina* (Some former *Chirita*), 'World checklist of Gesneriaceae' (Skog and Boggan, 2007) and 'The genera of Gesneriaceae' (Weber and Skog, 2007), and consulting Dr. Fang Wen based on his rich experience and knowledge about the identification and taxonomy of *Primulina*, we found the morphology of these newly discovered plants to be similar to *P. juliae* (Hance) Mich. Mich. & A. Weber (Wei *et al.*, 2010). However, they differ significantly in some characters. We have determined that they belong to a new species of *Primulina viz. P. wenii*, sp. nov.The new species *Primulina wenii* is described and illustrated below and compared with the morphologically similar *P. juliae* and their conservation status is evaluated.

Material and Methods

An overview of the genus *Primulina* from South, South-west and East China was made. All the available specimens of *Primulina* stored in the following herbaria in China (ANU, BJFC, CDBI, CTC, HGAS, HIB, IBK, IBSC, KUN, PE, SZ) were used as material of *Primulina* from recent fieldwork by the re-searching team of authors in South, South-west and East China (Thiers, 2015). All the morphological characters were studied under dissecting microscopes, and described using the terminology presented by Wang *et al.* (1990, 1998).

Leaf material of the new species was collected in the field and dried by silica gel for DNA extraction (Chase and Hills, 1991). The nuclear ribosomal internal transcribed spacer (ITS) region and the plastid *trn*L-F intron spacer (*trn*L-F) were used as molecular markers. The molecular methods and protocols followed Möller *et al.* (2009, 2011). GenBank accession numbers for ITS and *trn*L-F of the new species are KX985576 and KX985577, respectively (Table 1). To elucidate the phylogenetic affinities of the new species, ingroups (22 species of *Primulina*) and outgroups [*Ornithoboea wildeana* Craib, *Paraboea rufescens* (Franch.) B.L.Burtt] selection were chosen based on recent phylogenetic analyses (Li *et al.*, 2007; Möller*et al.*, 2011; Weber *et al.*, 2011; Kang *et al.*, 2014), and from which sequences were available from GenBank (Table 1).

Sequence data were edited and assembled using Lasergene Navigator 7.1 (DNAStar, Madison, Wisconsin, USA) and then aligned with the MEGA 5.1 (Tamura *et al.*, 2011) with additional manual refinements where necessary. Phylogenetic analyses were performed using maximum parsimony (MP) method implemented in PAUP* 4.0b10. Heuristic searches were performed using a starting tree built from stepwise in addition with TBR branch swapping and 1,000 random addition replicates. To assess confidence in clades, bootstrap analyses based on 1,000 replicates with 10 random additions per replicate were used.

Table 1. List of species along with accession number of *trn*L-F and ITS sequences used in this study.

Species name	Voucher number	*trn*L-F	ITS
Ornithoboea wildeana Craib	Jia-Mei Li Ljm-04-44	DQ872824	DQ865197
Paraboea rufescens(Franch.) Burtt.	Jia-Mei Li 0185	DQ872825	DQ865196
Primulina bipinnatifida (W.T. Wang) Y.Z. Wang	Guangxi NA/Li J.M.NA	DQ872806	DQ872842
Primulina danxiaensis	DXS04	KF498157	KF498050
Primulina dryas (Dunn) Mich. Möller & A. Weber	T.C. Godfrey 369	FJ501524	FJ501348
Primulina glandulosa (D. Fang, L. Zeng & D.H. Qin) Y.Z. Wang	Jia-Mei Li 054291	DQ872804	DQ872841
Primulina glandulosa var. *yangshuoensis* (Fang Wen, Yue Wang & Q.X. Zhang) Mich. Möller & A. Weber	M. Möller MMO 06-912	HQ632948	HQ633045
Primulina heterotricha (Merr.) Y.Z. Wang	Yin-Zheng Wang 06731	DQ872816	DQ872826
Primulina juliae	JXGF01	KF498228	KF498107
Primulina linearifolia (W.T. Wang) Y.Z. Wang	Jia-Mei Li 11121	DQ872810	DQ872834
Primulina longgangensis (W.T. Wang) Y.Z. Wang	A. Takhtajan & N. Aruzytov 1975	AJ492290	FJ501347
Primulina minutimaculata (D. Fang & W.T. Wang) Y.Z. Wang	Jia-Mei Li 067134	DQ872815	DQ872828
Primulina mollifolia (D. Fang & W.T. Wang) Y.Z. Wang	Jia-Mei Li 054281	DQ872802	DQ872847
Primulina ophiopogoides (D. Fang & W.T. Wang) Y.Z. Wang	Yin-Zheng Wang 06713‹	DQ872814	DQ872829
Primulina pinnata (W.T. Wang) Y.Z. Wang	Expedition Beijing 896526	FJ501526	FJ501349
Primulinapinnatifida (Hand.-Mazz) Y.Z. Wang	Q.J. Xie J-037	FJ501527	FJ501350
Primulina pteropoda (W.T. Wang) Y.Z. Wang	Yin-Zheng Wang 06731:	DQ872817	DQ872827
Primulina repanda var. *guilinensis*	ex Smithsonian Institute 94-083	AJ492292	FJ501351
Primulina spadiciformis (W.T. Wang) Mich. Möller & A. Weber	ex Smithsonian Institute 94-087	AJ492291	FJ501346
Primulina spinulosa (D. Fang & W.T. Wang) Y.Z. Wang	Yin-Zheng Wang 06713:	DQ872813	DQ872830
Primulina tabacum Hance	Q.J. Xie & C.X. Ye	AJ492300	FJ501352
Primulina weii Mich. Möller & A. Weber	Jia-Mei Li, Ljm-04-42	DQ872811	DQ872832
Primulina wenii Jian Li & L.J. Yan sp. nov.	Jian Li & F. Wen 20130412-01	KX985577	KX985576
Primulina wentsaii (D. Fang& L. Zeng) Y.Z. Wang	Jia-Mei Li 11630	DQ872812	DQ872831
Primulina xiuningensis (X.L. Liu & X.H. Guo) Mich. Möller & A. Weber	ZJJS01	KF498252	KF498124

Results and Discussion

Taxonomical treatment

Primulina wenii Jian Li & L. J. Yan, **sp. nov.** **(Figs 1 & 2).**

Diagnosis: *Primulina wenii* Jian Li & L. J. Yan differs from its congener, *P. juliae* (Hance) Mich. Möller & A. Weber, by the combination of the following characters: the lateral veins of leaf blade 3 or 4; peduncle densely covered with erectly spreading white eglandular pubescent and villous hairs; bracts 3, outside pubescent and villous, inside pubescent; corolla outside densely white glandular-pubescent, inside sparsely pubescent; filaments glabrous; stigma trapeziform, 2-lobed to irregularly lobed (Table 2)

Type: CHINA. Fujian province, Fuzhou city, Rixi town, 26.35139°N, 119.27056°E, 130 m, on moist tufa and the rocky surface of limestone cliff, 12 April 2013, *Jian Li & F. Wen* 20130412-01 (*Holotype*: IBK!; *Isotype*: IBK!).

Table 2. Morphological comparison of *Primulina wenii* sp. nov. and *P. juliae*.

Characters	*P. wenii*	*P. juliae*
Leaf blade indumentum	Lower surface densely appressed pubescent and veins villous, upper surface densely pubescent and villous	Both surfaces appressedpuberulent
Leaf blade margin	Irregularly serrated in number and size	Dentate to pinnately lobed basally or crenate, apex acute to rounded
Lateral veins of leaf blade	3 or 4, abaxially obviously prominent, adaxially conspicuously sunken	4 or 5, abaxially inconspicuously prominent, adaxially inconspicuous
Peduncle indumentum	Densely covered with erectly spreading white eglandular pubescent and villous hairs	Densely covered with spreading puberulent hairs
Bract number and indumentum	3; outside pubescent and villous, inside pubescent	2; outside puberulent, inside glabrous
Corolla indumentum	Outside densely white glandular-pubescent, inside sparsely pubescent	Outside sparsely puberulent, inside nearly glabrous
Filaments	Glabrous	Glabrous to glandular puberulent near apex
Pistil	Densely covered with white glandular puberulent hairs	Puberulent
Stigma	Trapeziform, 2-lobed to irregularly lobed	Narrowly oblong, 2-lobed

Herbs, perennial. Rhizome cylindric, 2–4 cm long, 0.6–1.0 cm in diam. Leaves 4–6, basal; petiole compressed, 1.9–2.5 × c. 0.8 cm, densely villous; leaf blade pale green to green, oblong to broadly elliptic, 10–20 × 7–14 cm, herbaceous but chartaceous when dried, lower surface densely appressed pubescent and along veins villous, upper surface densely pubescent and villous, base attenuate to cuneate, margin irregularly serrate, ciliate, apex obtuse; lateral veins 3 or 4 on each side of midrib, abaxially obviously prominent, adaxially conspicuously sunken. Cymes 3–4, axillary, 3(5)–7-flowered or more; peduncle green, 8–10 cm long, 2.5–3.0 mm in diam., densely covered with erectly spreading white eglandular pubescent and villous hairs; bracts 3, lateral ones

Fig. 1. *Primulina wenii* **sp. nov**. A. Habitat; B. Habit in flowering; C. Leaf blade – adaxial surface; D. Leaf blade – abaxial surface; E. Cyme, buds and bracts; F. bracts; G. Cymes and flowers; H. Peduncle and its indumentum; I. Front view of corolla; J. Upward view of corolla; K. Lateral view of corolla; L. Stigma with dissected lobes; M. Curved immature capsules. A and M: collected from type locality, 22 Jul 2013; B–L: photographed in nursery of Gesneriad Conservation Center of China, Guilin, 12 Apr 2014, by F. Wen.

Fig. 2. *Primulina juliae.* A. Habit; B. Leaf blade – adaxial surface; C. Front view of corolla; D. Lateral view of corolla; E. Cyme, calyx lobes and indumentum; F. Young straight capsules. A– B and E– F: photographed in type location by J. Li; C– D: photographed in nursery of Gesneriad Conservation Center of China by F. Wen.

conspicuously opposite, green, linear to lanceolate, 14–16 × 2.5–3.0 mm, the central one smaller, 10–12 × 1.3–1.6 mm, persistent at flowering; outside white pubescent and villous, inside white pubescent, margin entire and ciliate, apex acute. Calyx 5-partite to base; lobes lanceolate, 14–15 × c. 2.5 mm, nearly equal, outside densely erectly white villous and pubescent, inside sparsely shortly pubescent to nearly glabrous, margin entire, apex acute. Corolla outside pale bluish purple, throat with 2 purple stripes, upper portion of inside corolla surface with 2 dark purplish brown swollen spots, 2 swollen spots sparsely covered with short glandular hairs, inside of lobes pale purplish blue with slightly darker purple longitudinal stripes, corolla c. 3.2 cm long, outside densely white glandular-pubescent, inside sparsely pubescent; tube nearly tubular, c. 2.8 cm long,

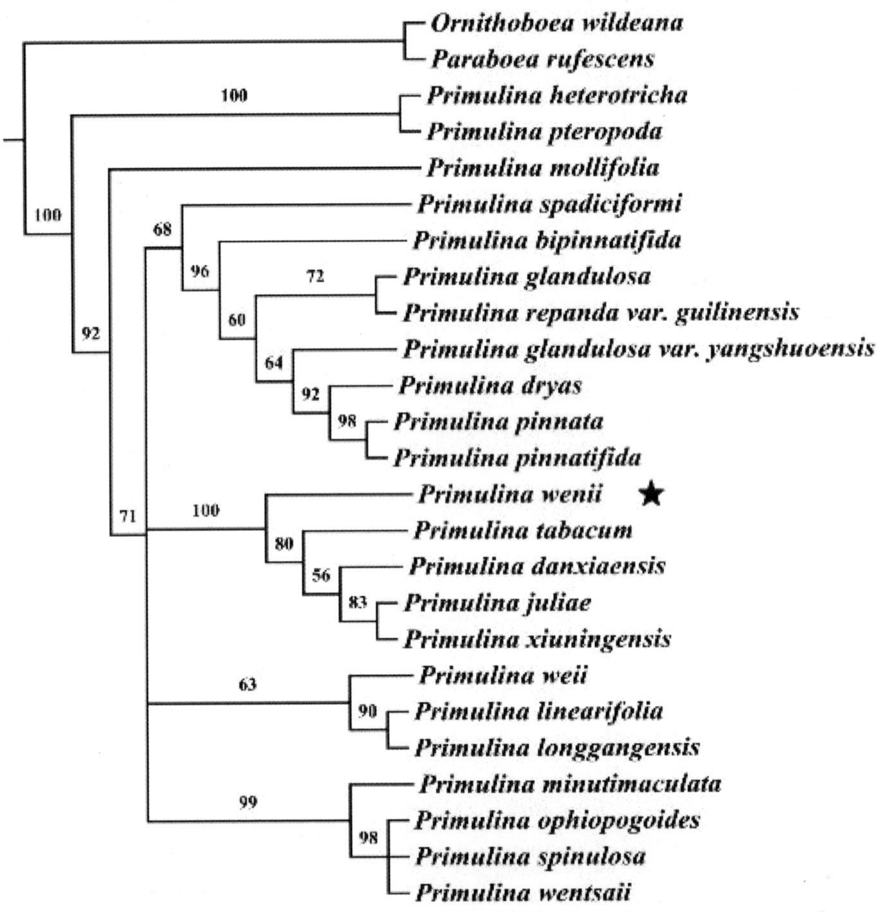

Fig. 3. The strict consensus tree of *Primulina* taxa based on a maximum parsimony (MP) analysis of combined ITS and *trn*L-F. Numbers of the branches indicate bootstrap values > 50% by MP analysis. * indicates the new species.

orifice c. 1.7 cm in diam.; limb distinctly 2-lipped; upper lip 2-lobed, lobes triangular to semicircle, 1.0–1.2 cm long; lower lip 3-lobed, lobes oblong to triangle, 1.0–1.3 cm long. Stamens 2, adnate to c. 1.2 cm above corolla base; filaments white, c. 1 cm long, geniculate c. 3 mm above insertion, glabrous; anthers reniform, slightly constricted at middle, c. 3.5 mm long; staminodes 3, lateral ones adnate to c. 1 cm above corolla base, white, linear, c. 7.5 mm long, glabrous, apex small capitate, central staminode adnate to c. 0.5 cm above corolla base, punctiform, inconspicuous, c. 0.5 mm long. Disk annular, c. 10 mm high; pistil c. 2.3 cm long, densely covered with white eglandular hairs; style linear, c. 1.3 cm, pubescent; stigma trapeziform, 2-lobed to irregularly lobed, c. 2.5 mm long. Fruit c. 5 cm long, straight to slightly curved, hairy.

Flowering and fruiting period: April – June.

Etymology: The species epithet '*wenii*' is used to commemorate a Chinese botanist, Fang Wen, who studied Gesneriaceae of China for many years, and is devoted to the conservation and taxonomy of Chinese Gesneriaceae plants.

Vernacular name: Chinese: Wēn Shì Bào Chūn JǜTái.

Habitat: *Primulina wenii* grows on shaded moist turf of limestone cliff in subtropical limestone evergreen broad-leaf forest on a North-facing slope of a limestone hill at an altitude of about 130 meters. The climate of Fuzhou, Fujian province is the monsoon of subtropical moist marine climate zone. The average annual temperature of Fuzhou is 20.1°C, the average annual precipitation is c. 1718.1 mm.

Notes: East China is not the area with high biodiversity of Gesneriaceae in China. For example, only six new species of Gesneriaceae, namely *Primulina xiziae* Fang Wen, Yue Wang & G. J. Hua (Li *et al.*, 2012), *P. chizhouensis* Xin Hong, S.B. Zhou & F.Wen (Hong *et al.*, 2012), *P. suichuanensis* X.L. Yu & J.J. Zhou (Zhou *et al.*, 2016), *Didymocarpus dissectus* Fang Wen, Y. L. Qiu, Jie Huang & Y.G. Wei (Wen *et al.*, 2013), *Oreocharis striata* Fang Wen & C.Z. Yang (Yang *et al.*, 2015) and *Beccarinda baolianis* Q.W. Lin (Lin, 2016) were discovered and described in the past decade. Further, Fujian province, belongs to E China, only possesses three species of *Primulina* before this new one was discovered. This also implies that more field investigations will help more discoveries.

Molecular analysis

The combined matrix had a length of 1,670 characters, 672 for ITS and 998 for *trn*L-F. Of the 254 (15.2%) varied, 257 (15.4%) were parsimoniously informative, including the indels. The maximum parsimony analysis on the combined matrix resulted in two trees of 937 steps in length, a consistency index (CI) of 0.7311, retention index (RI) of 0.6644 and homoplasy index (HI) of 0.2689. The strict consensus tree (Fig. 3) was highly resolved and tree topology was consistent with the previous phylogenetic study by Möller *et al.* (2011) and Kang *et al.* (2014).

DNA analyses of sequence data show that the new species is sister to a clade (BS = 100) comprising of *Primulina tabacum*, *P. danxiaensis*, *P. juliae* and *P. xiuningensis*. Of these species the undescribed species is morphologically most similar to *P. juliae* from which it can be distinguished by certain qualitative and quantitative characters in leaf blade, bract, indumentum, corolla, filament, stigma and pistil. For instance, it is obviously different that the peduncles of *P. wenii* are densely covered with erectly spreading white eglandular pubescent and villous hairs, the number of bracts is three, filaments are glabrous, and pistil is densely covered with white glandular puberulent hairs. The morphological differences between the two species (*P. wenii* and *P. juliae*) are depicted in Table 2.

Acknowledgements

The authors would like to thank Dr. Fang Wen and Prof. Yi-Gang Wei for identifying specimens and cultivating plants. We also thank some anonymous reviewers for their comments on an earlier draft of this paper. This study was supported by the Guangxi Natural Science Foundation (2015GXNSFBB139004) and the Key Research and Development Project of Guangxi (Guike AB16380053).

References

Chase, M.W. and Hills, H. 1991. Silica gel: an ideal material for field preservation of leaf samples for DNA studies. Taxon **40**: 215–220.

Fang, D. and Qin, D.H. 2004. *Wentsaiboea* D. Fang. & D.H. Qin, a new genus of the Gesneriaceae from Guangxi, China.ActaPhytotax. Sin. **42**: 533–536.

Geng, H.Q., Ma, Q.X., Yan, Y.H., Li, Y.F., Jiang, Q.Y., Lǘ, L.L., Wu, W.H., Ding, X. and Hou, X.L. 2014. New materials of plants in Fujian province (II).Subtrop. Plant Sci. **43**: 24–28.

Hance, H.F. 1883. New Chinese Cyrtandreae. J. Bot. **21**: 165–170.

Hong, X., Zhou, S.B. and Wen, F. 2012. *Primulina chizhouensis* sp. nov. (Gesneriaceae), a new species from a limestone cave in Anhui, China. Phytotaxa **50**: 13–18.

IUCN 2012. IUCN Red List categories and criteria. Version 3.1, ed. 2. Gland & Cambridge, IUCN.

Kang, M., Tao, J., Wang, J., Ren, C., Qi, Q., Xiang, Q.Y. and Huang, H. W. 2014. Adaptive and non-adaptive genome size evolution in Karst endemic flora of China. New Phytol. **202**: 1371–1381.

Li, J., Wang, Y., Hua, G.J. and Wen, F. 2012. *Primulina xiziae* sp. nov. (Gesneriaceae) from Zhejiang Province, China. Nordic J. Bot.**30**: 77–81.

Li, J.M. and Wang, Y.Z. 2007. Phylogenetic reconstruction among species of *Chiritopsis*and *Chirita* Sect. *Gibbosaccus* (Gesneriaceae) based on *nr*DNA ITS and *cp*DNA*trn* L-F sequences. Syst. Bot.**32**: 888–898.

Li, Z.Y. and Wang, Y.Z. 2004. *Primulina*, *Chirita* and *Chiritopsis*. *In*: Li, Z.Y. and Wang, Y.Z. (eds), Plants of Gesneriaceae in China. Henan Sci. & Technol. Publ. House, Zhengzhou, pp. 170–282.

Lin, Q.W. 2016.*Beccarinda baolianis*, a new species of Gesneriaceae from Fujian Province. Bull. Bot. Res. **36**: 650–652.

Liu, X.L. and Guo, X.H. 1989. A new species of *Chiritopsis* from Anhui. Bull. Bot. Res. **9**: 51–54.

Liu, Y., Xu, W.B. and Pan, B. 2010. *Wentsaiboea tiandengensis* sp. nov.and *W.luochengensis* sp. nov. (Gesneriaceae) from Karst caves in Guangxi, southern China. Nordic J. Bot **28**: 739–745.

Möller, M., Pfosser, M., Jang, C.G., Mayer, V., Clark, A., Hollingsworth, M.L., Barfuss, M.H.J., Wang, Y.Z., Kiehn, M.and Weber, A. 2009. A preliminary phylogeny of the 'Didymocarpoid Gesneriaceae' based on three molecular data sets: incongruence with available tribal classifications. Am. J. Bot. **96**: 989–1010.

Möller, M. and Clark, J.L. 2013. The state of molecular studies in the family Gesneriaceae: A review. Selbyana **31**: 95–125.

Möller, M., Forrest, A., Wei, Y.G. and Weber, A. 2011. A molecular phylogenetic assessment of the advanced Asiatic and Malesian didymocarpoid Gesneriaceae with focus on non-monophyletic and monotypic genera. Plant Syst. Evol. **292**: 223–248.

Möller, M., Wei, Y.G., Wen, F., Clark, J.L. and Weber, A. 2016. You win some you lose some: updated generic delineations and classification of Gesneriaceae-implications for the family in China. Guihaia**36**: 44–60.

Skog, L. E. and Boggan, J.K. 2007. World checklist of Gesneriaceae. Department of Botany, Smithsonian Institution, Washington, DC. http://botany.si.edu/Gesneriaceae/Checklist [accessed 28 April 2010].

Tamura, K., Peterson, D., Peterson, N., Stecher, G., Nei, M. and Kumar, S. 2011. MEGA5: molecular evolutionary genetics analysis using maximum likelihood, evolutionary distance, and maximum parsimony methods. Mol. Biol. Evol. **28:** 2731–2739.

The International Plant Names Index. 2015+. [continuously updated]: *Primulina*. Published at http://www.ipni.org/ [accessed 12 Feb 2015]

Thiers, B. 2015. Index Herbariorum: a global directory of public herbaria and associated staff. New York Botanical Garden: published at http://sweetgum.nybg.org/ih/ [accessed 3 Feb 2015].

Wang, W.T. 1981. Quinque genera nova gesneriacearum E Sina. Bull. Bot. Res. **1**: 21–51.

Wang, W.T. 1985. A revision of the genus *Chirita* (Gesneriaceae) in China (II). Bull. Bot. Res., Harbin **5**: 37–86.

Wang, W.T., Pan, K.Y. and Li, Z.Y. 1990. Gesneriaceae. *In*: Wang, W.T. (Ed.), Flora Reipublicae Popularis Sinicae**69**. Science Press, Beijing, pp. 125–581.

Wang W.T., Pan, K.Y. and Li, Z.Y. 1998. Gesneriaceae. *In*: Wu, Z.Y. and Raven, P.H. (eds), Flora of China Vol. **18**. Science Press, Beijing & Missouri Bot. Garden Press, St. Louis, 322 pp.

Wang, Y.Z., Mao, R.B., Liu, Y., LI, J.M., Dong, Y., LI, Z.Y. and Smith, J. F. 2011. Phylogenetic reconstruction of *Chirita*and allies (Gesneriaceae) with taxonomic treatments. J. Syst. Evol. **49**: 50–64.

Weber, A., Middleton, D.J., Forrest, A., Kiew, R., Lim, C.L., Rafidah, A.R., Sontag, S., Triboun, P., Wei, Y.G., .Yao, T.L. and Möller, M. 2011. Molecular systematics and remodelling of *Chirita* and associated genera (Gesneriaceae). Taxon **60**: 767–790.

Weber, A. and Skog, L.E. 2007. The genera of Gesneriaceae. Basic information with illustration of selected species. Ed. 2. Available from http://www.genera-gesneriaceae.at/.

Wei, Y.G., Wen, F., Möller, M., Monro, A., Zhang, Q., Gao, Q., Mou, H.F., Zhong, S.H. and Cui, C. 2010. Gesneriaceae of South China. Guangxi Science and Technology Publishing House, Nanning, 777 pp.

Wen, F., Qiu, Y.L., Huang, J., Zhao, B. and Wei, Y.G. 2013. *Didymocarpus dissectu s*sp. nov. (Gesneriaceae) from Fujian, eastern China. Nordic J. Bot. **31**: 316–320.

Wen, F., Lai, B.D., Zhao, Z.G., He, J.Y. and Jiang, B.S. 2015. *Primulina heterochroa* (Gesneriaceae), a new species from a tropical limestone area in Guangxi, China. Willdenowia **45**: 45–51.

Yang, C.Z., Cai, D.L. and Wen, F. 2015. *Oreocharis striata* (Gesneriaceae), a new species from Fujan, China. Ann. Bot. Fennici **52**: 369–372.

Xu, W.B., Zhang, Q., Wen, F., Liao, W.B., Pan, B., Chang, H. and Chung, K.F. 2012. Nine new combinations and one new name of *Primulina*. Phytotaxa **64**: 1–8.

Zhang, Y.T. 1993. Gesneriaceae. *In*: Lin, L.G. (ed.), Flora of Fujian **3**. Fuzhou: Fujian Science and Technology Publishing House.

Zhou, D.S., Zhou, J.J., Li, M. and Yu, X.L. 2016: *Primulina suichuanensis* sp. nov. (Gesneriaceae) from Danxia landform in Jiangxi, China. Nordic J. Bot. **34**: 148–151.

31

ELEOCHARIS NEGLECTA (CYPERACEAE), A NEW SPECIES FROM KONKAN REGION OF MAHARASHTRA, INDIA

D.B. BORUDE, A.N. CHANDORE[1], A.R. GHOLAVE[2] AND S.R. YADAV[2]

Department of Botany, Abasaheb Marathe Arts and New Commerce, Science College, Rajapur 416702, District Ratnagiri, Maharashtra, India

Keywords: Cyperaceae; *Eleocharis*; India; Konkan; New species.

Abstract

Eleocharis neglecta Borude, Chandore, Gholave & S.R. Yadav, a new species from Konkan region of Maharashtra, India is described and illustrated. This species belongs to *Eleocharis* subgen. *Limnochloa* (P. Beauv. *ex* Lestib.) Torr. and it is closely allied to *Eleocharis dulcis* (Burm. f.) Trin. *ex* Hensch. in having terete and septate culms.

Introduction

The genus *Eleocharis* R.Br. (Cyperaceae) is distributed worldwide with c. 200 species (Gonzalez-Elizondo and Peterson, 1997) however; recently Govaerts *et al.* (2016) have reported c. 300 species for the world. The genus is widely distributed from tropical to temperate regions of both hemispheres. In India, it is represented by about 21 species (Prasad and Singh, 2002; Wadoodkhan, 2015). Recently *Eleocharis wadoodii* S.R. Yadav, Lekhak & Chandore and *Eleocharis konkanensis* Chandore, Borude, Kambale & S.R. Yadav have been described from the state of Maharashtra. (Yadav *et al.*, 2009; Chandore *et al.*, 2016). Another species of *Eleocharis* i.e. *E. khandwaensis* Mujaffar, Chandore & S.R. Yadav has been described from state of Madhya Pradesh (Mujaffar *et al.*, 2014). Similarly, one more species of *Eleocharis* i.e. *E. setifolia* (A. Rich) Raynal has been recorded for India by Wadoodkhan (2015). Therefore the present account of *Eleocharis* in India is about 25 species including 3 new species and one addition.

During our floristic survey, some specimens of *Eleocharis* were collected along with *Eleocharis dulcis* from estuaries of Ansure, Dandaansure, Tulsunde and Vengurle of Konkan region of Maharashtra, India. Careful observations were made on the specimens in field as well as in laboratory under stereo-zoom microscope. After examination of relevant literature (Cook, 1996; Koyama, 1985; Wadoodkhan, 2015) and critical analysis of plant specimens revealed that it is to date undescribed species of *Eleocharis*. Hence, it is described and illustrated here as a new species.

Eleocharis neglecta Borude, Chandore, Gholave & S.R. Yadav, **sp. nov.** **(Figs 1-3)**.

Diagnosis: The new species is similar to *Eleocharis dulcis* (Burm. f.) Trin. *ex* Hensch., but differs in its subrigid culms (vs. soft culms); culms slender after pressing and drying (vs. culms flat after pressing and drying); spikelet angular, 15–35 flowered, distinctly wider than culms (vs. cylindrical, 60–120 flowered, not wider than culms); sterile glumes ovate, acute at apex, (vs. obovate, acuminate at apex); fertile glumes fairly imbricate, c. ½ portion overlap by other glumes, persistent (vs. closely imbricate, c. ¾ portion overlaps by other glumes, deciduous); epidermal cells on achene surface longitudinally oblong–hexagonal (vs. epidermal cells on achene surface isodiametrically hexagonal to pentagonal).

[1]Corresponding author. Email: arunchandore@gmail.com

[2]Department of Botany, Shivaji University, Kolhapur 416004, Maharashtra, India.

Types: INDIA. Maharashtra, Ratnagiri District, Rajapur, on the way of Dandansure road, 16°33'736"N, 73°21'412"E, altitude 13 m, 30 August 2015, *A.N. Chandore* 1951 (*Holotype*: CAL; *Isotypes*: BSI, K, SUK).

Fig. 1. ***Eleocharis neglecta*** Borude, Chandore, Gholave & S.R. Yadav, **sp. nov.** A. Habitat (network of young culms); B. Habitat; C. Spikelets; D. Roots & stolon; E. T. S. of culm; F. Flower; G. Achene; H. Achene (SEM) (all from A.N.Chandore 1951). Photo plate by A.N. Chandore.

Fig. 2. A. ***Eleocharis neglecta*** habitat; B. ***E. dulcis*** habitat; C. & E. ***E. neglecta*** spikelets; D. & F. ***E. dulcis*** spikelet; G. ***E. neglecta*** culm; H. ***E. dulcis*** culm; I. ***E. neglecta*** fertile glume; J. ***E. dulcis*** fertile glume; K. ***E. neglecta*** sterile glume; L. ***E. dulcis*** sterile glume; M. ***E. neglecta*** achene; N. ***E. dulcis*** achene; O. ***E. neglecta*** epidermal cell on achene surface; P. ***E. dulcis*** epidermal cell on achene surface. Photoplate by A.N. Chandore.

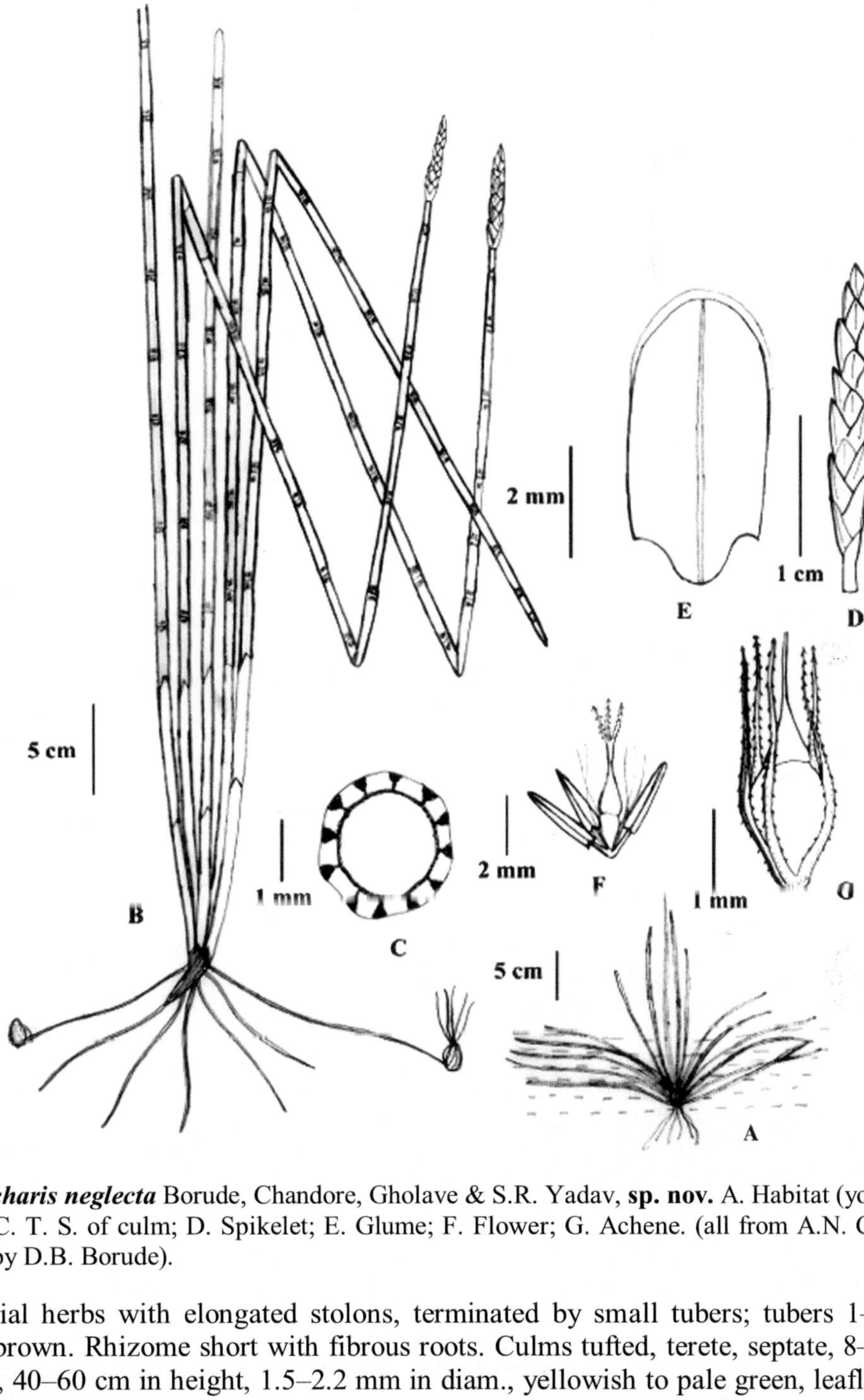

Fig. 3. ***Eleocharis neglecta*** Borude, Chandore, Gholave & S.R. Yadav, **sp. nov.** A. Habitat (young culms); B. Habit; C. T. S. of culm; D. Spikelet; E. Glume; F. Flower; G. Achene. (all from A.N. Chandore 1951, drawn by D.B. Borude).

Perennial herbs with elongated stolons, terminated by small tubers; tubers 1–3 mm thick, whitish to brown. Rhizome short with fibrous roots. Culms tufted, terete, septate, 8–15 culms per plant, erect, 40–60 cm in height, 1.5–2.2 mm in diam., yellowish to pale green, leafless; sheaths 2 per culm, outer sheath small, 4–5 cm long, dark brown, oblique, acute at apex; inner sheath membranous, 10–14 cm long, oblique at mouth, sheath apex acuminate. Spikelet solitary, terminal, straight, elongated, angular, 1.5–2.5 cm long, 1.5–2.5 mm wide, distinctly wider than the culms, acute, many flowered, straw brown; rachilla persistent. Glumes 20–30 per spikelet; lower glumes sterile, ovate, acute at apex, 5.0–6.0 × 3.0–4.0 mm, margin membranous; fertile glumes oblong-ovate to elliptic, 4.5–5.0 × 1.8–2.0 mm, faintly striations, imbricate, persistent, convolute when dry, obtuse at apex, scarious along margins towards the apex, mid-nerve prominent.

Hypogynous bristles 6 to 7, unequal, two to three times longer than achene, retrorsely barbed throughout, linear, gradually narrowed upwards, slightly connate at base, pale brown, c. 4 mm long. Stamens 3; anthers linear, c. 3 mm long, apiculate, yellowish brown; filament up to 7 mm long, translucent or whitish. Style 3-branched, c. 3.2 mm long; style base conical, persistent on the nut, c. 0.6 × 0.6 mm, as wide as the narrowed annular apex, brown to dark brown, flattened; stigmas 3, c. 3 mm long, scaberulous. Achene obovoid, biconvex, swollen at middle, 1.4–1.7 × 0.8–1.0 mm (excluding style-base), narrowed at base, yellowish to light brown, brown at maturity, surface appearing smooth; but under high magnification epidermal cell minute arranged in several vertical rows on each face, cells longitudinally oblong–hexagonal.

Phenology: Flowering and Fruiting from August to October.

Etymology: The specific epithet *neglecta* is given because it was a neglected species.

Distribution: India. Maharashtra, Konkan region, Sindhudurg District- Vengurle; Ratnagiri District- Ansure, Dandaansure, Tulsunde.

Habitat: *Eleocharis neglecta* Borude, Chandore, Gholave & S.R. Yadav grows along side of estuaries of Konkan and road side at an altitude of about 7-50 m in associations with *Cyperus rotundus* L., *Eleocharis dulcis* (Burm. f.) Trin. ex Hensch., *E. geniculata* (L.) Roem. & Schult., *Fimbristylis acuminata* Vahl, *F. dichotoma* (L.) Vahl, *F. tomentosa* Vahl, *Zoysia matrella* (L.) Merr., etc.

Additional specimens examined (Paratypes): INDIA. Maharashtra, Konkan, Ratnagiri District, Ansure, 30 August 2015, *A.N. Chandore* 1959 (SUK); Ratnagiri District, Tulsunde, 31 August 2015, *A.N. Chandore* 1960 (SUK); Sindhudurg District, Vengurle, 13 September 2015, *A N. Chandore* 1971 (SUK).

Note: *Eleocharis neglecta* Borude, Chandore, Gholave & S.R. Yadav generally grows in association with *Eleocharis dulcis*; therefore earlier workers might have neglected this species or wrongly identified it as *E. dulcis*. A comparative account of *E. neglecta* Borude, Chandore Gholave & S.R. Yadav with *E. dulcis* (Burm. f.) Trin. *ex* Hensch., is given in Table 1.

Table 1. Comparison between *Eleocharis dulcis* (Burm. f.) Trin. *ex* Hensch., and *E. neglecta* Borude, Chandore, Gholave & S.R. Yadav.

Characters	*E. dulcis*	*E. neglecta* **sp. nov.**
Culms	Soft, deep green, flat after pressing and drying	Subrigid, yellowish to pale green, slender after pressing and drying
Culm size	5.0–9.0 mm in diameter, 50–100 cm in height	1.5–2.2 mm in diameter; 40–60 cm in height
Inner sheath	15–25 cm long	10–14 cm long
Spikelets	Cylindrical; 3.0–5.5 cm long, 4.0 – 6.0 mm wide, not wider than culms	Angular; 1.4–2.8 cm long, 1.5–2.5 mm wide; distinctly wider than culms
Sterile glumes	Obovate, acuminate, 4.0–5.0 × 5.0–6.0 mm	Ovate, acute at apex, 5.0–6.0 × 3.0–4.0 mm
Fertile glumes	Closely imbricate, c. ¾ portion overlaps by other glumes, deciduous	Fairly imbricate, c. ½ portion overlap by other glumes, persistent
Flowers	15–35 flowers per spikelet	60–120 flowers per spikelet
Style	2-branched (rarely 3-branched)	3-branched
Achene	Obovoid to suborbicular, 1.8–2.0 × 1.0–1.2 mm	Obovoid, 1.4–1.7 × 0.8–1.0 mm
Surface of Achene (high magnification)	Epidermal cells isodiametrically hexagonal to pentagonal	Epidermal cells longitudinally oblong–hexagonal

Acknowledgements

We are thankful to the Head, Department of Botany, Shivaji University, Kolhapur for providing necessary facilities. The authors are grateful to Director, Botanical Survey of India for permission to consult *Eleocharis* herbarium. ANC & DBB are thankful to SERB, Department of Science and Technology (DST), New Delhi for financial assistance (File No.:-SR/FT/LS-82/2012), under DST Fast Track Young Scientist Scheme. Dr. ANC is thankful to the Principal, Abasaheb Marathe Arts & New Commerce Science College, Rajapur, for laboratory facilities.

References

Chandore, A.N., Borude, D.B., Kambale, S.S. and Yadav, S.R. 2016. *Eleocharis konkanensis,* a new species of Cyperaceae from the Konkan region of Western Ghats, India. Phytotaxa **252** (2): 154–158.

Cook, C.D.K. 1996. Aquatic and Wetland Plants of India. Oxford University Press, London, pp 126–131.

Gonzalez-Elizondo, M.S. and Peterson, P.M. 1997. A classification of and key to the supraspecific taxa in *Eleocharis* (Cyperaceae). Taxon **46**: 433–449.

Govaerts, R. 2016. World Checklist of Cyperaceae, <http://apps.kew.org/wcsp/qsearch.do>. Royal Botanic Gardens, Kew. Retrieved on 13 January 2016.

Koyama, T. 1985. *Eleocharis* R.Br. Cyperaceae *In*: Dassanayake M.D., Fosberg F.R. (Ed.) A revised Handbook to the Flora of Ceylon 5. Oxford & IBH, New Delhi, India, pp. 255–267.

Mujaffar, S., Chandore, A.N., and Yadav, S.R. 2014. *Eleocharis khandwaensis* sp. nov. (Cyperaceae) from the Madhya Pradesh, India. Nordic Journal of Botany **32**: 710–712.

Prasad, V.P. and Singh, N.P. 2002. Sedges of Karnataka (India). Reprinted from J. Econ. Taxon. Bot. Addl. Ser. No. 21. Scientific Publisher, Jodhpur, 137 pp.

Wadoodkhan, M.A. 2015. Cyperaceae of Western Ghats, West Coast and Maharashtra. Dattsons Publishers, Nagpur, pp. 126–148.

Yadav, S.R., Lekhak, M. and Chandore, A.N. 2009. A New Species of *Eleocharis* (Cyperaceae) from Western Ghats, India. Rheedea **19**: 37–40.

32

COLOCASIA HASSANII (ARACEAE), A NEW SPECIES OF AROID FROM BANGLADESH

HOSNE ARA[1]

Bangladesh National Herbarium, Chiriakhana Road, Mirpur-1, Dhaka-1216, Bangladesh

Keywords: New species; Araceae; *Colocasia hassanii*; Bangladesh.

Abstract

Colocasia hassanii, a new species of Araceae from Bangladesh, is described and illustrated. *C. hassanii* is closely related to *C. esculenta* (L.) Schott but differs from the later by the number of constriction (two) in spathe, presence of a sterile male zone in spadix, and bitter in taste. A morphological comparison between *C. hassanii* and *C. esculenta* is provided.

Introduction

The genus *Colocasia* Schott is represented by 13 species in the world (Yin, 2006) and eight species are found in Asia and Malay Archipelago (Mayo *et al.*, 1997). In Bangladesh, so far this genus was known to contain the following nine species: *C. affinis* Schott, *C. esculenta* (L.) Schott, *C. fallax* Schott, *C. gigantea* (Blume) Hook. f., *C. heterochroma* H. Li *et* Z.X. & Wei, *C. lihengiae* C.L. Long *et* K.M. Liu, *C. mannii* Hook. f., *C. oresbia* A. Hay and *C. virosa* Kunth. (Ara, 2007; Ara and Hassan, 2012).

During the revisionary work on Araceae since 1988 from Bangladesh the author has made an extensive field survey throughout the country and collected a few number of *Colocasia* Schott specimens which did not match with any species previously known. The flowering and fruiting specimens were collected and preserved for identification. The rhizomes were also collected and grown in the garden of Bangladesh National Herbarium (DACB), Dhaka and in the Botanical Garden of the University of Dhaka. The specimens were critically examined and compared with the identified specimens of *Colocasia* Schott available at BK, BKF, BM, CAL, DACB, K, DUSH (Dhaka University Salar Khan Herbarium), HCU (Herbarium of Chittagong University), BCSIRH (Herbarium, Bangladesh Council for Scientific and Industrial Research) and BFRIH (Herbarium, Bangladesh Forest Research Institute). Moreover, consultation of relevant literature (Wallich, 1829-1849; Roxburgh, 1832; Wight, 1843-1845; Hooker, 1893; Prain, 1903; Engler and Krause 1920; Haines, 1924; Heinig, 1925; Hill, 1939; Raizada, 1941; Sinclair, 1956; Mitra, 1958; Hu, 1968; Hotta, 1970; Rao and Verma, 1976; Nicolson, 1976, 1979, 1987; Madison, 1978; Nasir, 1978; Heng, 1979; Mayo, 1985; Karthikeyan *et al.,* 1989; Naithani, 1990; Heng and Wei, 1993; Noltie, 1994; Hay, 1996; Pullaiah, 1997; Toha, 2000; Long and Liu, 2001; Yin, 2006) on the family Araceae revealed that collected specimens differed from other described species of the genus *Colocasia* Schott. These specimens appeared to be closely related to *C. esculenta* Schott morphologically but differed with some important characters. After critical study these specimens were recognized as to belong a new species, *Colocasia hassanii* **sp. nov.**

[1]Corresponding author. Email: bnh_mirpur@yahoo.com

Results and Discussion

Colocasia hassanii H. Ara, **sp.nov.** **(Figs 1& 2)**.

Diagnosis: *Colocasia hassanii* H. Ara is closely related to *C. esculenta* (L.) Schott but can be easily differentiated by the number of constriction (two) present in spathe, presence of sterile male zone in spadix and bitter in taste.

Bengali name: *Tita Kachu.*

English name: Bitter Taro (proposed).

Holotype: Bangladesh, Bandarban district, on the way to Betchari, 22.9.2004, Hosne Ara HA 1215, 1216 (DACB).

Annual herb. Underground corm perennial, c. 2.5 cm in diam., and c. 2 cm high; stolons 1–2, trailing horizontally, non-branching, thin, pale green or pale purple, c. 20 cm long, c. 0.5 cm in diam. Leaves 4–6, peltate, 15–45×8–30 cm, ovate, acute, cordate, dark green above and light green beneath, base shallowly cordate, glaucous, venation pinnately reticulate, sinus 1–3 cm long; petiole 20–65 cm long, sheathing for about one-third at the base. Inflorescence solitary or paired. Peduncle 10–14 cm long, shorter than petiole. Spathe having two constrictions, 15.0–25.5 cm long, lower convolute part (tube) light green, 4.0–6.5 cm long, c. 2 cm in diam., nearly cylindrical; upper part lanceolate, light yellow, 11–19 cm long. Spadix sessile, shorter than spathe, 10–19 cm long. Female zone cylindrical, 2.8–3.0 cm long, c. 1 cm in diam.; staminode c. 1.8 cm long, creamy; sterile portion slender, c. 1.8 cm long, c. 0.4 cm broad at the middle, creamy; male portion 2.1–2.8 cm long, 0.4–0.5 cm in diam.; a sterile male portion present below the appendix, 0.4–0.5 cm long, c. 0.2 cm in diam.; appendix 6.5–10.0 cm long, 0.3–0.5 cm in diam. Ovary numerous, narrow, green, 3–5×2–3 mm; style very short, green; stigma c. 0.05 cm in diam., yellow, placentation parietal, ovules 6, c. 1.5×1.0 mm; staminode cream in colour, c. 0.15×0.2 cm; male flower 6-8 androus, c. 0.1×0.12 cm. Fruit a berry.

Flowering and fruiting period: June to October.

Ecology: Grows in the hilly areas as undergrowth.

Distribution: Eastern hilly areas of Bangladesh (Bandarban, Khagrachari and Rangamati hill districts), Cox's Bazar and Moulvibazar districts.

Specimens examined: Bandarban: On the way to Betchari, 22.9.2004, Hosne Ara HA 1215, 1216 (DACB); Meghla forest area, 23.9.2004, Hosne Ara HA 1291, 1292 (DACB); Cox's Bazar: Chota Inani, 30.9.2005, Hosne Ara HA 2432 (DACB); Bara Inani, 30.9.2005, Hosne Ara HA 2462 (DACB); Himchari area, 30.9.2005, Hosne Ara HA 2522 (DACB); Khagrachari: Alutilla, 11.7.2003, Hosne Ara and Sarder Nasir Uddin HA 461 (DACB); Jamtoli, 12.7.2003, Hosne Ara HA 484 (DACB); Moulvibazar: Adampur beat, Kawargola forest, 3.7.2005, Hosne Ara HA 1747 (DACB); Lawachara reserve forest, 4.7.2005, Hosne Ara HA 1777 (DACB); Adampur beat, Kawargola forest, 6.10.2005, Hosne Ara HA 2625 (DACB); Rangamati: Kaptai, Shilsori village, Velbapara, 8.7.2003, Hosne Ara and Sarder Nasir Uddin HA 391 (DACB); Kaptai, Sitapahar, 14.10.2003, Hosne Ara HA 711 (DACB); Rajbari area, 18.9.2004, Hosne Ara HA 1122 (DACB); Dhaka: Bangladesh National Herbarium garden (Cultivated), 18.6.2015, Hosne Ara HA 2881 (DACB) [Originally collected from Bandarban district].

Ethnobotanical information: The indigenous people of Bangladesh use its petioles in curries.

Etymology: This species is named in honour of Dr. Mohammad Abul Hassan, Professor of Botany, University of Dhaka, who has made outstanding contribution to the taxonomy of flowering plants in Bangladesh.

Note: The chromosome number has been determined for the new species *Colocasia hassanii*. Preliminary determination of 2n chromosome number appears as 28.

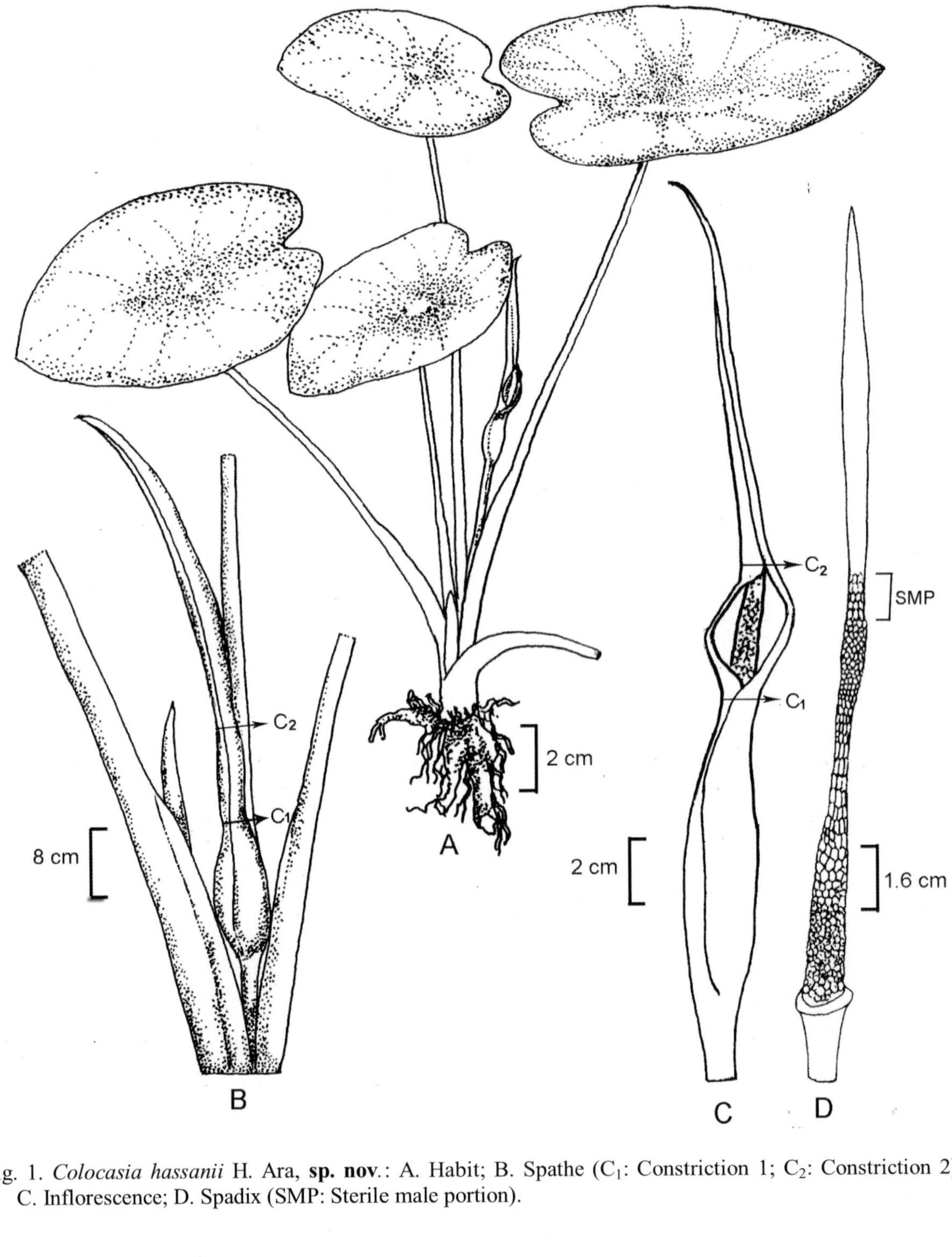

Fig. 1. *Colocasia hassanii* H. Ara, **sp. nov**.: A. Habit; B. Spathe (C_1: Constriction 1; C_2: Constriction 2); C. Inflorescence; D. Spadix (SMP: Sterile male portion).

Fig. 2. *Colocasia hassanii* H. Ara, **sp. nov**.: A. Tuber; B. Habit at BNH garden; C. Habit at home garden; D. Petiole; E.F.G. Inflorescence; H.I.J. Spadix; K. Appendix; L. Infructescence.

The major morphological differences between *C. hassanii* sp. nov. and *C. esculenta* are outlined in Table 1.

Table 1. Morphological comparison of *Colocasia hassanii* sp. nov. with *C. esculenta*.

Characters	*Colocasia hassanii* **sp. nov.**	*Colocasia esculenta*
Petiole	c. 62 cm long, green or purple, spotted	c. 100 cm long, green or purple, not spotted
Spathe	Two constrictions present between tube and limb	One constriction present between tube and limb
Tube and limb of the spathe	Both sides of the lower spathe light green and both sides of the limb light yellow and leathery	Both sides of the lower spathe green and both sides of the limb golden yellow and leathery
Spadix	Sterile male portion present	Sterile male portion absent
Tip of the appendix	Pointed	Obtuse (blunt)
Taste	Bitter	Not bitter

Conservation status: *Colocasia hassanii* sp. nov., according to IUCN Red list category (IUCN, 2017) is considered as Least Concern (LC) as the species is found in Bandarban, Cox's Bazar, Khagrachari, Moulvibazar and Rangamati and seems to be available to the local people.

Acknowledgements

The author is grateful to the authorities of the BK, BKF, BM, CAL, DACB, K, DUSH, HCU, BCSIRH and BFRIH for providing facilities to consult aroid materials and their libraries. The author expresses her gratitude to Professor Dr. M. Oliur Rahman, Department of Botany, University of Dhaka for his encouragement and help during preparation of the manuscript. Thanks are also due to Ms. Mahmuda Akter, Senior Artist-cum-illustrator, Bangladesh National Herbarium for drawing the illustrations, and the researchers at Cytogenetics Laboratory of the Department of Botany, University of Dhaka for their cooperation in chromosomal investigation of the new species *Colocasia hasanii*.

References

Ara, H. 2007. Araceae. *In*: Siddiqui, K.U., Ahmed, Z.U., Begum, Z.N.T., Hassan, M.A., Khondker, M., Rahman, M.M., Kabir, S.M.H., Ahmed, M., Ahmed, A.T.A., Rahman, A.A. and Haque, E.U. (Eds), Encyclopedia of Flora and Fauna of Bangladesh, Vol. **11**. Angiosperms: Monocotyledons (Agavaceae-Najadaceae). Asiatic Society of Bangladesh, Dhaka, pp. 19–98.

Ara, H. and Hassan, M.A. 2012. Five new records of Aroids of Bangladesh. Bangladesh J. Plant Taxon. **19**(1): 17–23.

Engler, A. and Krause, K. 1920. Additamentumad Araceas-Philodendroideas, Araceae-Colocasioideae. *In*: Engler, A. (Ed.), Das Pflanzenreich **71** (IV.23E): 3–132.

Haines, H.H. 1924 (Reprint 1978). Araceae. Botany of Bihar and Orissa, Part V-VI. Bishen Singh Mahendra Pal Singh, Dehra Dun, India, pp. 856–873.

Hay, A. 1996. A new Bornean species of *Colocasia* Schott (Araceae: Colocasieae) with a synopsis of the genus in Malesia and Australia. Sandakania **7**: 31–48.

Heinig, R.L. 1925. List of Plants of the Chittagong Collectorate and Hill Tracts. The Bengal Government Branch Press, Darjeeling, India, pp. 74–75.

Heng, Li. 1979. Araceae (excluding *Acorus*). *In*: Cheng, Wu and Heng, Li (Eds), Fl. Reipubl. Popularis Sin. **13**(2): 1–242.

Heng, Li and Wei, Z.X. 1993. *Colocasia heterochroma*, a new species of *Colocasia* from Araceae. Acta. Bot. Yunnanica **15**(1): 16–17.

Hill, A.F. 1939. The nomenclature of the taro and its varieties. Bot. Mus. Leafl. Harv. Univ. **7**(7): 113–118.

Hooker, J.D. 1893 (Reprint 1954). The Flora of British India, Vol. **6**. L. Reeve & Co. Ltd., Kent, England, pp. 490–555.

Hotta, M. 1970. A system of the family Araceae in Japan and adjacent areas. Mem. Fac. Sci. Kyoto Imp. Univ., Ser. Biol. **4**: 72–96.

Hu, S.Y. 1968. Araceae. Studies in the Flora of Thailand, no. 41. Dansk Bot. Arkiv **23**(4): 409–457.

IUCN 2017. IUCN Standards and Petitions Subcommittee Guidelines for Using the IUCN RedList Categories and Criteria. Version **13.** Prepared by the Standards and Petitions Subcommittee. Downloaded from http://www.iucnredlist.org/documents/RedListGuidelines.pdf. pp.1–108.

Karthikeyan, S., Jain, S.K., Nayar, M.P. and Sanjappa, M. 1989. Florae Indicae Enumeratio: Monocotyledonae. Flora of India Series **4**. Araceae. Botanical Survey of India, Calcutta, pp. 5–15.

Long, C.-L. and Liu, K.-M. 2001. *Colocasia lihengiae* (Araceae: Colocasieae), a new species from Yunnan, China. Bot. Bull. Acad. Sin. **42**: 313–317.

Madison, M. 1978. The genera of Araceae in the northern Andes. Aroideana **1**: 31–53.

Mayo, S.J. 1985. Araceae. *In*: Polhill, R.M. (Ed.), Flora of Tropical East Africa. Balkema, Rotterdam, the Netherlands, pp.1–71.

Mayo, S.J., Bogner, J. and Boyce, P.C. 1997. The Genera of Araceae. Royal Botanic Gardens, Kew, UK, pp. 1–370.

Mitra, J.N. 1958. Araceae. Flowering plants of Eastern India, Monocotyledons, Vol. **1**. The World Press Private Ltd, Calcutta, India, pp. 70–85.

Nasir, Y.J. 1978. Araceae. *In*: Nasir, E. and Ali, S.I. (Eds), Flora of West Pakistan, No. **120**, Agricultural Research Council, Islamabad, Pakistan, pp.1–17.

Naithani, H.B. 1990. Flowering Plants of India, Nepal and Bhutan (not recorded in Sir J. D. Hooker's Flora of British India). Rohini Printers and Publishers, Dehra Dun, India, pp. 451–455.

Nicolson, D.H. 1976 (Reprint 1978). Araceae. *In*: Saldanha, C.J. and Nicolson, D.H. (Eds), Flora of Hassan District, Karnataka, India. Amerind Publishing Co. Pvt. Ltd., New Delhi, Bombay, Calcutta and New York, pp. 781–790.

Nicolson, D.H. 1979. Araceae. *In*: Smith, A.C. (Ed.), Flora Vitiensis Nova, Vol. **1**. Pacific Tropical Botanical Garden, Hawaii, pp. 438–460.

Nicolson, D.H. 1987. Araceae. *In*: Dassanayake, M.D. (Ed.), A Revised Handbook to the Flora of Ceylon, Vol. **6**. Balkema, Rotterdam, the Netherlands, pp. 17–101.

Noltie, H.J. 1994. Araceae. *In:* Grierson, J.D. and Long, D.G. (Eds), Flora of Bhutan, Vol. **3**(1). Royal Botanic Garden, Inverleith Row, Edinburgh, UK, pp.121–158.

Prain, D. 1903. Bengal Plants **2**. Indian reprint (1963), Botanical Survey of India, Calcutta, pp. 830–840.

Pullaiah, T. 1997. Flora of Andhra Pradesh, Vol. **3**. Monocotyledons. Scientific Publishers, Jodhpur, India, pp. 1019–1030.

Raizada, M.B. 1941. On the Flora of Chittagong. Indian Forester **67**(5): 245–254.

Rao, A.S. and Verma, D.M. 1976. Materials towards a monocot flora of Assam-V. Bull. Bot. Surv. Ind. **18**(1-4): 9–34.

Roxburgh, W. 1832. Tetrandia Monogynia in Willium Carey's Flora Indica; or Descriptions of Indian Plants. Mission Press, Serampore, India, pp. 620–635.

Sinclair, J. 1956. The Flora of Cox's Bazar, East Pakistan. Bull. Bot. Soc. Beng. **9**(2): 110–111.

Toha, A.M.M. 2000. Biodiversity of aroids in southeast Bangladesh. M. Phil Thesis (unpublished). Aberdeen University, Scotland, U.K.

Wallich, N. 1829-1849. A numerical list of dried specimens of plants in the East Indian Company's Museum. Generally cited Wall. Cat., *ined.*

Wight, R. 1843-1845. Icones Plantarum Indiae Orientalis or Figures of Indian Plants, Vol. **3** (part 1). Messrs, Ostell, Lepage and Co., Calcutta. pp. 772–808.

Yin, J-T. 2006. *Colocasia tibetensis* (Araceae, Colocasieae), a new species from southeast Tibet, China. Ann. Bot. Fenn. **43**: 53–56.

33

THREE NEW SPECIES OF ARACEAE FROM BANGLADESH

HOSNE ARA[1] AND MD. ABUL HASSAN[2]

*Bangladesh National Herbarium, Chiriakhana Road, Mirpur-1
Dhaka-1216, Bangladesh*

Keywords: *Alocasia hararganjensis*; *Alocasia salarkhanii*; *Typhonium elatum*; New species; Araceae; Bangladesh.

Abstract

Three new species belonging to two genera of Araceae from Bangladesh, namely *Alocasia hararganjensis* H. Ara & M.A. Hassan, *Alocasia salarkhanii* H. Ara & M.A. Hassan and *Typhonium elatum* H. Ara & M.A. Hassan are described and illustrated. Diagnostic morphological characters of each of the new species are provided. A comparative morphological characteristic for each new species with their closest described species has also been provided.

Introduction

The family Araceae de Juss. is represented globally by 3,645 species under 144 genera (Boyce and Croat, 2011). The family is represented in Bangladesh by 27 genera and 109 species, of which 80 species are wild and 29 are cultivated (Ara, 2016).

During revisionary work of the family Araceae from Bangladesh the first author made an extensive field survey (137 field trips) throughout the country, especially in the hilly forest areas during 1988-2014 and collected 2,698 aroid specimens. Most of the collected specimens were identified by the first author. However, fifteen specimens remain distinct from any previously described and named species by different morphological characters. Twelve specimens identified belonging to the genus *Alocasia* (Schott) G. Don were not identical to any other described species. However, nine specimens of *Alocasia* (Schott) G. Don are closely related to *Alocasia fallax* Schott and remaining three are closely related to *A. fornicata* (Roxb.) Schott. Rest three specimens were identified as to belonged to the genus *Typhonium* Schott. These specimens are morphologically related to *Typhonium trilobatum* (L.) Schott. These specimens were critically examined and compared with the identified specimens of *Alocasia* (Schott) G. Don and *Typhonium* Schott available at BK, BKF, BM, CAL, DACB, K, DUSH (Dhaka University Salar Khan Herbarium), HCU (Herbarium of Chittagong University), BCSIRH (Herbarium, Bangladesh Council for Scientific and Industrial Research) and BFRIH (Herbarium, Bangladesh Forest Research Institute). Moreover, consultation of the relevent literatures (Wallich, 1829-1849; Roxburgh, 1832; Wight, 1843-1845; Hooker, 1893; Jackson, 1893-1955, Prain, 1903; Engler, 1920; Engler and Krause, 1920; Hu, 1968; Nicolson, 1976, 1979, 1987; Nasir, 1978; Nicolson and Sivadasan, 1981; Mayo, 1985; Karthikeyan *et al.,* 1989; Hay, 1991, 1993, 1998, 1999; Noltie, 1994; Sriboonma *et al.* 1994; Sookchaloem, 1995; Mayo *et al.*, 1997; Hetterscheid and Boyce, 2000; Toha, 2000; Boyce, 2007; Li *et al.*, 2010) on the family Araceae revealed that collected specimens differed from other described species of the genus *Alocasia* (Schott) G. Don and *Typhonium*

[1]Corresponding author, Email: bnh_mirpur@yahoo.com
[2]Department of Botany, University of Dhaka, Dhaka-1000, Bangladesh.

Schott. After critical study these specimens were recognized as new species, *Alocasia hararganjensis* sp. nov., *Alocasia salarkhanii* sp. nov. and *Typhonium elatum* sp. nov. The new species are described and illustrated below.

Results and Discussion

1. Alocasia hararganjensis H. Ara & M.A. Hassan, **sp. nov.** **(Figs 1 & 2).**

Diagnosis: *Alocasia hararganjensis* H. Ara & M.A. Hassan is closely related to *A. fallax* Schott but readily differentiated by the leaf shape which are narrowly to ovate sagittate with acute tip; tip of the cataphylls c. 8 cm long; and no sterile male zone above the sterile interstice.

Holotype: Bangladesh, Moulvibazar district, Gazipur beat, Hararganj reserve forest, 21.05.2005, Hosne Ara HA 1740 (DACB).

Massive sub-arborescent pachycaul herb, c 3.5 m high. Stem erect to decumbent, c. 10 cm in diameter, c 2.3 m long, clothed in the brown remains of old leaf bases. Leaves several together in terminal crown, held almost erect or slightly curved; petiole c. 1.2 m long, sheathing c. 1/2 from the base of petiole, eglandular, light green, wing of sheath out-rolled; blade narrowly sagittate to ovate-sagittate, slightly glossy, leathery, glabrous, dark-green adaxially, pale green abaxially, usually bullate, 40-50 cm long, base 40-48 cm in diameter, margin entire to slightly sinuate; anterior lobes 35-42 cm long with apiculate tip c. 1 cm long; anterior costa prominent on both surfaces, glabrous, primary veins 7-11 on each side, prominent on both surfaces, diverging at 45^0-70^0; secondary veins flush on both surfaces; interprimary collective veins well-defined; sub-marginal vein c. 3 mm from the margin; glands in the axils of primary veins absent or extremely inconspicuous; posterior lobes 30-35 cm long, rounded, peltate, c. 1.9 cm long; posterior costae straight. Inflorescences 10 in the centre of the leafy crown, bloom one after another, subtended by cataphylls; cataphylls 66-68 cm long, tip of the cataphylls pointed, c. 8 cm long, green; peduncle smooth, 50-64 cm long, 1.0-1.5 cm in diameter at the base, green. Spathe 18.5-21.0 cm long, constricted at level of sterile zone of spadix; lower spathe 4.5-5.0 cm long, light green, broadly ovoid-cylindric; limb 14-16 cm long, 6.5-7.0 cm in diameter, light greenish yellow, thinly leathery. Spadix shorter than spathe, 12-15 cm long, sessile. Female zone 1.8-2.0 cm long, 1.5-2.0 cm in diameter at the base, with 120-200 close-packed pistils; ovary green, ovoid to subglobose, 2-3 x 2.0-2.3 mm, unilocular, with basal placenta; style 0.5-0.8 mm long, 0.6 mm in diameter, cream coloured; stigma subglobose, shallowly 3-4 lobed, the lobes rounded, creamy; sterile interstice 2.0-2.5 cm long, with 7-8 whorls of rhombohexagonal synandrodia, the lowermost whorls isodiametric with female zone and resembling connate staminodes; sterile male zone absent above the sterile interstice; male zone creamy, 2.5-3.5 cm long, 1.0-1.5 cm in diameter; synandria creamy, rhombo-hexagonal, 2.0-2.3 mm in diameter, opening through apical slits; appendix ivory, 5.5-7.5 cm long, slightly thicker than the male zone at the base, then tapering gradually to a fine point, smooth to faintly rugose. Fruits yellowish (usually does not mature), rather small, 3-4 mm in diameter; fruiting peduncles 40-59 cm long, bend, fruiting spathe whitish, pendulous, 5-6 cm long, the spathe dehiscing longitudinally.

Flowering period: March to October.

Ecology: Grows on the hill slopes as undergrowth.

Distribution: North-Eastern part of Bangladesh, Habiganj and Moulvibazar districts.

Specimens examined: **Habiganj**: Kalenga beat, Kalenga, 03.07.2005, Hosne Ara HA 1771 (DACB); **Moulvibazar**: Gazipur beat, Hararganj reserve forest, 07.05.2003, Hosne Ara HA 315 (DACB); Sreemongal, Lawachara reserve forest, 15.05.2005, Hosne Ara HA 1468 (DACB); Madhabkundo, 20.05.2005, Hosne Ara HA 1707 (DACB); Gazipur beat, Hararganj reserve forest, 21.05.2005, Hosne Ara HA 1740 (DACB); Lawachara reserve forest, 04.07.2005, Hosne Ara HA

1779 (DACB); Gazipur beat, Hararganj forest, 05.07.2005, Hosne Ara HA 1804 (DACB); **Dhaka**: Bangladesh National Herbarium garden (Cultivated), 22.06.2015, Hosne Ara HA 2885 (DACB) [Originally collected from Hararganj reserve forest under Moulvibazar district].

Etymology: The species is named after the type locality-Hararganj in Moulvibazar district, from where the species was first collected.

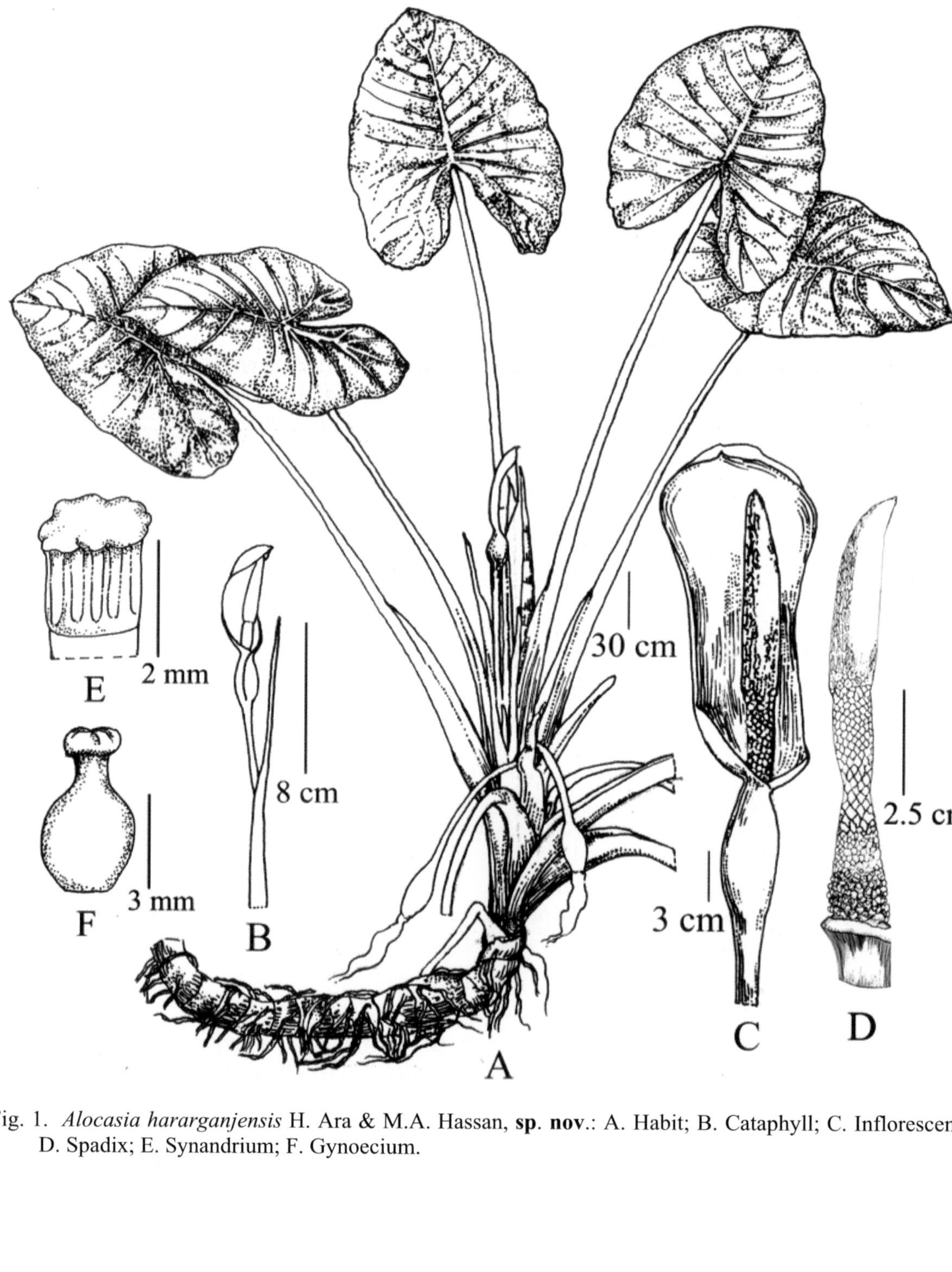

Fig. 1. *Alocasia hararganjensis* H. Ara & M.A. Hassan, **sp. nov.**: A. Habit; B. Cataphyll; C. Inflorescence; D. Spadix; E. Synandrium; F. Gynoecium.

Fig. 2. *Alocasia hararganjensis* H. Ara & M.A. Hassan, **sp. nov.**: A. Tuber; B. Habit-at BNH garden; C. Flower with cataphyll; D. Upper portion of the blade; E. Lower portion of the blade; F-H. Inflorescence; I. Spadix; J. Male zone; K. Synandrium; L. Sterile interstice; M. Lower portion of the spadix; N. Gynoecium, O. Appendix; P-Q. infructescence.

Note: The chromosome number has been determined for the new species *Alocasia hararganjensis*, and 2n chromosome number for this species appears to be 30.

The major morphological differences between the new species *Alocasia hararganjensis* H. Ara & M.A. Hassan and its closely allied species *Alocasia fallax* Schott are outlined in Table 1.

Table 1. Morphological comparison of *Alocasia hararganjensis* H. Ara & M.A. Hassan, sp. nov. with *Alocasia fallax* Schott.

Characters	*Alocasia hararganjensis* H. Ara & M.A. Hassan, **sp. nov.**	*Alocasia fallax* Schott
Leaf shape	Narrowly ovate to ovate sagittate, tip acute	Round and crisped, tip round
Cataphyll	Tip of the cataphylls long pointed, c. 8 cm long	Tip of the cataphylls shortly pointed, c. 2.8 cm long
Spadix	Male zone 2.5-3.5 cm long (not less than 2.5 cm long)	Male zone c. 1.8 cm long (not more than 2 cm long)
Sterile male zone	Absent above the sterile interstice	Present above the sterile interstice
Chromosome number (2n)	30	28 (Sultana *et al.*, 2011)

Conservation status: Near threatened (IUCN, 2017). Restricted distribution. Both *in situ* and *ex situ* conservation measures are suggested.

2. Alocasia salarkhanii H. Ara & M.A. Hassan, **sp. nov.** **(Figs 3 & 4).**

Diagnosis: *Alocasia salarkhanii* H. Ara & M.A. Hassan is closely related to *Alocasia fornicata* (Roxb.) Schott but can be easily differentiated by the colour of petiole and peduncle deep purple; shape of lamina long elliptic; number of inflorescences groups less than 4; length of style c. 0.5 mm long; male zone c. 2.3 cm long; and width of the appendix base and male zone are equal.

Holotype: Bangladesh, Moulvibazar district, Lawachara reserve forest, 15.05.20005, Hosne Ara HA 1467 (DACB).

Small to moderately robust herb. Stem erect to decumbent, up to 4 cm in diameter, c. 60 cm long, clothed in the brown remains of old leaf bases. Leaves several together, held almost erect or slightly curved; petiole 35-52 cm long, sheathing c. 1/3 from the base of petiole, eglandular, deep purple, wings of the sheath out-rolled; blade hastato-sagittate, rather narrowly triangular, margin entire to slightly undulate, glossy, leathery, dark green, glabrous adaxially, pale green and pubescent abaxially, 31-51 cm long, 14.0-21.5 cm in diameter at the base; anterior lobes 18-33 cm long with apiculate tip 0.6-1.0 cm long; anterior costa prominent on both surfaces, glabrous on both surfaces, primary veins 6 on each side, prominent on both surfaces, diverging at 45^0-90^0, secondary venation flush on both surfaces, mostly arising from the primary veins at a wide angle, then deflected towards the margin, forming variously well-defined interprimary collective veins or these absent, interprimary collective veins when present weakly undulating to strongly zig-zag at base broad acute angles; submarginal vein 0.8-1.0 cm from the margin; glands in the axils of primary veins absent or extremely inconspicuous; posterior lobes 13-18 cm long, acute, peltate, 2.3-3.0 cm long; posterior costae straight to incurved. Inflorescences 3 in the center of the leaf crown, bloom one after another, subtended by a cataphyll, cataphyll up to 25 cm long, purple; peduncle deep purple, smooth, 26-31 cm long, 0.7-1.3 cm in diameter at the base. Spathe c. 15 cm long; lower spathe convolute, c. 4.2 cm long, green; limb c. 11.7 cm long, c. 4.8 cm in diameter, light yellowish with violet or pink mixed on both sides. Spadix shorter than spathe, c. 11.5 cm long. Lower fertile female zone c. 1 cm long; sterile female zone c. 0.5 cm long, c. 1.3 cm in diameter at the base; pistil closely packed; ovary sub-globose, green, c. 2 x 2 mm, unilocular, with basal placenta; style distirct, stout, c. 0.5 mm long, c 0.8 mm in diameter, light yellow; stigma 3-4 lobed, lobes acute, light yellow; sterile interstice c. 2 cm long, narrower than the fertile zones,

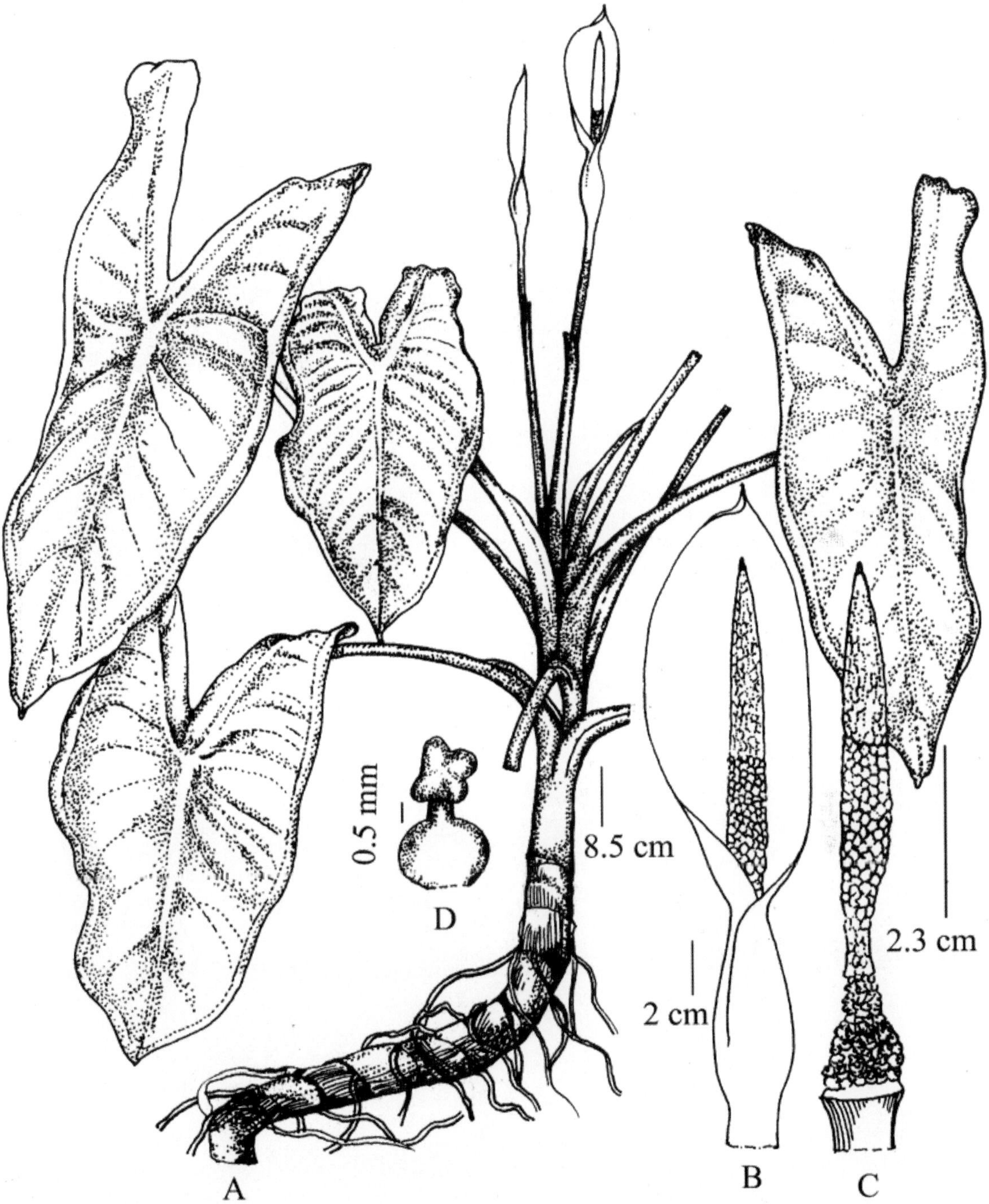

Fig. 3. *Alocasia salarkhanii* H. Ara & M.A. Hassan, **sp. nov**.: A. Habit; B. Inflorescence; C. Spadix; D. Gynoecium.

Fig. 4. *Alocasia salarkhanii* H. Ara & M.A. Hassan, **sp. nov.**: A. Wild habitat; B,C. Habit-at home garden; D,E. Inflorescence; F. Inside of the spathe; G. Outside of the spathe; H. Spadix; I,J. Lower portion of the spadix; K. Appendix.

corresponding with the spathe constriction; lower synandrodia often with incompletely connate staminodes, the rest elongate rhombo-hexagonal, flat-topped; male zone cyclindric, somewhat tapered at the base, c. 2.3 cm long, c. 0.8 cm thick, ivory in colour; synandria more or less hexagonal, c. 2 mm in diameter, androus; appendix c. 3.5 cm long, c. 0.8 cm thick at the middle, about the same thickness at the male zone, gradually tapering to a pointed tip, cream coloured. Fruit orange-red; fruiting peduncle 33-35 cm long, fruiting spathe ovoid, 4-6 cm long.

Flowering and fruiting period: May - August.

Chromosome number: 2n = 28 (Afroz *et al.*, 2013).

Ecology: Grows on the hilly area as undergrowth.

Distribution: North-Eastern part of Bangladesh, Moulvibazar district.

Specimens examined: **Moulvibazar**: Lawachara reserve forest, 15.05.2005, Hosne Ara HA 1467 (DACB); 04.07.2005, Hosne Ara HA 1781 (DACB); **Dhaka**: Khilgaon, Tilpapara (Cultivated), 07.05.2006, Hosne Ara HA 2630 (DACB); Bangladesh National Herbarium garden (Cultivated), 30.05.2006, Hosne Ara HA 2651 (DACB) [Originally collected from Lawachara reserve forest under Moulvibazar districts].

Etymology: The species is named in honour of Dr. Md. Salar Khan, known as the father of plant taxonomy in Bangladesh.

The major morphological and cytological differences between the new species *Alocasia salarkhanii* H. Ara & M.A. Hassan and its closely allied species *Alocasia fornicata* (Roxb.) Schott are outlined in Table 2.

Table 2. Morphological and cytological comparison of *Alocasia salarkhanii* H. Ara & M.A. Hassan, **sp. nov. with its closely related *Alocasia fornicata*** (Roxb.) Schott.

Characters	*Alocasia salarkhanii* H. Ara & M.A. Hassan, **sp. nov.**	*Alocasia fornicata* (Roxb.) Schott
Petiole and peduncle	Deep purple	Green
Leaf blade	Elliptic	Hastate
Inflorescence	In groups of up to 3	In groups of 12, up to 25
Style	c. 0.5 mm long	c. 1.0 mm long
Male zone	c. 2.3 cm long	c. 1.5 cm long
Appendix	Base of the appendix equal to the male zone	Base of the appendix wider than the male zone
Chromosome number	28 (22m+6sm)	28 (18m+10sm)
CMA and DAPI	CMA-band and DAPI band absent	CMA-band and DAPI band present

m = metacentric chromosome, sm = submetacentric chromosome.

Conservation status: Critically Endangered (IUCN, 2017). Located in a very restricted area, should be conserved both by *in situ* and *ex situ* methods.

3. Typhonium elatum H. Ara & M.A. Hassan, **sp. nov.** **(Figs 5 & 6).**

Diagnosis: *Typhonium elatum* H. Ara & M.A. Hassan differs from its closely related species *T. trilobatum* (L.) Schott by its green colour petiole which is more than 50 cm long; peduncle colour light green; length of the spathe c. 40 cm long and shape narrowly lanceolate; spadix length c. 29 cm long; base of the appendix multifurcate and colour of the staminode yellow.

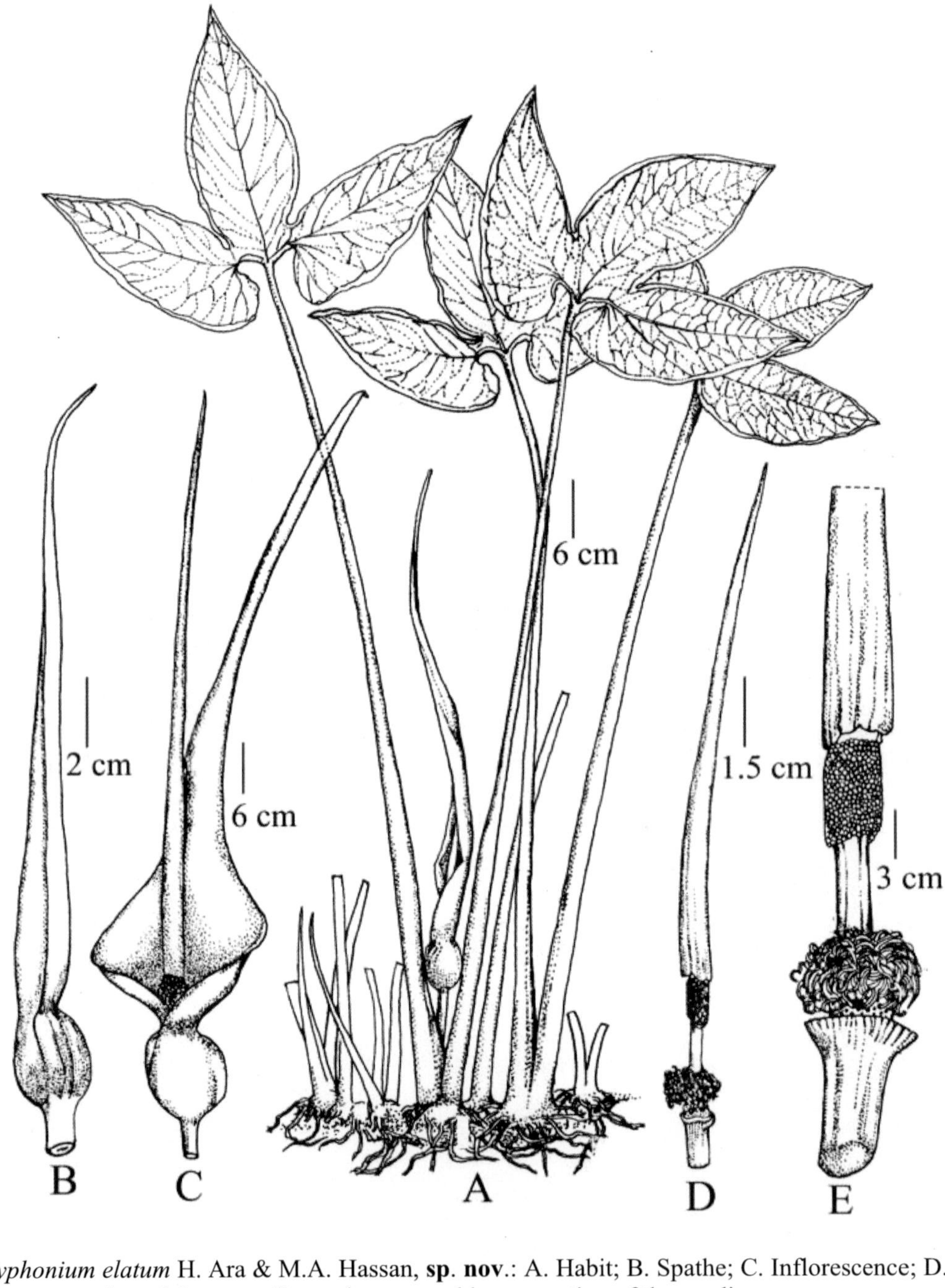

Fig. 5. *Typhonium elatum* H. Ara & M.A. Hassan, **sp. nov**.: A. Habit; B. Spathe; C. Inflorescence; D. Spadix; E. Lower portion of the appendix, male zone and lower portion of the sapdix.

Holotype: Bangladesh, Sherpur district, Samaschura beat, 10.10.2003, Hosne Ara HA 701 (DACB).

Tuber c. 5 cm long, c. 4 cm in diameter, developing several offsets. Leaf paired; petioles up to 64 cm long, c. 1 cm in diameter, pale green. Leaf blade usually deeply trilobed, anterior lobe elliptic-lanceolate, up to c. 25 cm long, c. 13 cm in diameter, posterior lobe c. 19.5 cm long, c. 9 cm in diameter, bright pale green. Inflorescence paired. Peduncle c. 10 cm long, c. 0.8 cm in diameter, pale green. Spathe c. 40 cm long, tube and blade separated by a strong constriction; tube

Fig. 6. *Typhonium elatum* H. Ara & M.A. Hassan **sp. nov.**: A,B,C. Habit - at home garden; D,E. Inflorescence, F. Spadix; G. Male zone and lower portion of the appendix; H,I. Lower portion of the spadix.

c. 3.5 cm long, outside pale green, inside glossy dark purple; blade c. 36.5 cm long, outside pale green, inside glossy dark purple. Spadix shorter than spathe, c. 29 cm long. Female zone c. 0.8 cm long, c. 0.8 cm in diameter, flowers congested; sterile zone between female and male zone, 2.7-4.0 cm long, the lower 0.5-2.2 cm with yellow staminodes, the remainder naked, light pink, longitudinally grooved; male zone cylindrical, c. 1.8 cm long, c. 1 cm in diameter, base and top oblique, flowers congested; appendix very shortly stipitate, narrowly lanceolate, 20-24 cm long, c. 1 cm in diameter at the base, top acute, base multifurcate, with surface shallowly and irregularly furrowed, dark purple, producing a strong unpleasant smell at female anthesis. Stamens 0.6-0.9 mm long, light pinkish. Staminodes filiform, c. 1.6 cm long, c. 0.5 mm in diameter, dark yellowish, curled. Ovary elongate, cylindric, c. 1.5 mm long, 1.1-1.2 mm in diameter, white with a faint pinkish flush near the top, unilocular, with one basal ovule; stigma sessile, depressed, hemispheric, c. 0.8 mm in diameter, c. 0.2 mm high, papillose, pink. Fruit not observed, usually die before fruit formation.

Flowering period: March - October.

Chromosome number: 2n = 18 (Huq *et al.*, 2007).

Ecology: Grows on the hilly area as undergrowth.

Distribution: Central parts of Bangladesh (Sherpur district).

Specimen examined: **Sherpur**: Samaschura beat, 10.10.2003, Hosne Ara. HA 701 (DACB); *ibid*, 23.06.2004, Hosne Ara HA 1060 (DACB); **Dhaka**: Khilgaon, Tilpapara (Cultivated), 02.07.2015, Hosne Ara HA 2889 (DACB) [Originally collected from Samaschura beat under Sherpur district].

Etymology: The species is named after tallness of its habit.

The major morphological and cytological differences between the new species *Typhonium elatum* H. Ara & M.A. Hassan and its closely allied species *Typhonium trilobatum* (L.) Schott are presented in Table 3.

Table 3. Morphological and cytological comparison of *Typhonium elatum* H. Ara & M.A. Hassan, **sp. nov. with *Typhonium trilobatum*** (L.) Schott.

Characters	*Typhonium elatum* H. Ara & M.A. Hassan, **sp. nov.**	*Typhonium trilobatum* (L.) Schott
Petiole	Green, ≥ 50 cm long	Purple, ≤ 45 cm long
Leaf shape	Elliptic lanceolate	Ovate, ovate-lanceolate
Peduncle	Light green	Purple
Spathe	c. 40 cm long, narrow lanceolate	15-18 cm long, ovate to broadly ovate
Spadix length	c. 29 cm long	c. 13.5 cm long
Appendix	Base of the appendix multifurcate	Base of the appendix truncate
Staminodes	Yellow	White
Ovary	c. 1.5 mm long, c. 1.2 mm in diameter, white with a faint pinkish flush near the top	c. 1.0 mm long, c. 0.7 mm in diameter, cream colour
Chromosome number	18 (12m + 6sm)	18 (16m + 2sm)
DAPI band	DAPI band 5	DAPI band 3

m = metacentric chromosome, sm = submetacentric chromosome.

Note : Although the two species bear same chromosome number, they differ by the karyotype formulae and DAPI characters.

Conservation status: Endangered (IUCN, 2017). Collected from a single locality, now conserved through *ex situ* in a home garden.

Acknowledgements

The authors are grateful to the authorities of the BK, BKF, BM, CAL, DACB, K, DUSH, HCU, BCSIRH and BFRIH for providing facilities to consult aroid materials and their libraries. The authors express their gratitude to Professor Dr. M. Oliur Rahman, Department of Botany, University of Dhaka for his encouragement and enormous help during preparation of the manuscript. Thanks are also due to Ms. Mahmuda Akter, Senior Artist-cum-illustrator, Bangladesh National Herbarium for drawing the illustrations, and the researchers at Cytogenetics Laboratory of the Department of Botany, University of Dhaka for their cooperation in chromosomal investigation of the new species.

References

Afroz, M., Sultana, S.S. and Alam, S.S. 2013. Karyotype and RAPD analysis of three morphological forms of *Alocasia fornicata* (Roxb.) Schott. Cytologia **78**(3): 269–275.

Ara, H. 2016. Taxonomic studies in the family Araceae from Bangladesh. Ph.D. Thesis (unpublished), Department of Botany, University of Dhaka, Bangladesh, pp. 1–524.

Boyce, P.C. 2007. Studies on the *Alocasia* Schott (Araceae-Colocasieae) of Borneo: I. Two new species from Sarawak, Malaysian Borneo. Gard. Bull. Singapore **58**(2): 141–154.

Boyce, P.C. and Croat, T.B. 2011 (onwards). The Überlist of Araceae, Totals for Published and estimated number of species in Aroid Genera. http://www.aroid.org/genera/180211 uberlist.pdf. (Accessed 05 November 2018).

Engler, A 1920. Araceae-Aroideae, Araceae-Pistiodeae. *In:* Engler, A. (Ed.), Das Pflanzenreich **73** (IV. 23F): 1-274.

Engler, A. and Krause, K. 1920. Additamentum ad Araceas-Philodendroideas, Araceae-Colocasioideae. *In*: Engler, A. (Ed.), Das Pflanzenreich **71** (IV.23E): 3–132.

Hay, A. 1991. The Genus *Alocasia* (Araceae) in Australasia. Blumea **35**: 499–545.

Hay, A. 1993. The genus *Typhonium* (Araceae-Areae) in Australasia. Blumea **37**(2): 345–376.

Hay, A. 1998. *Alocasia* in West Malesia and Sulawesi. Gard. Bull. Singapore **50**(2): 325–327.

Hay, A. 1999. *Alocasia* in the Philippines. Gard. Bull. Singapore **51**: 29–30.

Hetterscheid, W.L.A. and Boyce, P.C. 2000. A reclassification of *Sauromatum* Schott and new species of *Typhonium* Schott (Araceae). Aroideana **23**: 48–55.

Hooker, J.D. 1893 (Reprint 1954). The Flora of British India, Vol. **6**. L. Reeve & Co. Ltd., Kent, England, pp. 490–555.

Hu, S.Y. 1968. Araceae. Studies in the Flora of Thailand, no. 41. Dansk Bot. Arkiv **23**(4): 409–457.

Huq, S., Ara, H., Hassan, M.A. and Alam, S.S. 2007. Comparative karyotype analysis of *Typhonium trilobatum* L. and its two morphological forms. Cytologia **72**(2): 233–238.

IUCN 2017. Studies and Petitions Subcommittee Guidelines for Using the IUCN Red List http://www.incnredlist.org/documents/RedListGudelines.pdf. pp. 1−108.

Jackson, B.D. 1893-1955. Index Kewensis (An Eumeration of the Genera and Species of Flowering Plants). Vols. **1** & **2.** Oxford at the Clardon press.

Karthikeyan, S., Jain, S.K., Nayar, M.P. and Sanjappa, M. 1989. Florae Indicae Enumeratio: Monocotyledonae. Flora of India Series **4**. Araceae. Botanical Survey of India, Calcutta, pp. 5–15.

Li, H., Zhu, G., Boyce, P.C., Murata, J., Hetterscheid, W.L.A., Bogner, J. and Jacobsen, N. 2010. Araceae. *In*: Wu, Z.Y., Raven, P.H. and Hong, D.Y. (Eds). Flora of China, Vol. 23 (Acoraceae through Cyperaceae). Science Press, Beijing, and Missouri Botanical Garden Press, St. Louis. pp. 3–79.

Mayo, S.J. 1985. Araceae. *In*: Polhill, R.M. (Ed.), Flora of Tropical East Africa. Balkema, Rotterdam, the Netherlands, pp.1–71.

Mayo, S.J., Bogner, J. and Boyce, P.C. 1997. The Genera of Araceae. Royal Botanic Gardens, Kew, UK, pp. 1–370.

Nasir, Y.J. 1978. Araceae. *In*: Nasir, E. and Ali, S.I. (Eds), Flora of West Pakistan, No. **120.** Agricultural Research Council, Islamabad, Pakistan, pp.1–17.

Nicolson, D.H. 1976 (Reprint 1978). Araceae. *In*: Saldanha, C.J. and Nicolson, D.H. (Eds), Flora of Hassan District, Karnataka, India. Amerind Publishing Co. Pvt. Ltd., New Delhi, Bombay, Calcutta and New York, pp. 781–790.

Nicolson, D.H. 1979. Araceae. *In*: Smith, A.C. (Ed.), Flora Vitiensis Nova, Vol. **1**. Pacific Tropical Botanical Garden, Hawaii, pp. 438–460.

Nicolson, D.H. 1987. Araceae. *In*: Dassanayake, M.D. (Ed.), A Revised Handbook to the Flora of Ceylon, Vol. **6**. Balkema, Rotterdam, the Netherlands, pp. 17–101.

Nicolson, D.H. and Sivadasan, M. 1981. Four frequently confused species of *Typhonium* Schott (Araceae). Blumea **27**(2): 483–497.

Noltie, H.J. 1994. Araceae. *In*: Grierson, J.D. and Long, D.G. (Eds), Flora of Bhutan, Vol. **3**(1). Royal Botanic Garden, Inverleith Row, Edinburgh, UK, pp. 121–158.

Prain, D. 1903. Bengal Plants, Vol. **2**. Indian reprint (1963), Botanical survey of India, Calcutta, pp. 830–840.

Roxburgh, W. 1832. Tetrandia Monogynia in Willium Carey's Flora Indica; or Descriptions of Indian Plants. Mission Press, Serampore, India, pp. 620–635.

Sookchaloem, D. 1995. *Typhonium (*Araceae*)* in Thailand. Thai For. Bull. (Bot.) **23**: 18–39.

Sriboonma, D., Murata, J. and Iwatsuki, K. 1994. A revision of *Typhonium* (Araceae). J. Fac. Sci. Univ. Tokyo, Sect. III. **15**: 255–313.

Sultana, S.S., Ara, H. and Alam, S.S. 2011. Karyotype analysis with Orcein and CMA in two species of *Alocasia* (Schott) G. Don (Araceae). Bangladesh J. Bot. **40**(1): 53–56.

Toha, A.M.M. 2000. Biodiversity of aroids in southeast Bangladesh. M. Phil Thesis (unpublished). Aberdeen University, Scotland, U.K.

Wallich, N. 1829-1849. A numerical list of dried specimens of plants in the East Indian Company's Museum. Generally cited Wall. Cat., *ined.*

Wight, R. 1843-1845. Icones Plantarum Indiae Orientalis or Figures of Indian Plants, Vol. **3** (Part 1). Messrs, Ostell, Lepage and Co., Calcutta, pp. 772–808.

34

A NEW SPECIES OF *GLYPHIS* ACH. AND THREE NEW RECORDS OF *GRAPHIS* ADANS. (GRAPHIDACEAE) FROM BALI ISLAND, INDONESIA

JUNITA HARDINI[1], RINA SRI KASIAMDARI[2], SANTOSA AND PURNOMO

Department of Tropical Biology, Faculty of Biology, Universitas Gadjah Mada, Jl. Teknika Selatan, Sekip Utara Yogyakarta 55281, Indonesia

Keywords: *Glyphis batuana* sp. nov.; Graphidaceae, New records; Taxonomy; Bali Island.

Abstract

Glyphis batuana Hardini, Kasiamdari & *Purnomo* sp. nov. is a new species of lichenized fungus found on the bark of the Frangipani tree (*Plumeria* sp.). The new species from Batuan village (Gianyar districts), Bali Island, Indonesia is described and illustrated. It is characterized by its lirelliform, unbranched ascomata, entire labia, black, open disc with brown pruina, completely carbonized excipulum, 8-spored asci with 8-10 locular ascospores, and lack of secondary substances. A key to species of *Glyphis* Ach. in Indonesia is provided. Three new records of *Graphis* Adans., namely *G. conferta* Zenker, *G. immersella* Mull.Arg. and *G. nilgiriensis* Adaw. & Makhija are also reported.

Introduction

The Island of Bali is located in the central part of Indonesia, corresponding to the province of the same name. The lowland area of Bali has many Frangipani trees (*Plumeria* spp., mostly *P. rubra* L.), native in the Neotropics but cultivated world-wide and in Indonesia these are planted in open areas for religious ceremonies. Frangipani trees have a thin and smooth skin and are often overgrown with crustose lichens, particularly of the family Graphidaceae. Graphidaceae is the largest family of tropical crustose lichens, and consists of 79 genera, including *Glyphis* Ach. and *Graphis* Adans. (Lücking *et al.*, 2014, 2017). During a survey in the lowland areas of Bali (around Batuan Village in Gianyar District), a new species of *Glyphis* was discovered. This genus is characterized by brown-pruinose ascomata and carbonized excipula, together with hyaline, distoseptate, amyloid ascospores. The genus currently includes seven species, namely *G. atrofusca* (Mull. Arg) Lücking, *G. cicatricosa* Ach., *G. dictyospora* Staiger, *G. duriuscula* Stirt., *G. latissima* (Vain.) Zahlbr., *G. substriatula* (Nyl.) Staiger, and *G. scyphulifera* (Ach.) Staiger (Archer, 2009; Lücking *et al.* 2014, 2017).

Glyphis cicatricosa Arc. is the most common and widespread species, with a pantropical distribution. It was reported from Singapore (Sipman, 2003), the Philippines (Tabaquero, 2013), and also from Hong Kong, China (Aptroot, 1999), Australia (Archer, 2004), India (Singh and Sinha, 2010), Bolivia (Flakus *et al.*, 2013; Kukwa *et al.*, 2013), and Venezuela (Fuenmayor, 2013). In Indonesia, *Glyphis cicatricosa* is reported from Bogor and Cibodas (Sipman, 2003) and from Java (Groenhart, 1936). Other species of *Glyphis* found in Indonesia were also reported by Groenhart (1936), which are *G. verrucosa* Mont. *et* Bosch (Java), *G. javanica* Mull.Arg. (Java), *G. labyrinthica* Ach. (Bogor, Gunung Gede, Java), G. *lactea* Mull.Arg (Bogor), *G. heterostycha* Hepp. (Bogor, Cibodas, Gunung Salak, Java), *G. heteroclyta* Mont. (Bogor, Cibodas, Gunung

---

[1]Department of Biology, Faculty of Mathematics and Natural Science, Udayana University, Jl. Raya Kampus UNUD, Bukit Jimbaran, Kuta Selatan, Badung, Bali 80361, Indonesia.
[2]Corresponding author. Email: rkasiamdari@ugm.ac.id

Salak, Java), G. *Leprieurii* Mont (Bogor, Cibodas, Gunung Salak, Java), and *G. tricosa* Ach. (Java).

Graphis is one of the largest genera in Graphidaceae. Groenhart (1936) reported 30 species of *Graphis* found in Indonesia, of which four species were recorded from Bogor (*G. bataviana* A. Zahlbr., *G. bogoriensis* A. Zahlbr., *G. karsteni* A. Zahlbr., *G. schiffneri* A. Zahlbr.), four species from Cibodas (*G. curtiuscula* A. Zahlbr., *G. inamoena* A. Zahlbr., *G. overimii* A. Zahlbr., *G. treubii* A. Zahlbr.) and 22 species from Java (*G. afzelii* Ach., *G. angustata* Eschw., *G. aphanes* Mont. *et* Bosch., *G. cinerea* Fee, *G. duplicata* Ach., *G. elegans* Ach., *G. flavens* Müll.Arg., *G. grammitis* Fee, *G. intricata* Fee, *G. javanica* A. Zahlbr., *G. lineola* Ach., *G. ovata* Mass., *G. radiata* Nyl., *G. regularis* Müll.Arg., *G. rimulosa* Trev., *G. schizograpta* Müll.Arg., *G. scripta* Ach., *G. stenospora* Müll.Arg, *G. subassimilis* Müll.Arg, *G. tenella* var. *flavicans* Müll.Arg, *G. vittata* Müll.Arg, *G. zollingeri* A. Zahlbr.). Whereas, Lücking *et al.,* (2009) reported 20 species of *Graphis*, which were three species from Borneo (*G. dupaxana* Vain., *G. marginata* Raddi, *G. sarawakensis* Hale *ex* Lücking), one species from Celebes (*G. rustica* Kremp.), 12 species from Java (*G. assimilis* Nyl., *G. chlorotica* A. Massal*, G. crassilabra* Müll.Arg., *G. curtiuscula* Zahlbr., *G. duplicata* Ach., *G. flavens* Müll.Arg., *G. hossei* Vain., *G. leptoclada* Müll.Arg., *G. regularis* Müll.Arg., *G. subassimilis* Müll.Arg., *G. submarginata* Lücking, *G. vittata* Müll.Arg.), one species from Krakatau (*G. filiformis* Adaw. & Makhija), one species from Malang Java (*G. leucaenae* Aptroot), and also reported were *G. schiffneri* Zahlbr., and *G. japonica* (Mull.Arg.) A.W. Archer & Lücking. Lücking *et al.* (2014) also states that there were still many lichen family of Graphidaceae that not yet known and predicted to be mostly found in Southeast Asia one of which is Indonesia.

This study aimed to explore lichen Graphidaceae in Batuan village (Gianyar district) and Bunutbolong village (Jembrana district) and reported lichen taxa which were a new species and a new record for Indonesia. Further detail observations of morphological characters, anatomy, chemistry in samples of lichen Graphidaceae, resulted in a conclusion for new species, *Glyphis batuana* Hardini, Kasiamdari & Purnomo. and newly recorded *Graphis conferta* Zenker, *Graphis immersella* Mull. Arg*.,* and *Graphis nilgiriensis* Adaw. & Makhija, which were described in this paper.

Material and Methods

Study area

This research was conducted on the Bali Island with two locations: Gianyar district and Jembrana district (Fig. 1). Gianyar district is located at an altitude of 113 m a.s.l., temperature of 40□C, humidity of 40%, and Jembrana district is located at an altitude of 491 m a.s.l., temperature 25□C, humidity of 70%.

Sample collection and identification

The specimens were collected in July through December 2014 from the Gianyar District and Jembrana District, Bali Island, and deposited in the herbarium of Biology Museum, Faculty of Biology Universitas Gadjah Mada, Yogyakarta, Indonesia. The stereomicroscope (Olympus CX22) at magnifications of 7x to 45x were used for observation of morphological features. The anatomical features were studied under a light microscope XSZ-107BN at magnifications of 40x to 1000X with Optilab microscope digital camera. Thin hand-cut sections of thalli and ascomata were observed in water, 10% KOH and Iodine solutions. Lichen substances were identified by thin layer chromatography (TLC) following standard methods in solvent C and color spot tests as described by Orange *et al.* (2010). Microcrystals test (Hale, 1974; Huneck and Yoshimura, 1996) was performed by crystallizing reagents: GE (glycerine-acetic acid), GAW (glycerine alcohol water), GAoT (glycerine alcohol o-toluidin), GAAn (glycerine alcohol aniline). The photographs were taken using a Nikon Coolpix S220 digital camera.

Morphological work was carried out on the unknown collected specimen and were done by comparison and assessment of voucher specimen of the new taxa with that of the holotypes, isotypes and specimens the herbaria of images the lichen with The Field Museum (Robert Lücking, 2008, www.discoverlife.org), images the lichens with the Tropical Plant Guides and Discover Life. Specimens were also identified with the relevant literature, *viz.* Hayward (1977), Archer (1999, 2000, 2004, 2005, 2007), Sipman (2003) and Lücking *et al.* (2009, 2012, 2014), and then were consulted for identification and confirmation of specimens diagnosed as new taxa with curator for lichens.

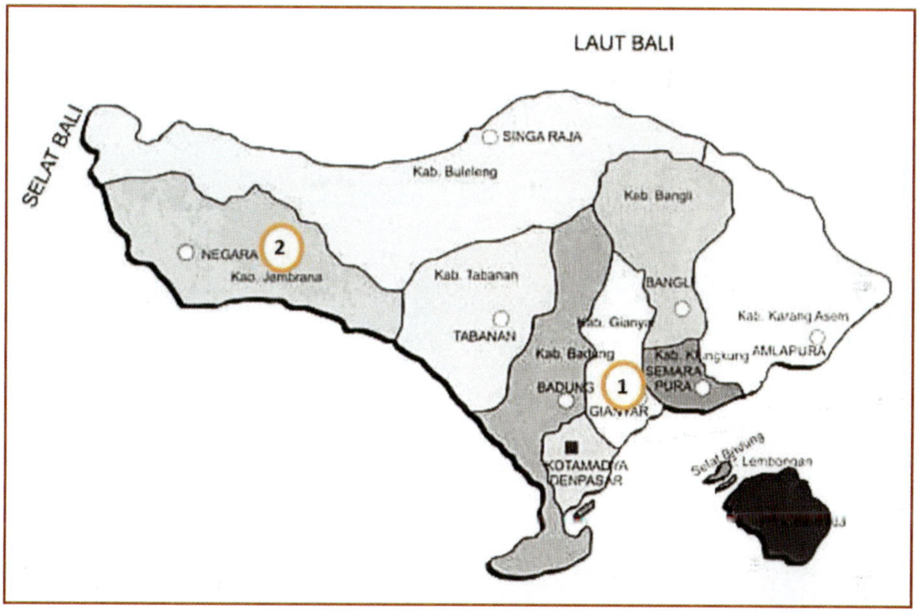

Fig. 1. Study area at the Bali Island. 1. Batuan village (Gianyar districts), 2. Bunutbolong village (Jembrana district).

Results and Discussion

Glyphis batuana Hardini, Kasiamdari & Purnomo, **sp. nov.** **(Fig. 2)**.

Myco Bank No.: 816969

Diagnosis: The new species, *Glyphis batuana* differing from *G. atrofusca* (Mull.Arg.) Lücking in the transversely septate ascospores and completely carbonized excipulum and from *G. cicatricosa* in the non-pseudostromatic lirellae and smaller ascospores.

Type: INDONESIA, Bali Island, Gianyar district, Batuan village, on the tree bark of Frangipani (*Plumeria* sp.), house plants and plants by the road, altitude 113 m a.s.l., 8°34'59"S - 115°16'2"E, July 2014, Jun-BG5 (Holotype, MBY).

Thallus corticolous, crustose, 3-5 cm, continuous, whitish green, surface smooth to uneven, rough; crystalline, 100-150 thick in cross-section. Cortex indistinct. Photobiont layer densely interspersed with calcium oxalate crystals, 50-70 um thick. Crystalline layer with cluster of calcium oxalate crystals scattered on the thallus surface and the bottom of the algae layers. Medulla indistinct.

Ascomata numerous, lirelliform, erumpent. Lirellae lacking thalline margin, short to elongate, 1-4 mm long and 0.15 – 0.30 mm broad, sinuous, unbranched to sparsely branched. Labia entire, black. *Disc* exposed, reddish brown with chocolate-brown pruina. Excipulum completely carbonized, in upper part 15-20 µm thick, laterally 15-30 µm thick. Hymenium hyaline, clear, 85-95µm high. Paraphyses unbranched, filiform, hyaline, sometimes pale brown at the tips. Asci clavate, 8-spored, 75-100 x 7-20 µm, I- blue. Ascospores hyaline, fusiform, distoseptate, transversely 7-9 septate, about 19-26 x 5-7 µm, halo not seen, I- blue.

Etymology: The specific epithet of *Glyphis batuana* refers to the location where the new taxon was discovered, Batuan village.

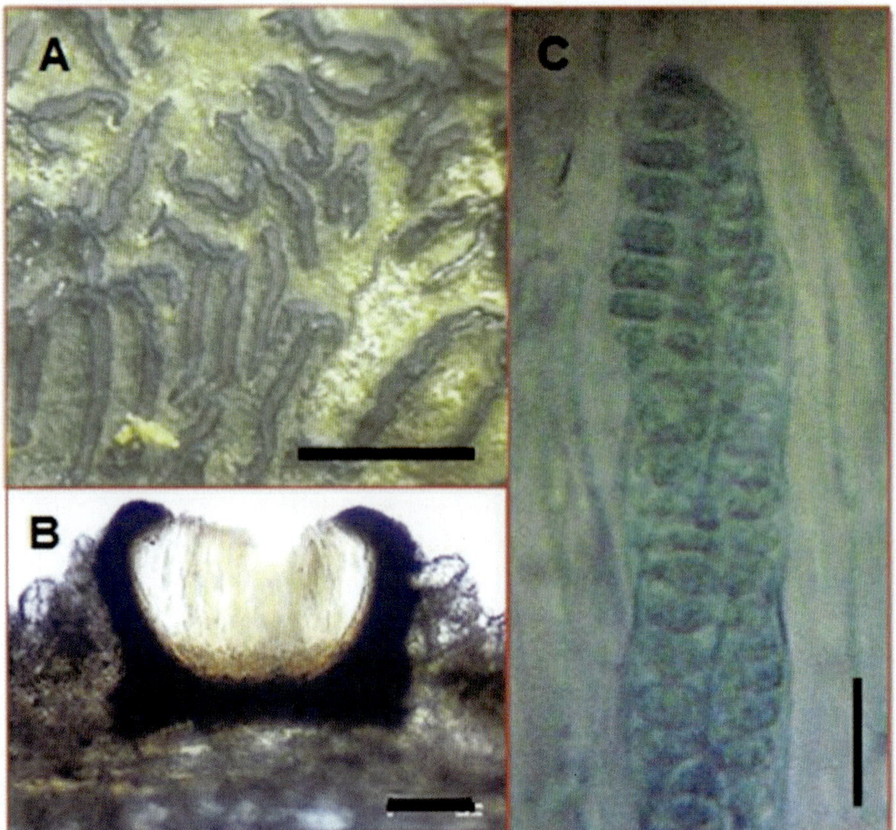

Fig. 2. A-C. *Glyphis batuana* Hardini, Kasiamdari & Purnomo **sp. nov.**. A. Habit; B. Cross section of ascomata; C. Ascospores. Scale bars: A = 2 mm, B = 100 µm; C = 10 µm.

Chemistry: Thallus and medulla K–, P–, C–, KC–, no substances detected in thin-layer chromatography.

Distribution and habitat: The type locality is situated in the lowland area of the Bali Island (113 m a.s.l.). The region has many Frangipani trees for religious ceremonies purposes, are planted in open ground with temperature of 38°C, humidity 40%. The new species is found in the young Frangipani tree that has smooth skin and thin. The new species is dispersed in small patches on thin tree bark with other species of *Graphis* and collected at a height of 1-1.5 m from tree base in exposed condition. The region is considered as eastern paleotropics, similar to other regions in Singapore, The Phillipines, India, Vietnam and Cambodia.

Notes: The new species *Glyphis batuana* is characterized by brownish green thallus, lirellae erumpent, elongate, sinuous and sparsely branched, thalline margin absent or lacking, labium entire and black, disc pruinose brown and exposed, completely carbonized exciple, transversely septate ascospores of about 19-26 x 5-7 µm, 7-8 septate, 8 spored per ascus, hymenium clear and 85-95 µm thick, no lichen substances found (Fig. 2). The organism resembles *G. atrofusca* (Mull.Arg.) Lücking in having ascomata lirelliform, open, scattered, sometimes branched. All species of *Glyphis* have this pruina and generally a carbonized excipulum and mostly lack substances. *G. atrofusca* basically differs in the muriform ascospores and laterally carbonized excipulum. Another similar species of the group is *G. cicatricosa* Ach., which has disc dark reddish brown, completely carbonized, 8 spores asci, that are comparable to *G. batuana*, but is distinct in having an ascomata immersed in conspicuous raised white pseudostromata, stromata rounded, richly branched, crowded and covering the surface of the stromata, thicker hymenium of about 120–160 µm, larger ascospores of 32–55 × 8–12 µm (Staiger, 2002; Archer, 2004). *G. scyphulifera* (Ach.) Staiger, has ascomata reddish brown, exciple completely carbonized, 8 per ascus, but differs by its rounded ascomata and muriform ascospores (Staiger, 2002). *G. substriatula* has elongate lirellae with distinct brown pruina along slit and lacking thalline margin (Straiger, 2005; Lücking *et al.*, 2014).

Key to the genus *Glyphis* in Indonesia

1.	Ascospores transversely septate	2
-	Ascospores muriform	3
2.	Ascomata pseudostromatic; ascospores 32-55 x 8-12 µm	*G. cicatricosa*
-	Ascomata lirellate, not in pseudostromata; ascospores 19-26 x 5-7 µm	*G. batuana* **sp. nov.**
3.	Ascomata rounded, sessile; ascospores 12-15 µm broad	*G. scyphulifera*
-	Ascomata lirellate	4
4.	Excipulum laterally carbonized	*G. atrofusca*
-	Excipulum completely carbonized	*G. substriatula*

New records of *Graphis* Adans.

Three species of *Graphis* (*G. conferta* Zenker, *G. immersella* Mull.Arg. and *G. nilgiriensis* Adaw. & Makhija) have been reported for the first time from Bali, which are new records for Indonesia. There are no previous reports about the collection of these three species of *Graphis* from Bali or any other part of Indonesia.

Graphis conferta Zenker, Pharmaceutische Waarenkunde (Eisenach) 1(3): 166 (1829). (**Fig. 3A**).

Thallus corticolous, crustose, brown, smooth surface shiny; ascomata lirellate, black, short, 0.2-3.0 mm, unbranched, sessile, flat labium, disc covered; basal thalline margin, excipulum completely carbonized, yellow hymenium, 100-110 µm high, clear; ascospores hyaline, 8 spored ascus, 20-28 x 5-7 µm, 6-10 septate, I + blue; No subtances.

Specimen examined: Indonesia: Bali Island: Jembrana district: Bunutbolong village, 491 m a.s.l., on bark of Frangipani (*Plumeria* sp.), July 2014, MBY.

Graphis immersella Mull.Arg. Bull. Herb. Boissier 3: 319 (1895). T: Cairns, Qld, 1893, J. F. Shirley 1793; lecto: *G. fide* A.W. Archer, Telopea 8: 281 (1999). (**Fig. 3B**).

Synonym: *Graphis leptalocarpa* A.W. Archer, (*Holotype:* Solomon Islands) Mycotaxon 83: 364 (2002); *G. manhaviensis* Zahlbr. (*Holotype:* China, Handel-Mazzetti, H. 1930. Symbolae Sinica 3:1-254).

Thallus corticolous, crustose, whitish-grey, surface of thallus uneven; ascomata lirellate, immersed, black, short to elongate, 2-5 mm, sparsely branched, lirellae variable; labia entire, non pruinose; disc exposed; lateral thalline margin; excipulum laterally carbonized; I –ve, yellowish to brownish hymenium, 90-115 µm high, clear; ascospores hyaline, 8 spored ascus, 5-7 septate, I + blue, 19-25 x 5-7 µm; K + yellow, thallus containing stictic acid.

The materials examined similar to *G. immersella* which has been described by Archer (1999, 2005) and Lücking *et al.* (2009), but this species has longer lirellae (2-5 mm), some branched dichotom at the end or middle. It has been previously reported in Australia (Archer 1999), and Solomon Islands (Archer, 2007).

Specimen examined: Indonesia: Bali Island: Jembrana district, Bunutbolong village, 491 m a.s.l., on bark of Frangipani (*Plumeria* sp.), July 2014, Jun-BJ12, MBY.

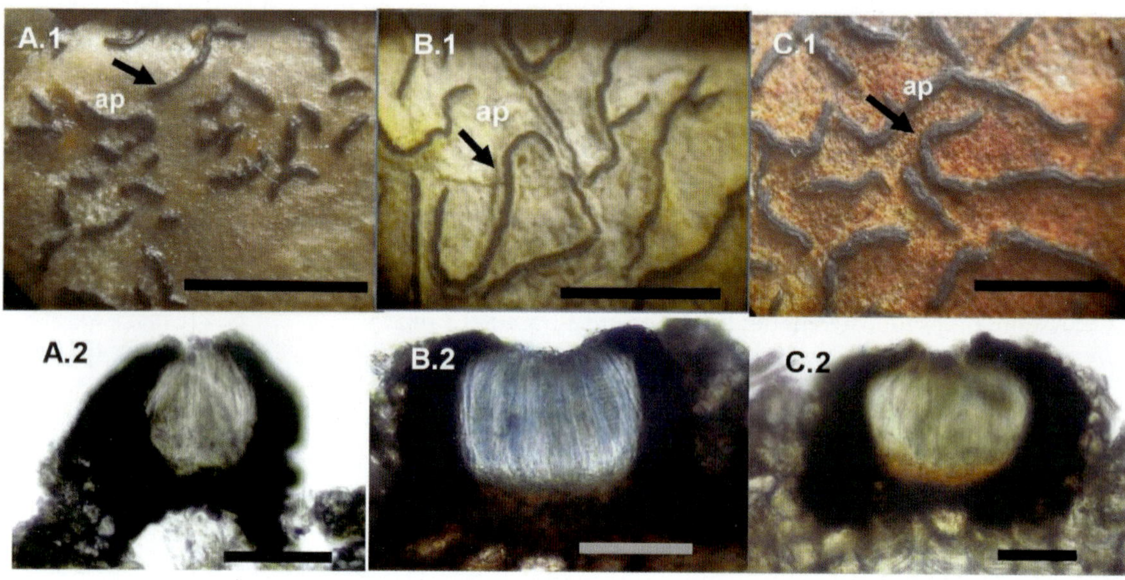

Fig. 3. A. *Graphis conferta* Zenker: A1. Thallus with apothecia (scale=2 mm), A2. Section of apothecium with hymenium, ascus and ascospores (scale=50 µm); B. *Graphis immersella* Mull.Arg.: B1. Thallus with apothecia (scale=2 mm), B2. Section of apothecium with hymenium, ascus and ascospores (scale = 50 µm); C. *Graphis nilgiriensis* Adaw. & Makhija: C1. Thallus with apothecia (scale = 2 mm), C2. Section of apothecium with hymenium, ascus and ascospores (scale = 50 µm); ap = apothecium.

Graphis nilgiriensis Adaw. & Makhija, Mycotaxon 96: 59 (2006). **(Fig. 3C)**.

Thallus corticolous, crustose, yellowish-orange, the surface uneven; ascomata lirellate, sessile, black, elongated, irregularly curved, striate, 1- 5 mm long, unbranched; labium entire; disc closed; basal thalline margin; excipulum completely carbonized; yellowish to brownish hymenium, of 74-99 µm high, clear; ascospores hyaline, 8 spored ascus, 5-9 septate, I + blue, 20-50 x 5-8 µm; K + yellow, containing stictic acid.

This species has been reported previously from India (Adawadkar and Makhija, 2006). Thallus colour of this species is yellowish orange, may be experiencing discoloration due to decomposition of chlorophyll (Lücking *et al.*, 2009) and ascomata length 1-5 mm.

Specimen examined: Indonesia: Bali Island: Jembrana district: Bunutbolong village, 491 m a.s.l., on bark of Frangipani (*Plumeria* sp.), July 2014, Jun-BJ12, MBY.

Acknowledgments

We express our thanks to Dr. Robert Lücking (Curator for cryptogams: lichens, fungi, bryophytes), Botany Department, The Field Museum, Chicago, Illinois, USA, for many discussions about the taxa of *Glyphis* and *Graphis* (Graphidaceae).This study is part of a Doctoral dissertation at the Biological Science Program, Faculty of Biology UGM, and funded by the Directorate General of Higher Education (DGHE), Ministry of Education and Culture, Republic of Indonesia for the Scholarship of National Graduate Study Program.

References

Aptroot, A. 1999. Annotated checklist of Hong Kong Lichens. Trop. Bryol. **17**: 57–101.

Archer, A.W. 1999. The lichen genera *Graphis,* and *Graphina* (Graphidaceae) in Australia 1: Species based on Australian type specimens. Telopea **8**: 273–285.

Archer, A.W. 2000. The lichen genera *Phaeographis* and *Phaeographina* (Graphidaceae) in Australia 1: Species based on Australian type specimens. Telopea **8**: 461–475.

Archer, A.W. 2004. Additions to the lichen flora of Fiji and Vanuatu based on Graphidaceae in the F.R.M. Wilson collection at the National Herbarium of New South Wales. Telopea **10**: 771–776.

Archer, A.W. 2005. New combinations and synonymies in the Australian Graphidaceae. Telopea **11**: 59–78.

Archer, A.W. 2007. Key and checklist for the lichen family Graphidaceae (lichenized Ascomycota) in the Solomon Islands. Syst. Biodivers. **5**: 9–22.

Archer, A.W. 2009. Graphidaceae. *In*: McCarthy, P.M. (Ed.), Flora of Australia. Vol. **57**. Lichens 5. Melbourne: ABRS/CSIRO.

Flakus, A., Sipman, H.J.M., Bach, K., Flakus, P.R., Knudsen, K., Ahti, T., Schiefelbein, U., Palice, Z., Jablonska, A., Oset, M., Meneses, Q.R.I. and Kukwa, M. 2013. Contribution to the knowledge of the lichen biota of Bolivia. Pol. Bot. J. **58**: 697–733.

Fuenmayor, G.L. 2013. Preliminary inventory of corticolous lichens of the Sierra de San Luis, Venezuela. Rev. Biodivers. Neotrop. **3**: 98–105.

Groenhart, P. 1936. Beiträge zur Kenntnis der javanischen Flechten I-III. Nederlandsch kruidkundig archief. **46**: 690–784.

Hale, M.E. 1974. The Biology of Lichens. London: Edward Arnold, pp. 119–129.

Hayward, G.C. 1977. Taxonomy of the lichen families Graphidaceae and Opegraphaceae in New Zealand. New Zealand J. Bot. **15**: 565–584.

Huneck, S. and Yoshimura, I. 1996. Identification of lichens substances. Springer,Berlin, Heidelberg, pp. 13–14, 47.

Kukwa, M., Schiefelbein, U. and Flakus, A. 2013. A contribution to the lichen family Graphidaceae (Ostropales, Ascomycota) of Bolivia. Herzogia **26**: 231–252.

Lücking, R., Archer, A.W. and Aptroot, A. 2009. A world-wide key to the genus *Graphis* (Ostropales, Graphidaceae). The Lichenologist **41**: 363–452.

Lücking, R., Sutjaritturakan, J. and Kalb, K. 2012. Validation of three species names and description of a new species in the genus *Graphis* (Ascomycota: Ostropales: Graphidaceae). The Lichenologist **44**: 391–394.

Lücking, R., Hodkinson, B.P. and Leavitt, S.D. 2017. The 2016 classification of lichenized fungi in the Ascomycota and Basidiomycota–Approaching one thousand genera. The Bryologist **119**: 361–416.

Lücking, R., Johnston, M.K., Aptroot, A., Kraichak, E., Lendemer, J.C., Boonpragob, K., Caceres, M.E.S., Ertz, D., Ferraro, L., Jia, Z.F., *et al.* 2014. One hundred and seventy-five new species of Graphidaceae: closing the gap or a drop in the bucket. Phytotaxa **189**: 7–38.

Orange, A., James, P.W. and White, F.J. 2010. Microchemical methods for the identification of lichens. London: British Lichen Society, pp. 1-101.

Singh, K.P. and Sinha, G.P. 2010. Indian Lichens: Annotated Checklist. Botanical Survey of India, Kolkata.

Sipman, H. 2003. Key to the lichen genera of Bogor, Cibodas and Singapore. <http:/www. bgbm.org/sipman/keys/Javagenera.htm>. Retrieved on 5 June 2013.

Staiger, B. 2002. Die Flechtenfamilie Graphidaceae: Studien in Richtungeiner Natürlichen Gliederung. Biblioth. Lichenol. **85**: 15–26.

Staiger, B. 2005. How to arrange the diversity of a tropical lichen family? Systematics and generic concepts in the lichen family Graphidaceae. Arch. Sci. **58**: 53–62.

Tabaquero, A.L., Bawingan, P.A. and Lücking, R. 2013. Key and checklist of Graphidaceae lichens in the Kalahan forest reserve, Nueva Vizcaya, Philippines. Philippine J. Syst. Biol. **7**: 22–38.

Permissions

All chapters in this book were first published in BJPT, by Bangladesh Association of Plant Taxonomists; hereby published with permission under the Creative Commons Attribution License or equivalent. Every chapter published in this book has been scrutinized by our experts. Their significance has been extensively debated. The topics covered herein carry significant findings which will fuel the growth of the discipline. They may even be implemented as practical applications or may be referred to as a beginning point for another development.

The contributors of this book come from diverse backgrounds, making this book a truly international effort. This book will bring forth new frontiers with its revolutionizing research information and detailed analysis of the nascent developments around the world.

We would like to thank all the contributing authors for lending their expertise to make the book truly unique. They have played a crucial role in the development of this book. Without their invaluable contributions this book wouldn't have been possible. They have made vital efforts to compile up to date information on the varied aspects of this subject to make this book a valuable addition to the collection of many professionals and students.

This book was conceptualized with the vision of imparting up-to-date information and advanced data in this field. To ensure the same, a matchless editorial board was set up. Every individual on the board went through rigorous rounds of assessment to prove their worth. After which they invested a large part of their time researching and compiling the most relevant data for our readers.

The editorial board has been involved in producing this book since its inception. They have spent rigorous hours researching and exploring the diverse topics which have resulted in the successful publishing of this book. They have passed on their knowledge of decades through this book. To expedite this challenging task, the publisher supported the team at every step. A small team of assistant editors was also appointed to further simplify the editing procedure and attain best results for the readers.

Apart from the editorial board, the designing team has also invested a significant amount of their time in understanding the subject and creating the most relevant covers. They scrutinized every image to scout for the most suitable representation of the subject and create an appropriate cover for the book.

The publishing team has been an ardent support to the editorial, designing and production team. Their endless efforts to recruit the best for this project, has resulted in the accomplishment of this book. They are a veteran in the field of academics and their pool of knowledge is as vast as their experience in printing. Their expertise and guidance has proved useful at every step. Their uncompromising quality standards have made this book an exceptional effort. Their encouragement from time to time has been an inspiration for everyone.

The publisher and the editorial board hope that this book will prove to be a valuable piece of knowledge for researchers, students, practitioners and scholars across the globe.

List of Contributors

Yi-Gang Wei, Fang Wen and Long-Fei Fu
Guangxi Institute of Botany, Guangxi Zhuang Autonomous Region and Chinese Academy of Sciences, Guilin 541006, China

Wen-Tsai Wang
College of Life Sciences, Guangxi Normal University, 541004, Guilin, China

M. K. Ratheesh Narayanan and M. Sivadasan
Department of Botany, Payyanur College, Kannur 670 327, Kerala, India
Department of Botany & Microbiology, College of Science, King Saud University, Riyadh 11451, Kingdom of Saudi Arabia

A. H. Alfarhan
Department of Botany & Microbiology, College of Science, King Saud University, Riyadh 11451, Kingdom of Saudi Arabia

C. N. Sunil
Department of Botany, S.N.M. College, Maliankara, Kerala, India

M. K. Nandakumar
M.S. Swaminathan Research Foundation, Puthoorvayal Kerala, India

T. Shaju
Jawaharlal Nehru Tropical Botanic Garden and Research Institute, Thiruvananthapuram 695 562, Kerala, India

A. S. M. Amal Tamimi
Department of Biology, College of Science, Princess Nora bint Abdulrahman University, Riyadh 11652, Kingdom of Saudi Arabia

C. Murugan
Botanical Survey of India (BSI), Andaman and Nicobar Regional Centre, Port Blair 744 102, Andaman and Nicobar Islands, India

Joju P. Alappatt
Forest Training Institute, Wimberligunj, Port Blair-744 206, South Andaman, Andaman and Nicobar Islands, India

S. Prabhu
Botanical Survey of India (BSI), Andaman and Nicobar Regional Centre, Port Blair-744 102, Andaman and Nicobar Islands, India

W. Arisdason
Botanical Survey of India, CGO Complex, Salt Lake City, Kolkata-711 064, West Bengal, India

S. Karuppusamy, M. Ajmal Ali and Fahad M. A. Al-Hemaid
Department of Botany, The Madura College, Madurai-625011, Tamil Nadu, India

K. M. Rajasekaran, Joongku Lee and Soo-Yong Kim
International Biological Material Research Center, Korea Research Institute of Bioscience and Biotechnology, Daejeon-305806, South Korea

Arun K. Pandey
Department of Botany, University of Delhi, Delhi-110007, India

Nasrin Jahan Diba and Sabrina Naz
Department of Botany, University of Rajshahi, Rajshahi-6205, Bangladesh

Hendrik Schubert
Institute of Bioscience, University of Rostock, Albert-Einstein-Str. 3, D-18059, Rostock, Germany

Jai Prakash Keshri and Prasant Mallick
Centre for Advanced Studies in Botany, University of Burdwan, Golapbag, Burdwan 713 104, West Bengal, India

Cheng-Hua Yang, Zheng-Guo Xie and Yong-Fu Yu
Forestry Academy of Guizhou Province, Guiyang 550005, Guizhou, People's Republic of China

Zhi-Rong Yang
State Key Laboratory of Systematic and Evolutionary Botany, Institute of Botany, Chinese Academy of Sciences, Beijing 100093, People's Republic of China

Özgür Eminagaoglu
Department of Forest Engineering, Faculty of Forestry, Artvin Coruh University, 08000 Artvin-Turkey

Melahat Özcan
Department of Biology, Faculty of Science and Arts, Artvin Coruh University, 08000 Artvin-Turkey

Nery Sofiyanti, Dyah Iriani, Fitmawati and Andesba A. Roza
Laboratory of Botany, Department of Biology, Faculty of Mathematics and Natural Sciences, University of Riau, Kampus Bina Widya Panam, Pekanbaru, Riau, Indonesia

Tong-Jun Liang and Bo Li
Lushan Botanical Garden, Jiangxi Province and the Chinese Academy of Sciences, Jiujiang, 332900, China

Min-Rong Luo and Liang Zhao
College of Life Sciences, Northwest A&F University, Yangling, Shaanxi, 712100, China

Amena Kibria and K. S. Hossain
Department of Botany, Jagannath University, Dhaka 1100, Bangladesh

N. Akhtar, M. A. A. Jahan, Md. A. M. Sarker and Mst. N. Begum
Biological Research Division, Bangladesh Council of Scientific and Industrial Research (BCSIR), Dhaka-1000, Bangladesh

C. N. Sunil and M. K. Ratheesh Narayanan
Department of Botany, S.N.M. College, Ernakulam–683 516, Kerala, India

Prajeesh Parameswaran
M. S. Swaminathan Research Foundation, Kalpetta, Wayanad–673 121, Kerala, India

M. Sivadasan and A. H. Alfarhan
Department of Botany & Microbiology, College of Science, King Saud University, Riyadh–11451, Kingdom of Saudi Arabia

Yun Lin and Lin Dong Duan
Hunan Food and Drug Vocational College, Changsha 410014, P. R. China

Hai Yan Bi
Beijing Museum of Natural History, Beijing 100050, P. R. China

Shou-Biao Zhou
College of Life Sciences, Anhui Normal University, CN-241000, Wuhu, China

Xin Hong
Gesneriad Conservation Center of China, Guangxi Institute of Botany, Guangxi Zhuang Autonomous Region and the Chinese Academy of Sciences, CN-541006, Guilin, China

Wen Ma and Fang Wen
Gesneriad Conservation Center of China, Guangxi Institute of Botany, Guangxi Zhuang Autonomous Region and the Chinese Academy of Sciences, CN-541006, Guilin, China

Akeem Babalola Kadiri
Department of Botany, University of Lagos, Akoka, Nigeria

Ming-Tai An and Yun Lin
Forestry College, Guizhou University, Guiyang 550025, Guizhou, P. R. China Hunan Medication Vestibule School, Changsha 410208, Hunan, P. R. China

Jia-Guo Wang and Jiang-Hua Wu
Guizhou Normal College, Guiyang 550018, Guizhou, P. R. China

Min Meng
Yinjiang Forestry Bureau, Yinjiang 555200, Guizhou, P. R. China

Khandakar Kamrul Islam and Naimur Rahman
Bangladesh National Herbarium, Chiriakhana Road, Mirpur-1, Dhaka-1216, Bangladesh

Bachan Lal Bhellum
Department of Botany, Government College for Women, Parade, Jammu, J & K State, India

Bikarma Singh
CSIR-Indian Institute of Integrative Medicine, Canal Road, Jammu-Tawi, India

M. Assadi and S. M. M. Hamdi
Research Institute Rangelands and Forests, Tehran, Iran

F. Fajani
Garmsar Branch, Islamic Azad University, Garmsar, Iran

Hong-Rui Lv, Jie Wang, Jia Feng, Jun-Ping Lv, Qi Liu and Shu-Lian Xie
School of Life Science, Shanxi University, Taiyuan 030006, China

K. Prasad
Botanical Survey of India, Deccan Regional Centre, Hyderabad 500048, India

P. V. Prasanna
Central National Herbarium, Botanical Survey of India, Acharya Jagadish Chandra Bose Indian Botanic Garden, Howrah 711 103, India

P. Biju
Department of Botany, Government College, Kasaragod, Kasaragod - 671123, Kerala, India

E. J. Josekutty
Department of Botany, Government Brennan College Thalassery, Kannur 670106, Kerala, India

Augustine Jomy
Department of Botany, St. Thomas College, Pala, Pala 686574, Kerala, India

C. N. Sunil, V. Nithya Madhanan, C. R. Remya Krishnan, V. V. Naveen Kumar, M. S. Simi and K. J. Jyothi
Post Graduate & Research Department of Botany, S.N.M. College, Maliankara, Ernakulam District, Kerala, India

Huseyin Inceer and Nursen Aksu Kalmuk
Karadeniz Technical University, Faculty of Science, Department of Biology, 61080 Trabzon, Turkey

Dong-Ping Zhao, Tong-Rui Zhang, Xue-Liang Bai and Dong-Mei Ren
Department of Biology, School of life Science, Inner Mongolia University, West College Road 235, Hohhot 010021, China

M. S. Kiran Raj
Department of Botany, Sree Narayana College, Cherthala, S. N. Alappuzha–688 582, Kerala, India

M. Sivadasan and A. H. Alfarhan
Department of Botany & Microbiology, College of Science, King Saud University, Riyadh–11451, Kingdom of Saudi Arabia

P. Dileep
Department of Botany, Regional Institute of Education (RIE), Mysore-570 006, Karnataka, India

M. Nazim Uddin
Department of Botany, Rangamati Government College, Rangamati, Bangladesh

M. Oliur Rahman
Department of Botany, University of Dhaka, Dhaka 1000, Bangladesh

M. Atiqur Rahman
Department of Botany, University of Chittagong, Chittagong 4331, Bangladesh

Shashi Kumar and Sushil Kumar Singh
Botanical Survey of India, Eastern Regional Centre, Shillong-793003, India

Anoop P. Balan and Harikrishnan Shanmugam
Indian Cardamom Research Institute, Spices Board, Myladumpara, Idukki, Kerala 685553, India

Jian Li, Xin Chen and Shu Li
Institute of Ecology, College of Life Sciences, Zhejiang University, Hangzhou, Zhejiang province, China, CN-315008, P. R. China

Li-Jiao Yan
Gesneriaceae Conservation Center of China (GCCC), Guilin Botanical Garden, Guangxi Inst. of Botany, Guangxi Zhuang Autonomous Region and Chinese Academy of Sciences, CN-541006 Guilin, P. R. China

D. B. Borude and A. N. Chandore
Department of Botany, Abasaheb Marathe Arts and New Commerce, Science College, Rajapur 416702, District Ratnagiri, Maharashtra, India

A. R. Gholave and S. R. Yadav
Department of Botany, Shivaji University, Kolhapur 416004, Maharashtra, India

Hosne Ara
Bangladesh National Herbarium, Chiriakhana Road, Mirpur-1, Dhaka-1216, Bangladesh

Md. Abul Hassan
Department of Botany, University of Dhaka, Dhaka-1000, Bangladesh

Junita Hardini and Santosa and Purnomo Rina
Department of Tropical Biology, Faculty of Biology, Universitas Gadjah Mada, Jl. Teknika Selatan, Sekip Utara Yogyakarta 55281, Indonesia

Sri Kasiamdari
Department of Biology, Faculty of Mathematics and Natural Science, Udayana University, Jl. Raya Kampus UNUD, Bukit Jimbaran, Kuta Selatan, Badung, Bali 80361, Indonesia

Index